Thermodynamics: Diverse Concepts and Applications

Thermodynamics: Diverse Concepts and Applications

Edited by **Lucy Flynn**

Willford Press
New York

Published by Willford Press,
118-35 Queens Blvd., Suite 400,
Forest Hills, NY 11375, USA
www.willfordpress.com

Thermodynamics: Diverse Concepts and Applications
Edited by Lucy Flynn

International Standard Book Number: 978-1-68285-112-8 (Hardback)

Printed in the United States of America.

Contents

Preface

Thermodynamics as a discipline aims to understand the correlation between heat, energy and work done and is widely applied in various scientific fields. The chapters in this book discuss some of the significant concepts and applications related to thermodynamics such as chemical thermodynamics, thermophysics, and heat mass transfer. There has been rapid progress in this field and its applications are finding their way across multiple industries. It is an essential guide for both professionals and those who wish to pursue this discipline further.

This book is a comprehensive compilation of works of different researchers from varied parts of the world. It includes valuable experiences of the researchers with the sole objective of providing the readers (learners) with a proper knowledge of the concerned field. This book will be beneficial in evoking inspiration and enhancing the knowledge of the interested readers.

In the end, I would like to extend my heartiest thanks to the authors who worked with great determination on their chapters. I also appreciate the publisher's support in the course of the book. I would also like to deeply acknowledge my family who stood by me as a source of inspiration during the project.

Editor

1

Effects of induced magnetic field and slip condition on peristaltic transport with heat and mass transfer in a non-uniform channel

Najma Saleem[1], T. Hayat[2]* and A. Alsaedi[3]

[1]Department of Mathematics, University of Management and Technology, CII Johar Town Lahore-54770, Pakistan.
[2]Department of Mathematics, Quaid-i-Azam University Islamabad 44000, Pakistan.
[3]Department of Mathematics, Faculty of Science, King Abdulaziz University, P. O. Box 80257, Jeddah 21589, Saudi Arabia.

We study the effects of induced magnetic field and slip on the peristaltic flow of Jeffrey fluid in a non-uniform channel. Flow analysis is discussed in the presence of heat and mass transfer. Main emphasis is given to the study of stream function, the longitudinal pressure gradient, the magnetic force function, the axial induced magnetic field, the current density, the temperature and concentration. Numerical integration is carried out for pressure rise per wavelength. The flow quantities of interest are discussed by graphical illustrations.

Key words: Jeffrey fluid, induced magnetic field, slip, heat and mass transfer.

INTRODUCTION

The study of peristaltic flow is of special interest for several applications in industry and physiology. Especially the peristaltic transport of non-Newtonian fluids (Ellahi, 2009) is a topic of major interest of the researchers in the physiological world. Such interest is stimulated because of its occurrence in several physiological processes including chyme movement in the gastrointestinal tract, urine transport from kidney to bladder, movement of ovum in the female fallopian tube, transport of spermatozoa in the ductus efferentes of the male reproductive tract, transport of bile in bile duct, in roller and finger pumps, in vasomotion of small blood vessels and many others. It is now a well accepted fact that the peristaltic flows of magnetohydrodynamic (MHD) fluids are important in medical sciences and bioengineering. The MHD characteristics are useful in the development of magnetic devices, cancer tumor treatment, hyperthermia and blood reduction during surgeries. Hence several scientists having in mind such importance extensively discussed the peristalsis with magnetic field effects (Reddy and Raju, 2010; Tripathi, 2011; Hayat et al., 2011a; Abd elmaboud and Mekheimer, 2011; Hayat et al., 2010a). Further, Singh and Rathee (2010, 2011) discussed the blood flow in the presence of an applied magnetic field. The pulsatile blood flow in the presence of applied magnetic field and body acceleration is examined by Sanyal et al. (2007). It is noticed that although ample literature on MHD peristaltic flow in the presence of applied magnetic field is available, very little attention is paid to the influence of induced magnetic field in peristalsis (Hayat et al., 2008a, b; Hayat et al., 2011b; Kotkandapani and Srinivas, 2008; Ali et al., 2008). In continuation, the induced magnetic field effect on the peristaltic motion in couple stress and micropolar fluids is addressed by Mekheimer (2008a, b). Hayat et al. (2010b, c) extended such discussion to third order and Carreau fluids. Abd elamboud (2011) examined induced magnetic field effect on peristaltic flow in an annulus.

The goal of the present study is to discuss the effect of induced magnetic field on peristaltic flow of Jeffrey fluid in non-uniform channel. The heat transfer and slip effects are also considered. The heat transfer analysis in such

*Corresponding author. E-mail: pensy_t@yahoo.com.

flow is important because of hemodialysis and oxygenation process. Further, there is always small amount of slippage in the real systems. Two different types of fluids exhibit slip effect. One case consists of fluids with very elastic properties and other rarefied gases. The slip effect appears in concentrated polymer solutions, molten polymers and non-Newtonian fluids. In the flow of dilute suspensions of particle, a clear layer is noticed next to a wall. Poiseuille observed such a layer using a microscope in the flow of blood through capillary vessels (Coleman et al., 1966). Few very recent contributions dealing with peristaltic flows subject to slip and heat transfer effects may be mentioned in the researches (Hayat et al., 2010d; Srinivas et al., 2009; Nadeem and Akram, 2010; Akbar et al., 2011).

ANALYSIS

We consider the MHD peristaltic flow of an incompressible Jeffrey fluid in a symmetric but non-uniform channel. The fluid is electrically conducting in the presence of constant magnetic field with strength H_0 applied normal to the flow. This gives rise to an induced magnetic field $H'(h'_{X'}(X',Y',t'), h'_{Y'}(X',Y',t'), 0)$ and ultimately the total magnetic field is $H'^{+}(h'_{X'}(X',Y',t'), H_0' + h'_{Y'}(X',Y',t'), 0)$. The flow generation is possible because of travelling wave on the channel walls.

The relevant equations for the present flow problem are given by:

$$\nabla.\mathbf{H}' = 0, \quad \nabla.\boldsymbol{\epsilon}' = 0, \quad \nabla\times\boldsymbol{\epsilon}' = -\mu_e\frac{\partial \mathbf{H}'}{\partial t}, \tag{1}$$

$$\nabla\times\mathbf{H}' = \mathbf{J}', \quad \mathbf{J}' = \sigma\{\boldsymbol{\epsilon}' + \mu_e(\mathbf{V}'\times\mathbf{H}'^{+})\}, \tag{2}$$

$$\nabla.\bar{\boldsymbol{V}}' = 0, \tag{3}$$

$$\rho\left[\frac{\partial}{\partial t} + (\boldsymbol{V}'.\nabla)\right]\bar{\boldsymbol{V}}' = div T - \mu_e\left\{(\mathbf{H}'^{+}.\nabla) - \frac{1}{2}(\mathbf{H}'^{+})^2\nabla\right\}, \tag{4}$$

$$\rho C_p\frac{dT'}{dt'} = \kappa\nabla^2 T' + \boldsymbol{T}.\boldsymbol{L}, \tag{5}$$

with $L = grad\boldsymbol{V}'$ and Cauchy stress tensor $(\mathbf{T})$ and extra stress tensor $(\mathbf{S}')$ in Jeffrey fluid are

$$T = -p'\bar{I} + \mathbf{S}', \tag{6}$$

$$\mathbf{S}' = \frac{\mu}{1+\lambda_1}(\dot{\gamma} + \lambda_2\ddot{\gamma}). \tag{7}$$

In the aforementioned expressions, p' denotes the pressure, $\mathbf{J}$ the current density, I the identity tensor, μ_e the magnetic permeability, ρ the fluid density, σ the electrical conductivity, ϵ' an induced magnetic field, C_p the specific heat at a constant volume, κ the thermal conductivity, T the temperature, μ the dynamic viscosity of fluid, γ the shear rate, λ_1 the ratio of relaxation to retardation times, and λ_2 the retardation time and dots characterize material differentiation.

Induction equation in view of combination of Equations (1) and (2) becomes

$$\frac{\partial \mathbf{H}^{+}}{\partial t} = \nabla\times(\boldsymbol{V}\times\mathbf{H}^{+}) + \frac{1}{\xi}\nabla^2\mathbf{H}^{+}, \tag{8}$$

where $\xi = 1/\mu\sigma$ is the magnetic diffusivity.

We perform the flow analysis in wave frame (x', y'). Hence the coordinates and velocities in the laboratory (X', Y') and wave (x', y') frames can be related through the following transformations:

$$x' = X' - ct', \quad y' = Y', \; u'(x', y') = U' - c, \; v'(x', y') = V' \tag{9}$$

where (U', V') and (u', v') are the respective velocities in the laboratory and wave frames.

The velocity V is chosen as follows:

$$\bar{V}' = (u', v', 0). \tag{10}$$

Introducing the non-dimensional quantities

$$x = \frac{x'}{\lambda}, \; y = \frac{y'}{a}, \; u = \frac{u'}{c}, \; v = \frac{v'}{c}, \; t = \frac{t'c}{\lambda}, \; p = \frac{a^2 p'}{c\lambda\mu}, \; S = \frac{aS'}{\mu c},$$

$$h = \frac{h'}{a}, \; \Psi = \frac{\Psi'}{ca}, \; \Phi = \frac{\Phi'}{H_0 a}, \; \delta = \frac{\lambda}{a}, \; R_e = \frac{\rho c a}{\mu}, \; R_e = \sigma\mu_e ac,$$

$$S_1 = \frac{H_0}{c}\sqrt{\frac{\mu_e}{\rho}}, \; p_m = p + \frac{1}{2}R_e\delta\frac{\mu_e(\mathbf{H}'^{+})^2}{\rho c^2}, \; Br = E_r P_r, \; \theta' = \frac{T - T_0}{T_1 - T_0},$$

$$P_r = \frac{\rho v C_p}{\kappa}, \; E_r = \frac{c^2}{C_p(T_1 - T_0)}, \; Sc = \frac{\mu}{\rho D}, \; Sr = \frac{\rho T_0 D K_T}{\mu T_m C_0}, \tag{11}$$

$$u = \frac{\partial\Psi}{\partial y}, \quad v = -\delta\frac{\partial\Psi}{\partial x}, \quad h_x = \frac{\partial\Phi}{\partial y}, \quad h_y = -\delta\frac{\partial\Phi}{\partial x}, \tag{12}$$

the incompressibility condition is satisfied whereas the other equations give

$$R_e\delta\left(\frac{\partial\Psi}{\partial y}\frac{\partial^2\Psi}{\partial x\partial y} - \frac{\partial\Psi}{\partial x}\frac{\partial^2\Psi}{\partial y^2}\right) = -\frac{\partial p_m}{\partial x} + \delta\frac{\partial}{\partial x}(S_{xx}) + \frac{\partial}{\partial y}(S_{xy}) + R_e S_1^2\frac{\partial^2\Phi}{\partial y^2} + R_e S_1^2\delta\left(\frac{\partial\Phi}{\partial y}\frac{\partial^2\Phi}{\partial x\partial y} - \frac{\partial\Phi}{\partial x}\frac{\partial^2\Phi}{\partial y^2}\right), \tag{13}$$

$$R_e\delta^3\left(\frac{\partial\Psi}{\partial x}\frac{\partial^2\Psi}{\partial x\partial y}-\frac{\partial\Psi}{\partial y}\frac{\partial^2\Psi}{\partial x^2}\right)=$$
$$-\frac{\partial p_m}{\partial y}+\delta^2\frac{\partial}{\partial x}(S_{yx})+\delta\frac{\partial}{\partial y}(S_{yy})-R_e\delta^2S_1^2\frac{\partial^2\Phi}{\partial x\partial y}-R_eS_1^2\delta^3\left(\frac{\partial\Phi}{\partial y}\frac{\partial^2\Phi}{\partial x^2}-\frac{\partial\Phi}{\partial x}\frac{\partial^2\Phi}{\partial x\partial y}\right), \tag{14}$$

$$\frac{\partial\Psi}{\partial y}-\delta\left(\frac{\partial\Psi}{\partial y}\frac{\partial\Phi}{\partial x}-\frac{\partial\Psi}{\partial x}\frac{\partial\Phi}{\partial y}\right)+\frac{1}{R_m}\left(\frac{\partial^2\Phi}{\partial y^2}+\delta^2\frac{\partial^2\Phi}{\partial x^2}\right)=E, \tag{15}$$

$$R_e\delta\left(\frac{\partial\Psi}{\partial y}\frac{\partial\theta}{\partial x}-\frac{\partial\Psi}{\partial x}\frac{\partial\theta}{\partial y}\right)=\frac{1}{P_r}\left(\delta^2\frac{\partial^2\theta}{\partial x^2}+\frac{\partial^2\theta}{\partial y^2}\right)+E_r\left\{\begin{array}{c}\delta S_{xx}\frac{\partial^2\Psi}{\partial x\partial y}+S_{xy}\\ \left(\frac{\partial^2\Psi}{\partial y^2}-\delta^2\frac{\partial^2\Psi}{\partial x^2}\right)-\delta S_{yy}\frac{\partial^2\Psi}{\partial y\partial x}\end{array}\right\}, \tag{16}$$

$$R_e\delta\left(\frac{\partial\Psi}{\partial y}\frac{\partial\varphi}{\partial x}-\frac{\partial\Psi}{\partial x}\frac{\partial\varphi}{\partial y}\right)=\frac{1}{Sc}\left(\delta^2\frac{\partial^2\varphi}{\partial x^2}+\frac{\partial^2\varphi}{\partial y^2}\right)+Sr\left(\delta^2\frac{\partial^2\theta}{\partial x^2}+\frac{\partial^2\theta}{\partial y^2}\right), \tag{17}$$

where Ψ, Φ, E_r, P_r, Sc, Sr, δ, R_e, R_m, and S_1 are respectively the stream function, magnetic force function, Eckert, Prandtl, Schmidt, Soret, wave, Reynolds, magnetic Reynolds and Strommer's numbers. Here p_m shows the total pressure which is sum of ordinary and magnetic pressures. Further, T_0, C_0 and T_1, C_1 are the temperatures and concentrations at the upper and lower walls respectively and

$$S_{xx}=\frac{2\delta}{1+\lambda_1}\left(1+\frac{\lambda_2c\delta}{d_1}\left(\frac{\partial\Psi}{\partial y}\frac{\partial}{\partial x}-\frac{\partial\Psi}{\partial x}\frac{\partial}{\partial y}\right)\right)\frac{\partial^2\Psi}{\partial x\partial y},$$

$$S_{xy}=\frac{1}{1+\lambda_1}\left(1+\frac{\lambda_2c\delta}{d_1}\left(\frac{\partial\Psi}{\partial y}\frac{\partial}{\partial x}-\frac{\partial\Psi}{\partial x}\frac{\partial}{\partial y}\right)\right)\left(\frac{\partial^2\Psi}{\partial y^2}-\delta^2\frac{\partial^2\Psi}{\partial x^2}\right),$$

$$S_{yy}=-\frac{2\delta}{1+\lambda_1}\left(1+\frac{\lambda_2c\delta}{d_1}\left(\frac{\partial\Psi}{\partial y}\frac{\partial}{\partial x}-\frac{\partial\Psi}{\partial x}\frac{\partial}{\partial y}\right)\right)\frac{\partial^2\Psi}{\partial x\partial y},$$

with the boundary conditions given following

$$\Psi=0,\ \frac{\partial^2\Psi}{\partial y^2}=0,\ \frac{\partial\Phi}{\partial y}=0,\ \frac{\partial\theta}{\partial y}=0,\ \frac{\partial\varphi}{\partial y}=0,\ at\ y=0, \tag{18}$$

$$\Psi=F,\ \frac{\partial\Psi}{\partial y}=-\beta S_{xy}-1,\ \Phi=1,\ \theta=1,\ \varphi=0,\ at\ y=h(x), \tag{19}$$

where the dimensionless slip parameter $\beta(=L/a)$.

In view of long wavelength and low Reynolds number analysis, one has from Equations (13) to (19) as

$$-\frac{\partial p}{\partial x}+\frac{\partial}{\partial y}\left(\frac{1}{1+\lambda_1}\frac{\partial^2\Psi}{\partial y^2}\right)+R_eS_1^2\frac{\partial^2\Phi}{\partial y^2}=0, \tag{20}$$

$$-\frac{\partial p}{\partial y}=0, \tag{21}$$

$$\frac{\partial^2\Phi}{\partial y^2}=R_m\left(E-\frac{\partial\Psi}{\partial y}\right). \tag{22}$$

Equations (21) and (22) after eliminating the pressure give

$$\frac{\partial^2}{\partial y^2}\left(\frac{1}{1+\lambda_1}\frac{\partial^2\Psi}{\partial y^2}\right)+R_eS_1^2\frac{\partial^3\Phi}{\partial y^3}=0. \tag{23}$$

Now Equations (17) and (18) are presented in the forms

$$\frac{\partial^2\theta}{\partial y^2}+Br\left\{\frac{1}{1+\lambda_1}\left(\frac{\partial^2\Psi}{\partial y^2}\right)^2\right\}=0, \tag{24}$$

$$\frac{1}{Sc}\frac{\partial^2\varphi}{\partial y^2}+Sr\frac{\partial^2\theta}{\partial y^2}=0, \tag{25}$$

with the boundary conditions as follows

$$\Psi=0,\ \frac{\partial^2\Psi}{\partial y^2}=0,\ \frac{\partial\Phi}{\partial y}=0,\ \frac{\partial\theta}{\partial y}=0,\ \frac{\partial\varphi}{\partial y}=0,\ at\ y=0, \tag{26}$$

$$\Psi=F,\ \frac{\partial\Psi}{\partial y}=-\frac{\beta}{1+\lambda_1}\frac{\partial^2\Psi}{\partial y^2}-1,\ \Phi=1,\ \theta=1,\ \varphi=0,\ at\ y=h(x). \tag{27}$$

Our interest in this study is to perform the analysis for the following wave forms.

(1) Sinusoidal wave $h(x)=1+Kx+\phi\sin(2\pi(x-t))$.

(2) Triangular wave

$$h(x)=1+Kx+\phi\left[\frac{8}{\pi^3}\sum_{m=1}^{\infty}\frac{(-1)^{m+1}}{(2m-1)^2}\sin\{2(2m-1)\pi(x-t)\}\right].$$

(3) Square wave

$$h(x)=1+Kx+\phi\left[\frac{4}{\pi}\sum_{m=1}^{\infty}\frac{(-1)^{m+1}}{(2m-1)}\cos\{2(2m-1)\pi(x-t)\}\right].$$

(4) Trapezoidal wave

$$h(x)=1+Kx+\phi\left[\frac{32}{\pi^2}\sum_{m=1}^{\infty}\frac{(-1)^{m+1}\sin\left\{\frac{\pi}{8}(2m-1)\right\}}{(2m-1)^2}\sin\{2(2m-1)\pi(x-t)\}\right],$$

with $\phi=b/a$ (amplitude ratio) and $K=\lambda k/a(k\ll 1)$. Total number of terms in the series that are incorporated in the analysis are 50. Note that the expressions for triangular, square and trapezoidal waves are derived from Fourier series.

The pressure rise per wavelength is

$$\Delta p_\lambda=\int_0^1\frac{dp}{dx}dx. \tag{28}$$

EXACT SOLUTION

From Equations (20), (22) and (23) we have

$$\frac{\partial p}{\partial x} = \frac{\partial}{\partial y}\left(\frac{1}{1+\lambda_1}\frac{\partial^2\Psi}{\partial y^2}\right) + M^2\left(E - \frac{\partial\Psi}{\partial y}\right), \tag{29}$$

$$\frac{\partial^4\Psi}{\partial y^4} - (1+\lambda_1)M^2\frac{\partial^2\Psi}{\partial y^2} = 0. \tag{30}$$

The aforementioned equations along with the corresponding boundary conditions give

$$\Psi = \frac{1}{L^2}\{(\cosh(Ly) - \sinh(Ly))(C_2 + C_1\cosh(2Ly) + C_1\sinh(2Ly))\} + C_3 + C_4 y, \tag{31}$$

$$\frac{dp}{dx} = \frac{(C_1 - C_2)L}{(1+\lambda_1)} - \frac{\{(C_1 - C_2)L + L^2(C_4 - E)\}}{L}, \tag{32}$$

$$C_1 = -\frac{(F+h)(1+\lambda_1)L^2}{2L_1},\quad C_2 = \frac{(F+h)(1+\lambda_1)L^2}{2L_1},\quad C_3 = 0,$$

$$C_4 = \frac{FL(1+\lambda_1)\cosh(hL) + (1 + F\beta L^2 + \lambda_1)\sinh(hL)}{L_1},$$

where $M^2 = R_m R_s S_1^2$ (is the Hartman number), $L = M\sqrt{(1+\lambda_1)}$;
$L_1 = hL(1+\lambda_1)\cosh(hL) + (-1 + hL^2\beta - \lambda_1)\sinh(hL).$
The corresponding systems for Φ and θ after using Equation (31) finally give

$$\Phi = \frac{1}{2LM^2(1+\lambda_1)}\Big(2L^3(C_5 - C_6 y) - \big(L^3y^2(C_4 - E) + 2(C_1 - C_2)\cosh(Ly) + 2(C_1 + C_2)\sinh(Ly)\big)R_m\Big), \tag{33}$$

$$\theta' = -\frac{1}{4L^4}\Big(-4\big(-BrC_1C_2L^4y^2 + L^4(C_7 \mp y)\big) + Br(C_1^2 + C_2^2)L^2\cosh(2Ly) + Br(C_1^2 - C_2^2)L^2\text{sinhh}(2Ly)\Big), \tag{34}$$

$$C_5 = \frac{(hL(2(C_1+C_2) + hL^2(C_4 - E) + 2(C_1 - C_2)\cosh(hL) + 2(C_1 + C_2)\sinh(hL))R_m}{2L^3},$$

$$C_6 = \frac{(C_1 + C_2)R_m}{L^2},\quad C_8 = \frac{Br(C_1^2 - C_2^2)}{2L},$$

$$C_7 = \frac{Br(2hL(-C_1^2 + C_2^2 + 2C_1C_2hL) + (C_1^2 + C_2^2)\cosh(2hL) - (C_1^2 - C_2^2)\sinh(2hL))}{4L^2}.$$

Now the results for axial induced magnetic field h_x, current density distribution J_z and concentration distribution φ are given by

$$h_x = \frac{C_6L^2 - (L^2y + (C_1 + C_2)\cosh(Ly) + 2(C_1 - C_2)\sinh(Ly))R_m}{L^2}, \tag{35}$$

$$J_z = \frac{(2L^3(C_4 - E) + 2(C_1 - C_2)L^2\cosh(hL) + 2(C_1 + C_2)L^2\sinh(Ly))R_m}{2L^3}, \tag{36}$$

$$\varphi = \frac{1}{4L^4}\Big(4\big(C_9L^4 + y(BrC_1C_2L^4yScSr + C_{10}L^4)\big) + Br(C_1^2 + C_2^2)ScSrL^2\cosh(2Ly) + Br(C_1^2 - C_2^2)ScSrL^2\sinh(2Ly)\Big), \tag{37}$$

$$C_9 = \frac{BrL^2ScSr(2hL(-C_1^2 + C_2^2 + 2C_1C_2hL) + (C_1^2 + C_2^2)\cosh(2hL) + (C_1^2 - C_2^2)\sinh(2hL)}{2L^4},$$

$$C_{10} = \frac{Br(C_1^2 - C_2^2)ScSr}{2L},$$

RESULTS AND DISCUSSION

In order to predict the effects of pertinent parameters on various quantities such as pressure rise per wavelength (Δp_λ), the streamlines (Ψ), the axial induced magnetic field (h_x), the current density distribution (J_z), the temperature (θ') and concentration (φ) distributions, the Figures 1-7 are displayed for different wave forms. This analysis mainly focuses for the effects of slip parameter (β), Hartman number (M), the ratio of relaxation to retardation times (λ_1), Brinkman number (Br), Schmidt number (Sc) and Soret number (Sr).

Figures 1(a-d) characterize the pumping mechanism for different values of λ_1. There are four types of regions regarding pumping. When the dimensionless mean flow rate (θ) and pressure rise per wavelength (Δp_λ) are positive it defines the peristaltic pumping region. For $\theta > 0$ and $\Delta p_\lambda < 0$, we have augmented pumping and for $\theta = 0$, $\Delta p_\lambda > 0$ is a free pumping region. One also has retrograde pumping when $\theta > 0$ and $\Delta p_\lambda < 0$. Figure 1a shows clearly that for sinusoidal wave, the Δp_λ increases for small values of λ_1 in retrograde pumping region whereas a reverse situation is noticed in the augmented region, that is Δp_λ increases by increasing λ_1. The other wave forms also show the similar behavior as that of sinusoidal wave. It is also observed that Δp_λ is maximum for square wave and minimum for triangular wave. The variation of M on axial induced magnetic field (h_x) with y over a cross section $x = 0.1$ is displayed in the Figures 2(a-d). It is revealed from the Figure 2a that magnitude of axial induced magnetic field (h_x) increases for M. Interestingly, the axial induced magnetic field (h_x) in the upper half region is in one direction while

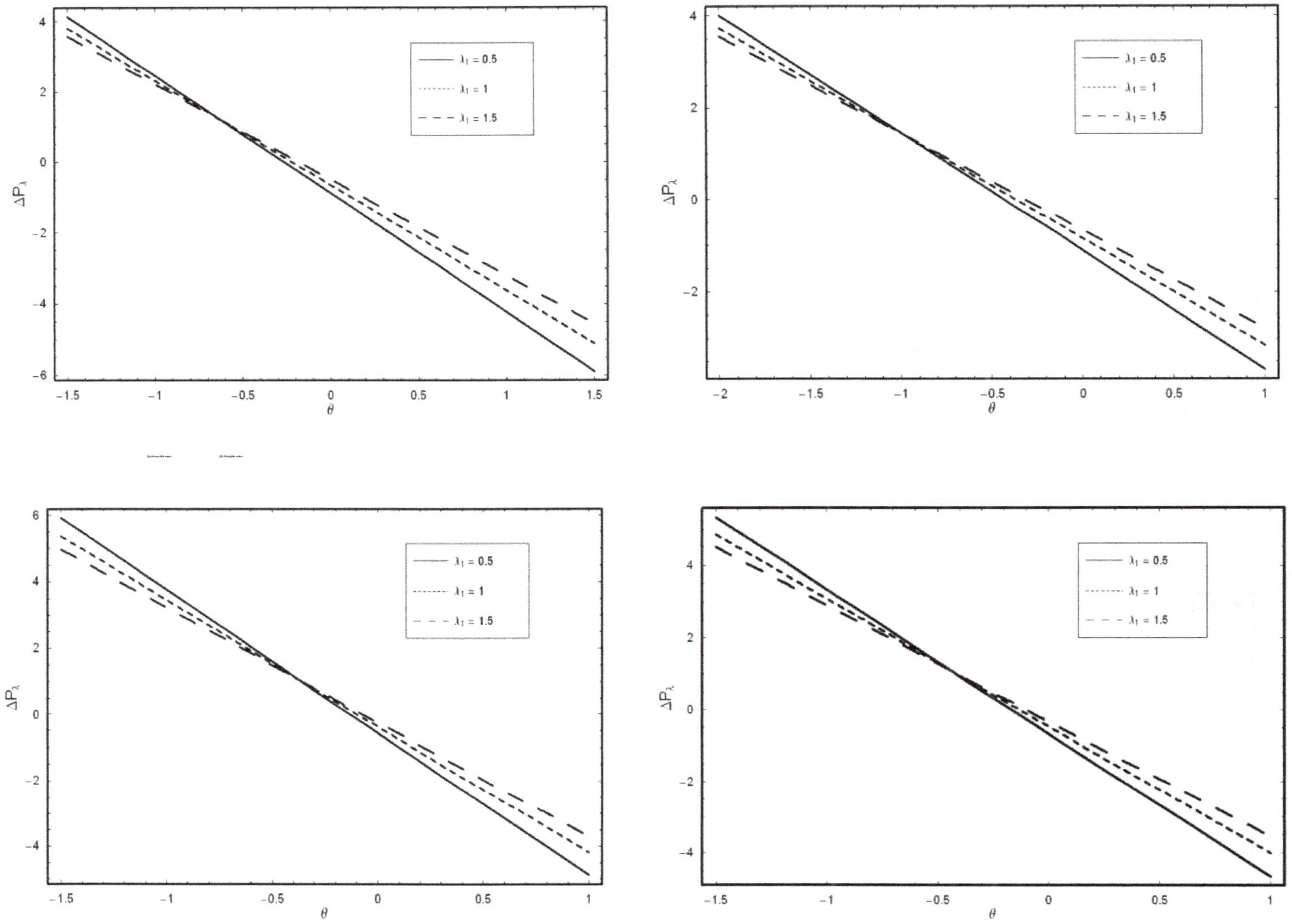

(c) square wave (d) trapezoidal wave

Figure 1. Plot for Δp_{λ} versus x. The other parameters are $K = 0.005,\ \phi = \frac{\pi}{6},\ E = 0.4,\ t = 0.1,\ \beta = 0.2,\ R_e = R_m = S_1 = M = 1.$

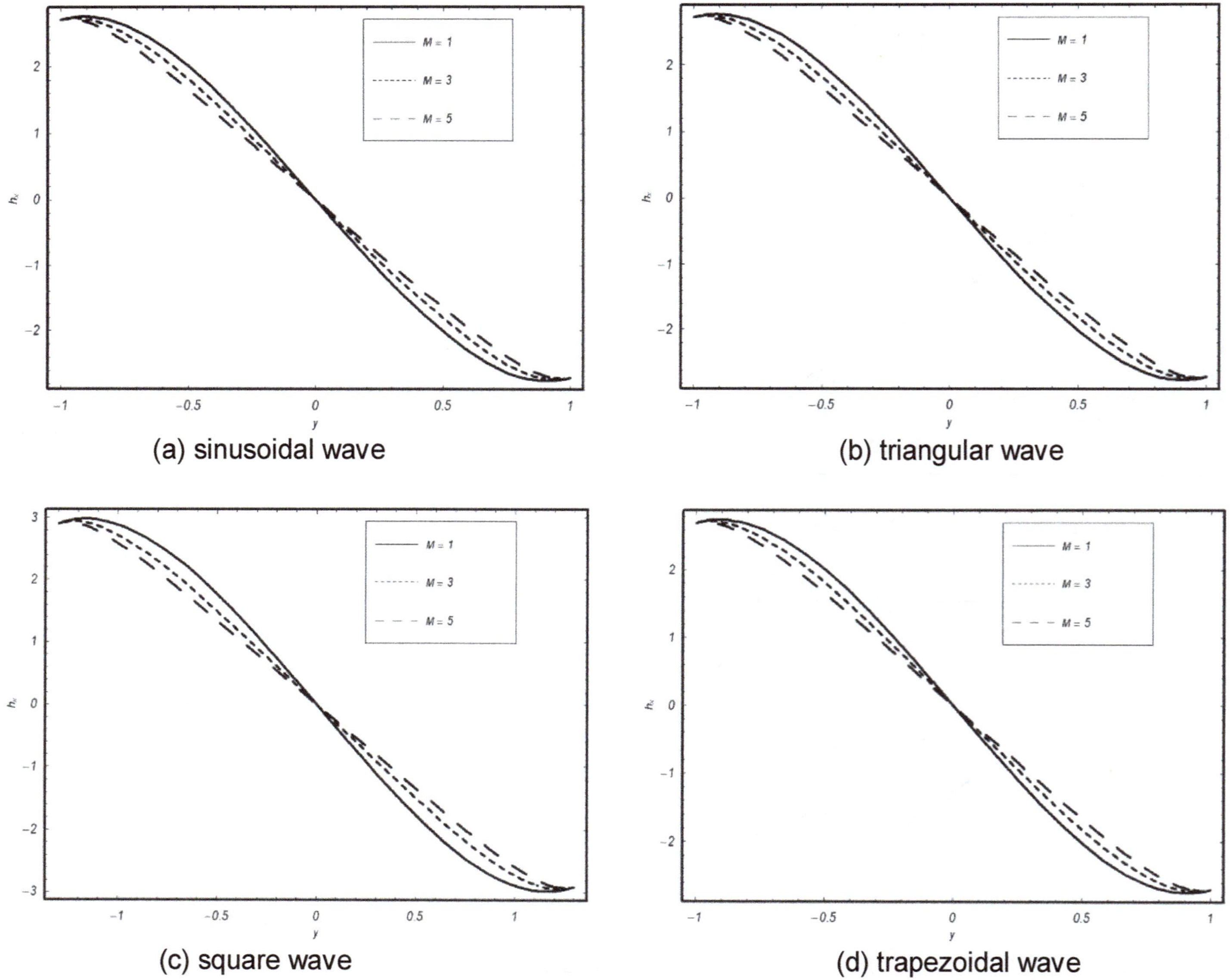

Figure 2. Variation of h_x versus y. Here $K = 0.005,\ \phi = \frac{\pi}{6},\ E = 0.3,\ t = 0.1,\ \beta = 0.03,\ \lambda_1 = 0.05,\ R_m = 1, \theta = 3$ and $x = 0.1$.

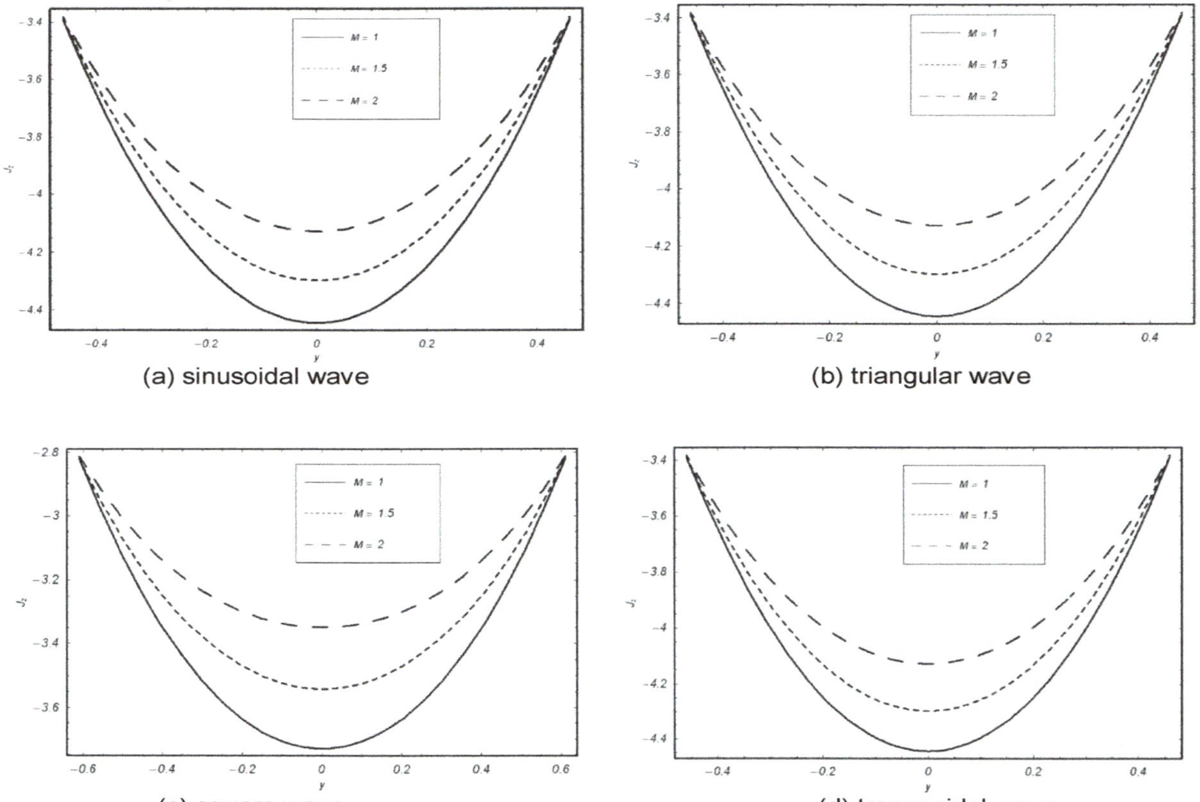

(a) sinusoidal wave (b) triangular wave

(c) square wave (d) trapezoidal wave

Figure 3. Variation of J_z versus y. Here $K = 0.005,\ \phi = \frac{\pi}{6},\ E = 0.3,\ t = 0.1,\ \beta = 0.03,\ \lambda_1 = 0.05,\ R_m = 1, \theta = 3$ and $x = 0.1$.

it is in the opposite direction in the lower half region. Further, h_x is equal to zero at $y = 0$. These results also hold for all the other considered wave forms in Figure 2(c-d). A comparative study further indicates that h_x is maximum for square wave and minimum for trapezoidal wave. To see the variation of current density distribution (J_z) versus y over a cross section $x = 0.1$, we have plotted the Figures 3(a-d). Here we found that J_z is an increasing function of M. Trapping is an interesting phenomenon in theory of peristaltic transport. The formation of an internally circulating bolus of the fluid by closed streamlines is called trapping and this trapped bolus is pushed ahead along the peristaltic wave with the speed of wave. Here Figures 4-5 have been

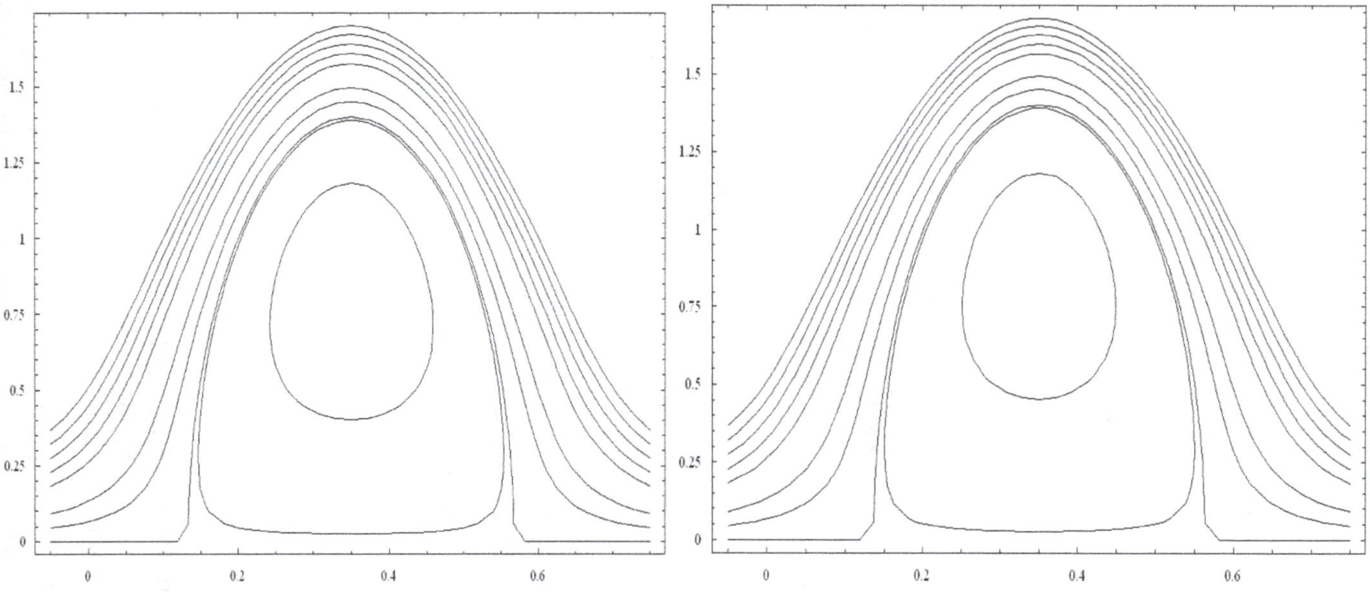

Figure 4a. Streamlines (sinusoidal wave) for $\lambda_1 = 0.5, \lambda_1 = 1.5$.

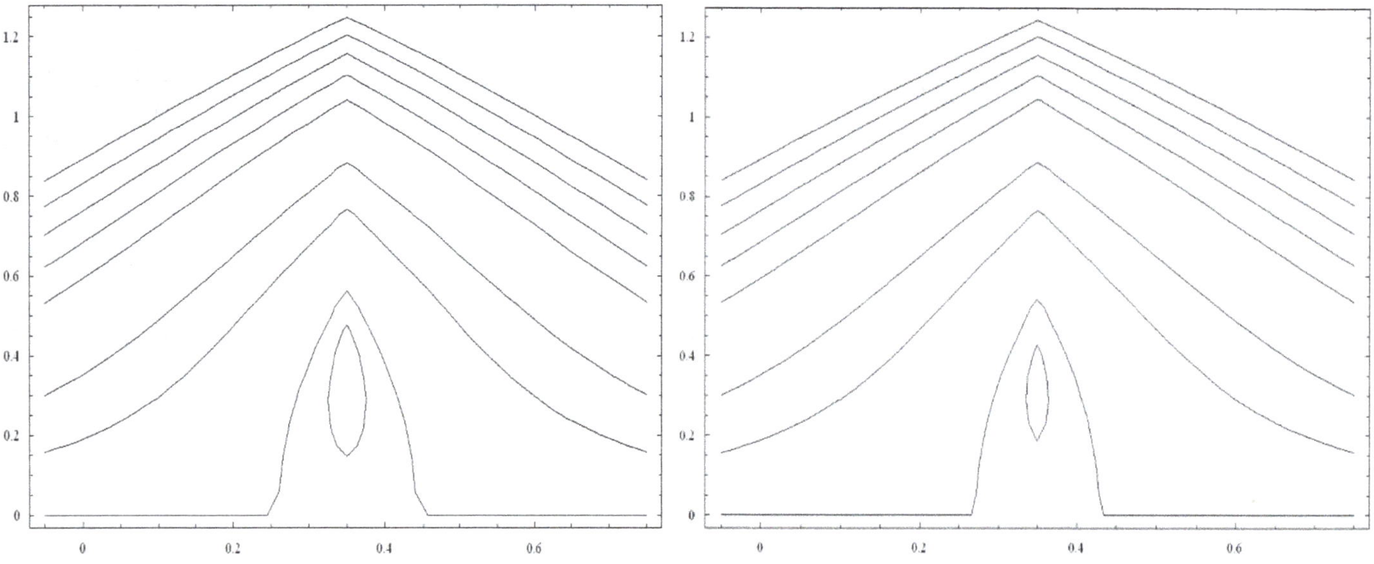

Figure 4b. Streamlines (triangular wave) for $\lambda_1 = 0.5, \lambda_1 = 1.5$.

plotted for the description of trapping when $M = 1, \beta = 0.03, \varphi = \frac{\pi}{6}, a = b = 0.4, d = 1.1$ and $\theta=0.61$. The streamlines for different values of λ_1 are shown in Figure 4. It is noticed that the size of the trapped bolus decreases from $\lambda_1 = 0.5$ to $\lambda_1 = 1.5$. In Figure 5 we have sketched the streamlines for different values of M. This Figure depicts that trapping reduces for large values of M. That is size of the trapped bolus is going to squeeze from hydrodynamic to magneto-hydrodynamic situations for all the considered wave forms.

In the Figures 6(a-d) and 7(a-d), the temperature and concentration profiles have been illustrated. As expected, the temperature is an increasing function of Br (Figure 7). In fact Brinkman number is a measure of the importance of viscous heating relative to the conductive heat transfer.

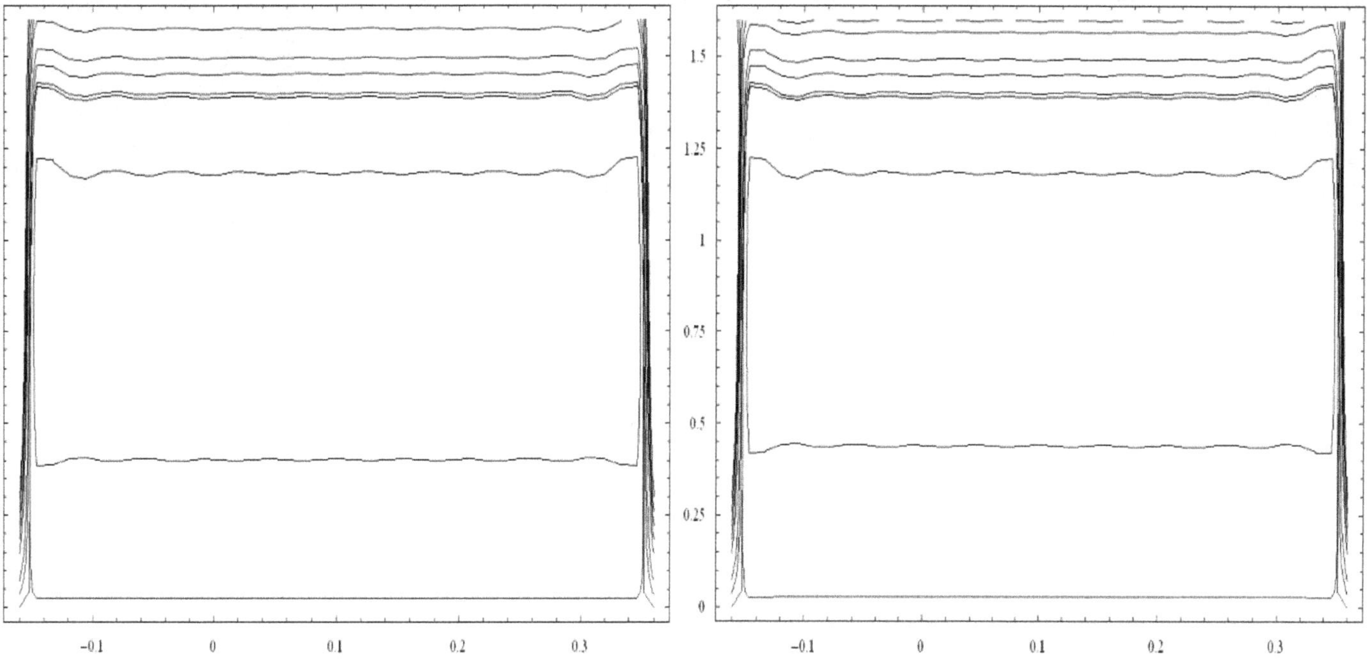

Figure 4c. Streamlines (square wave) for $\lambda_1 = 0.5, \lambda_1 = 1.5$.

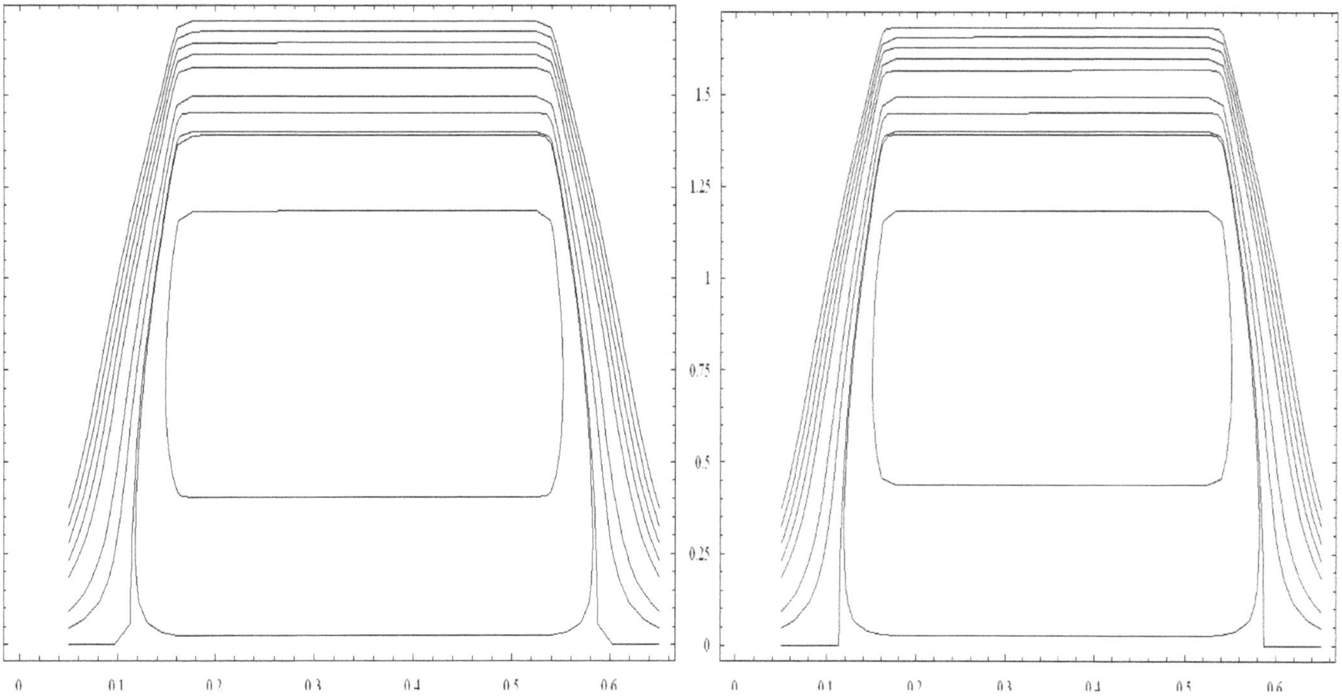

Figure 4d. Streamlines (trapezoidal wave) for $\lambda_1 = 0.5, \lambda_1 = 1.5$.

An increase in the Brinkman number increases the energy in the molecules and consequently the temperature increases. Here it is also observed that the temperature profile looks almost parabolic. The temperature is maximum for the sinusoidal and trapezoidal waves.

The Schmidt number is defined as the ratio of momentum diffusivity and mass diffusivity and is used to characterize the fluid flows in which there are simultaneous

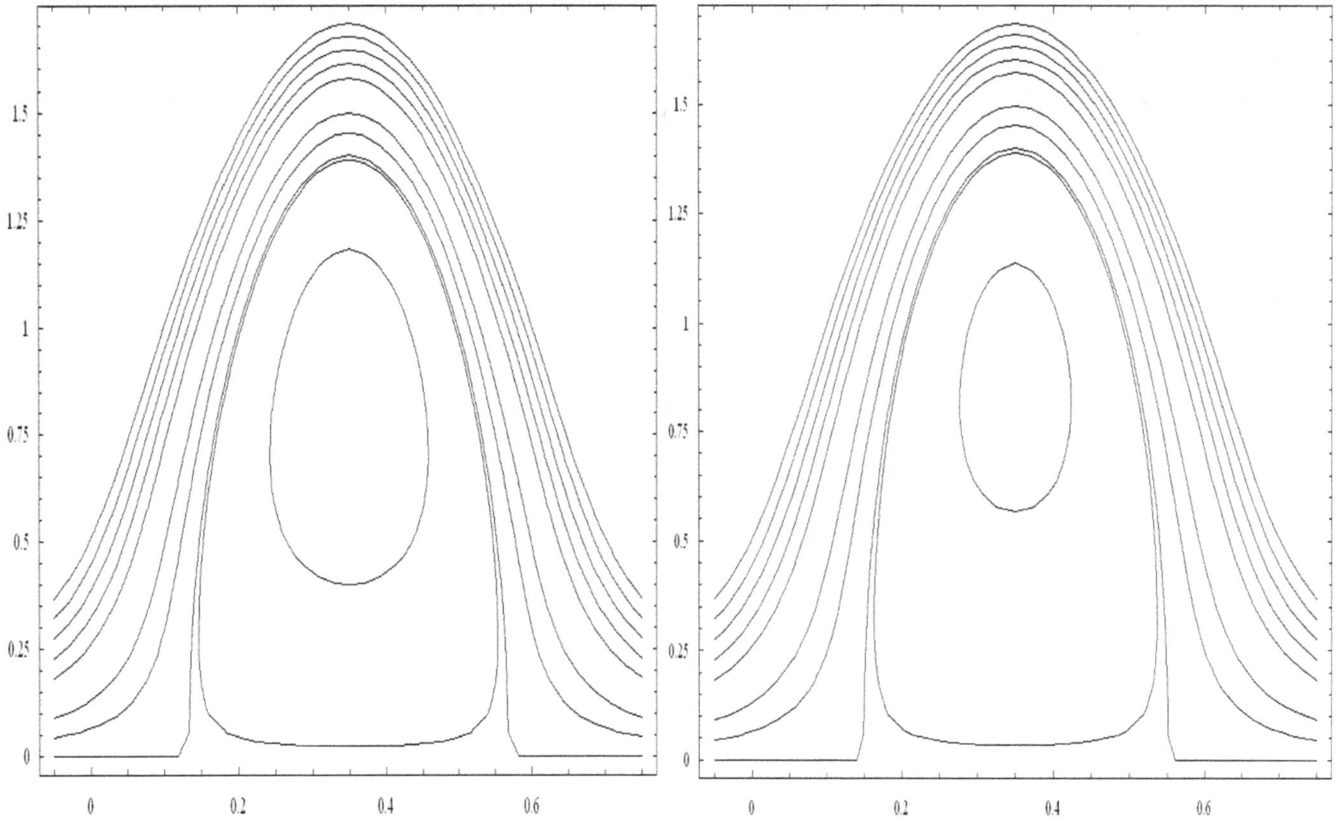

Figure 5a. Streamlines (sinusoidal wave) for $M = 1, M = 2$.

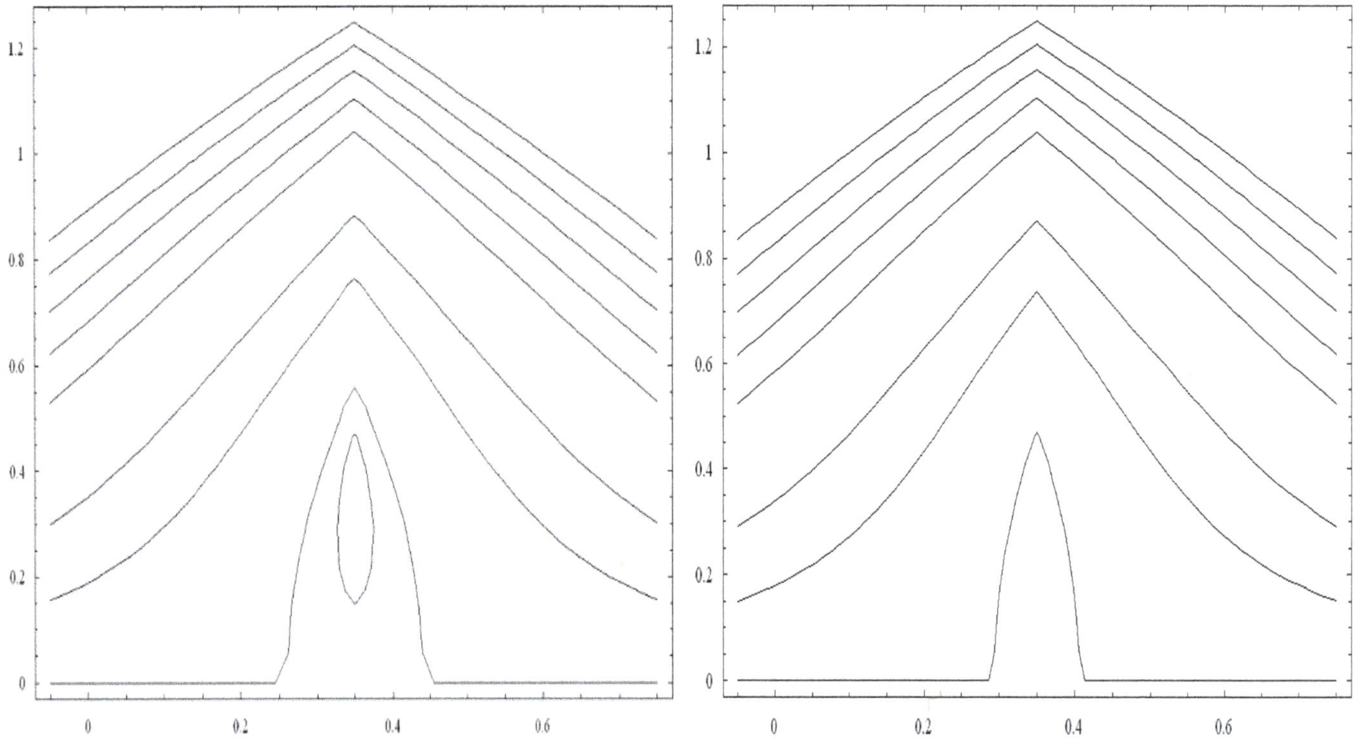

Figure 5b. Streamlines (triangular wave) for $M = 1, M = 2$.

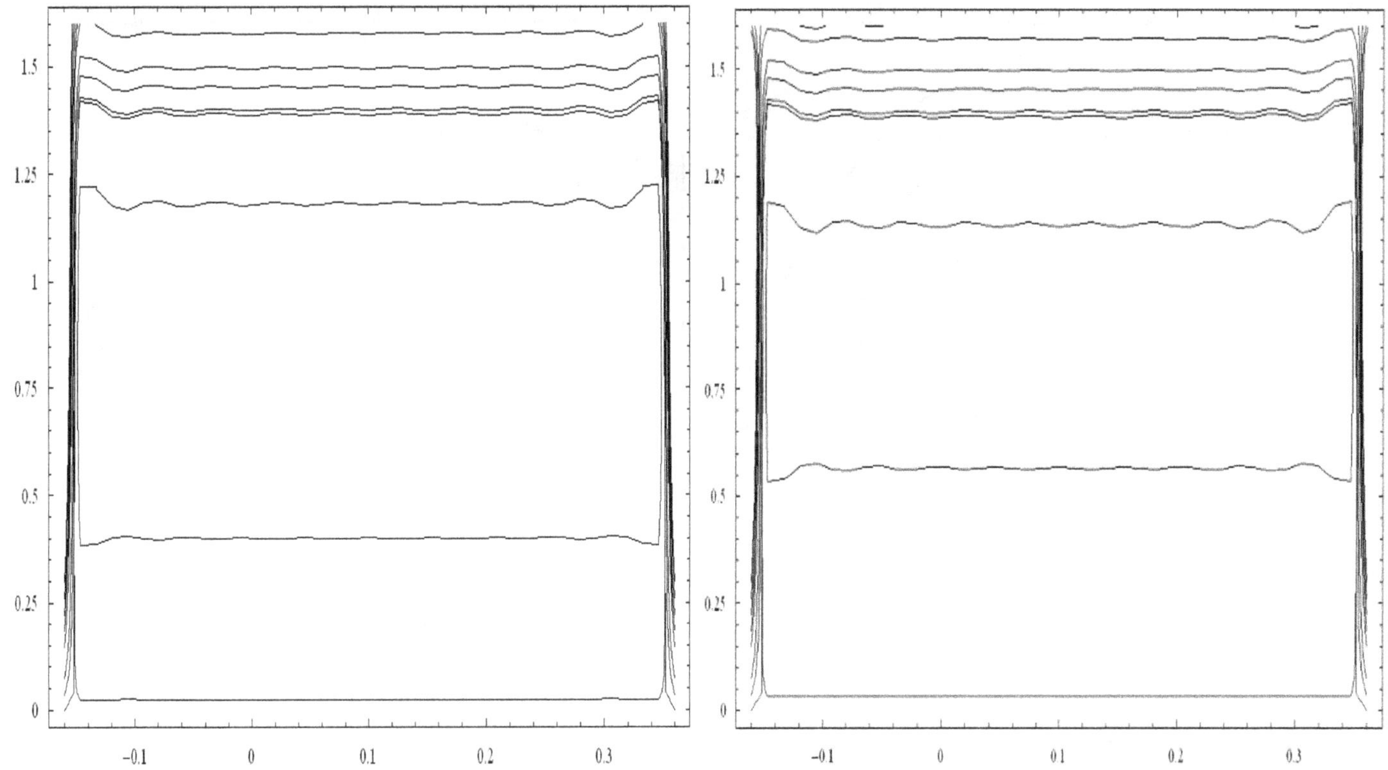

Figure 5c. Streamlines (square wave) for $M = 1, M = 2$.

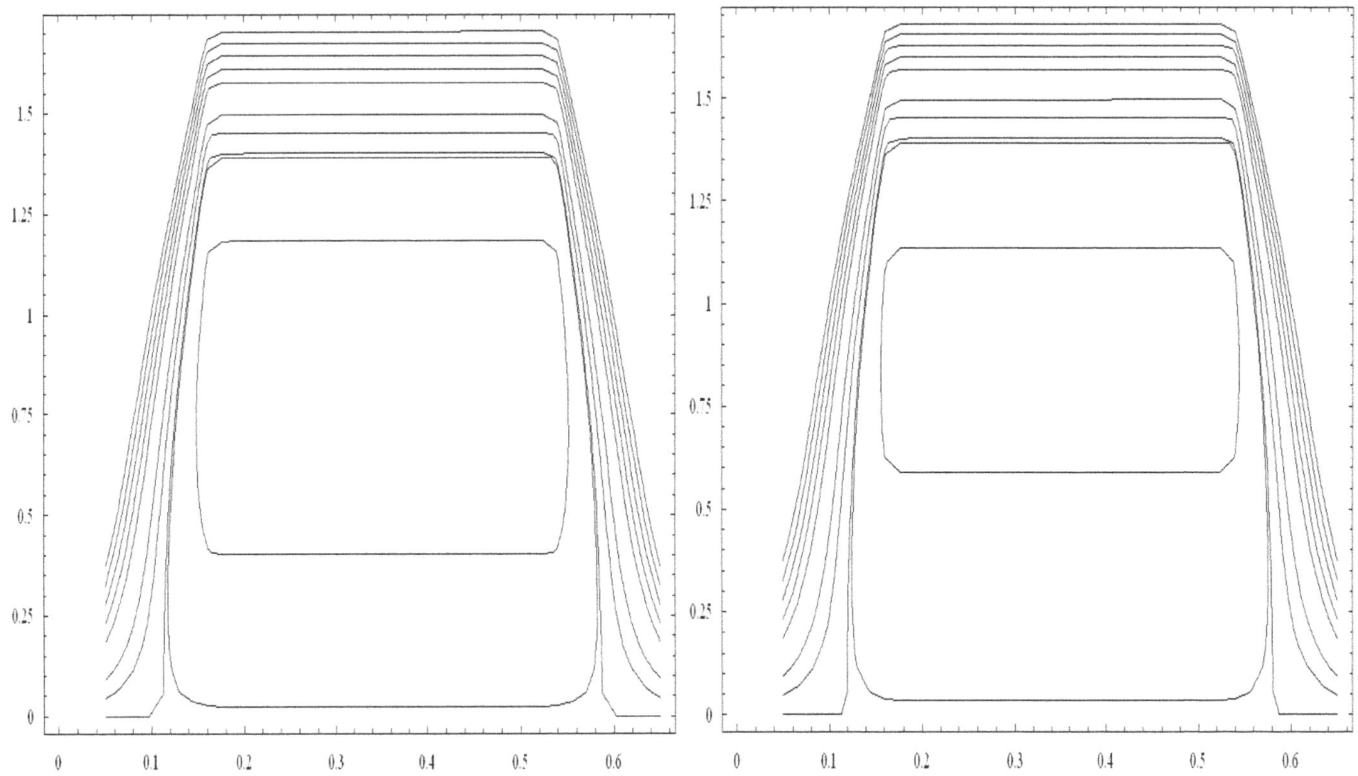

Figure 5d. Streamlines (trapezoidal wave) for $M = 1, M = 2$.

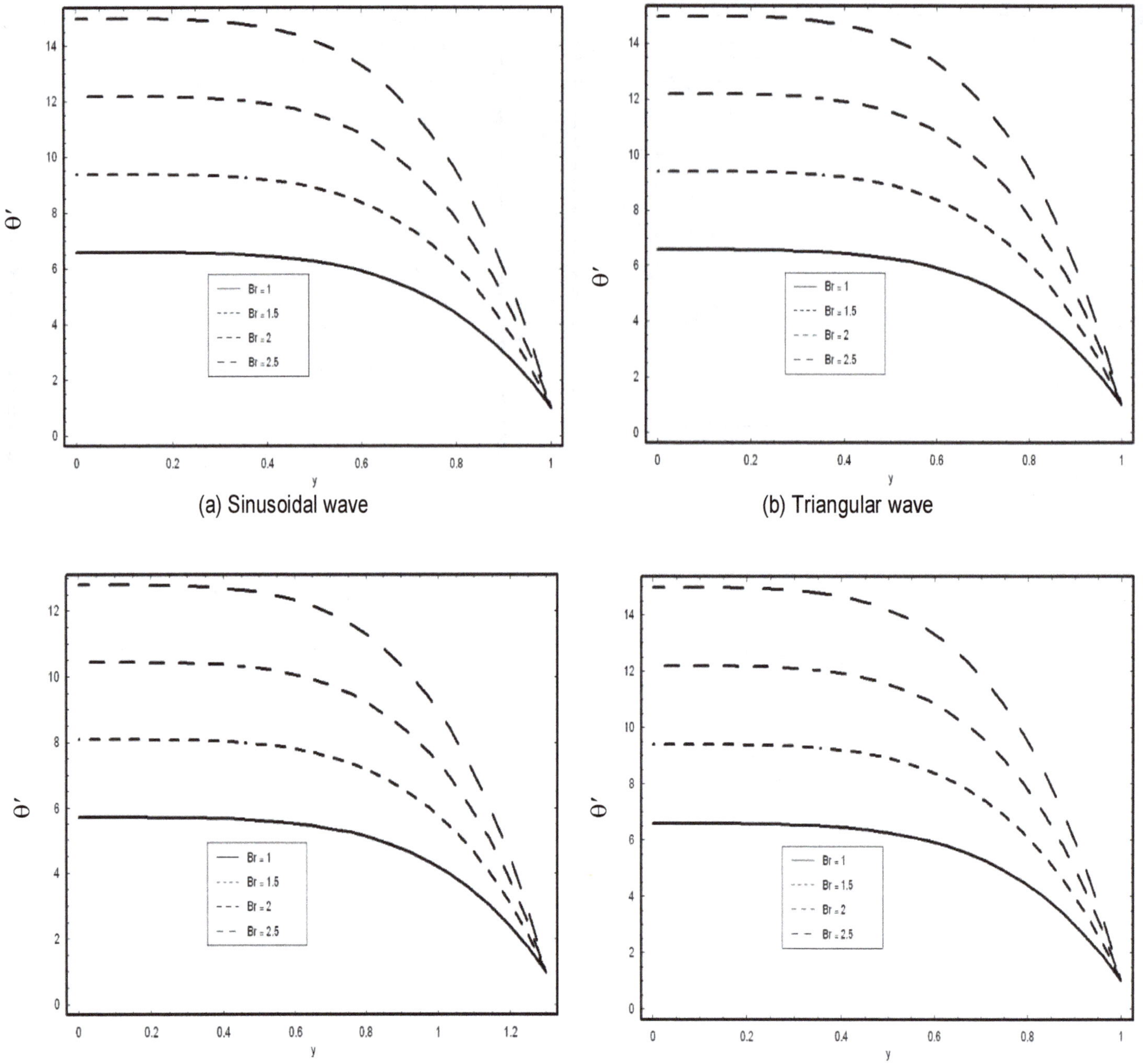

(a) Sinusoidal wave (b) Triangular wave

(c) Square wave (d) Trapezoidal wave

Figure 6. Variation of θ' versus y. Here $K = 0.005,\ \phi = \frac{\pi}{6},\ E = 0.03,\ t = 0.1,\ \beta = 0.03,\ \lambda_1 = 0.4, M = 0.8,\ R_m = 1, a = b = 0.4, d = 1.1$ and $x = 0.1$

momentum and mass diffusion convection processes. Since Sc has a direct relation with mass diffusion rate therefore increases for small values of Sc. It is observed that such decrease is maximum for square wave form.

ACKNOWLEDGEMENT

The research of Ahmed Alsaedi was partially supported by Deanship of Scientific Research (DSR), King Abdulaziz University, Jeddah, Saudi Arabia.

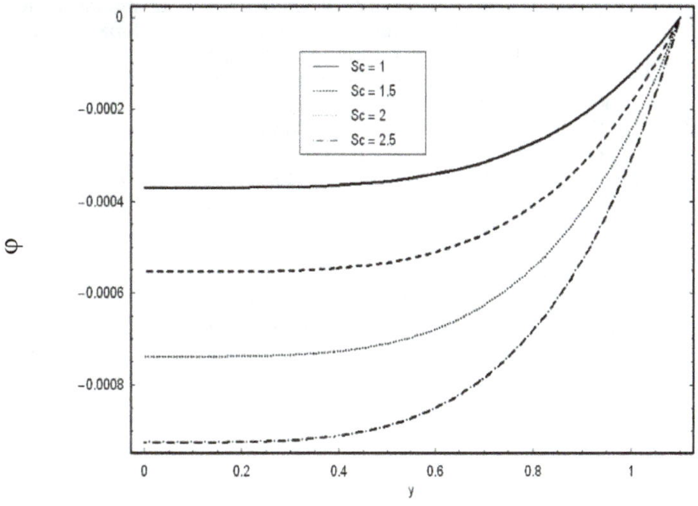

(a) Sinusoidal wave

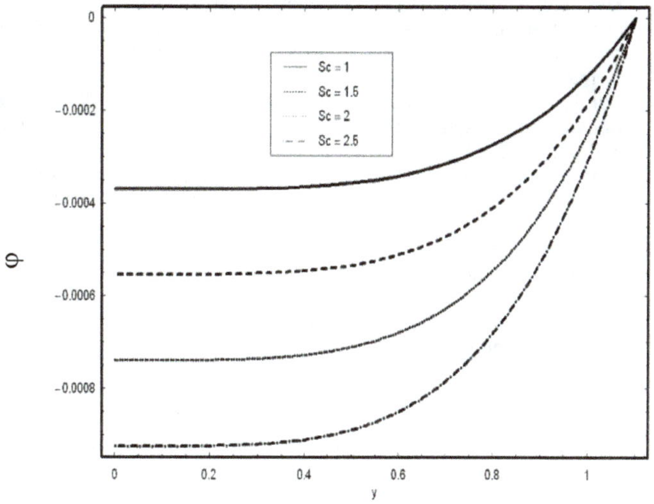

(b) Triangular wave

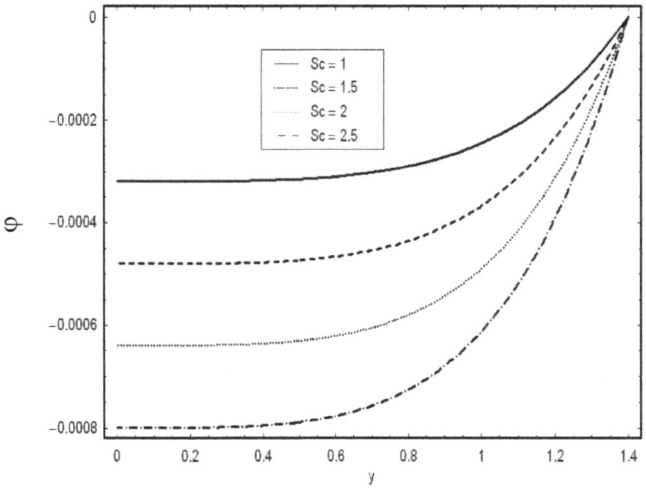

(c) Square wave

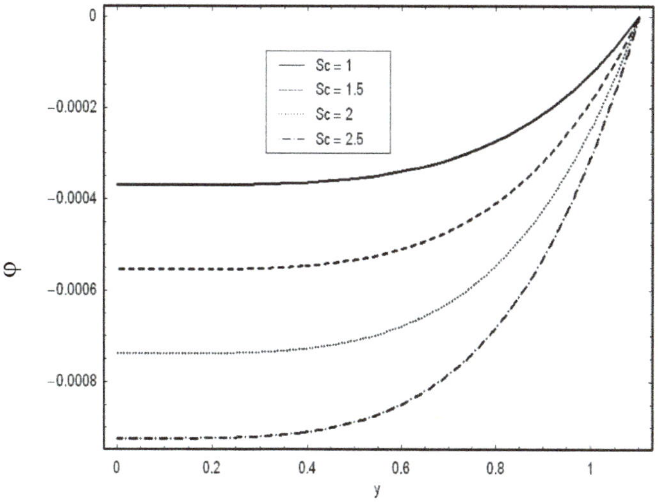

(d) Trapezoidal wave

Figure 7. Variation of φ versus y. Here $K = 0.005,\ \phi = \frac{\pi}{6},\ E = 0.03,\ t = 0.1,\ \beta = 0.03,\ \lambda_1 = 0.4, M = 0.8,\ R_m = 1, \theta = 2, Sc = Sr = 1,\ a = b = 0.4, d = 1.1$ and $x = 0.1$

REFERENCES

Abd elamboud Y, Mekheimer KhS (2011). Nonlinear peristaltic transport of a second order fluid through a porous medium. Appl. Math. Modelling, 35: 2695-2710.

Ali N, Hussain Q, Hayat T, Asghar S (2008). Slip effects on the peristaltic transport of MHD fluid with variable viscosity. Phys. Lett. A, 372: 1477-1489.

Abd elamboud Y (2011). Influence of induced magnetic field on peristaltic flow in an annulus. Comm. Nonlinear Sci. Numer. Simulation, in press.

Akbar NS, Hayat T, Nadeem S, Hendi AA (2011). Effects of slip and heat transfer on peristaltic flow of a third order fluid in an inclined asymmetric channel. Int. J. Heat and Mass Transfer, 54: 1654-1664.

Coleman BD, Markovitz H, Noll W (1966). Viscometric flows of non-Newtonian fluids. Springer- Verlag, Berlin, Heidelberg, New York.

Ellahi R (2009). Steady and unsteady flow for Newtonian and non-Newtonian fluids: Basics, concepts and methods. VDM, Germany.

Hayat T, Javid M, Hendi AA (2011a). Peristaltic transport of viscous fluid in a curved channel with compliant walls. Int. J. Heat Mass Transf., 54: 1615-1621.

Hayat T, Saleem N, Ali N (2010a). Peristaltic flow of a Carreau fluid in a channel with different wave forms. Numerical Methods for Partial Differential Equations, 26: 519-534

Hayat T, Hussain Q, Ali N (2008a). Influence of partial slip on the peristaltic flow in a porous medium. Physica A, 387: 3399-3409.

Hayat T, Qureshi MU, Ali N (2008b). The influence of slip on the peristaltic motion of a third order fluid in an asymmetric channel. Phys. Lett. A, 372:2653-2664.

Hayat T, Saleem N, Mesloub S, Ali N (2011b). MHD flow of a Carreau fluid in a channel with different wave forms, Zeitscrift fur Naturforschung A, 66a: 15-222.

Hayat T, Khan Y, Ali N, Mekheimer KhS (2010b). Effect of an induced magnetic field on the peristaltic flow of a third order fluid. Numerical

Methods for Partial Differential Equations, 26:345-366.

Hayat T, Saleem N and Ali N (2010c). Effect of induced magnetic field on peristaltic transport of a Carreau fluid, Commun. Nonlinear Sci. Numer. Simul., 15: 2407-2423.

Hayat T, Hina S, Ali N (2010d). Simultaneous effects of slip and heat transfer on peristaltic flow. Commun. Nonlinear Sci. Numer. Simul., 15: 1526-1537.

Kotkandapani M, Srinivas S (2008). Peristaltic transport of a Jeffrey fluid under the effect of magnetic field in an asymmetric channel. Int. J. Non-Linear Mech., 43: 915-924.

Mekheimer KhS (2008a). Effect of the induced magnetic field on peristaltic flow of a couple stress fluid. Phys. Lett. A, 372: 4271-4278.

Mekheimer KhS (2008b). Peristaltic flow of a magneto-micropolar fluid: Effect of induced magnetic field. J. Appl. Math., 2008: 1-23.

Nadeem S, Akaram S (2010). Heat transfer in peristaltic flow of MHD fluid with partial slip. Commun. Nonlinear Sci. Numer. Simul., 15: 312-321.

Reddy MS, Raju GSS (2010). Nonlinear peristaltic pumping of Johnson-Segalman fluid in an asymmetric channel under effect of a magnetic field. European J. Sci. Res., 46:147-164.

Sanyal DC, Das K, Debnath S (2007). Effect of magnetic field on pulsatile blood flow through an inclined circular tube with periodic body acceleration. Int. J. Phys. Sci. 11: 43-56.

Singh J, Rathee R (2010). Analytical solution of two-dimensional model of blood flow with variable viscosity through an indented artery due to LDL effect in the presence of magnetic field. Int. J. Phys. Sci., 5: 1857-1868.

Singh J, Rathee R (2011). Analysis of non-Newtonian blood flow through stenosed vessel in porous medium under the effect of magnetic field. Int. J. Phys. Sci., 6: 2497-2506.

Srinivas S, Gayathri R, Kothandapani M (2009). Influence of slip condition, wall properties and heat transfer on MHD peristaltic transport. Computer Physics Communications, 180: 2115-2122.

Tripathi D (2011). A mathematical model for the peristaltic flow of chyme movement in small intestine. Math. Biosciences, in press.

2

Synthesis and use of hydrotalcites as heat stabilisers in thermally processed powdered polyvinylchloride (PVC)

A. Bissessur[1] and M. Naicker[2]

[1]School of Chemistry and Physics, University of KwaZulu-Natal, Private Bag X54001, Durban, 4000 South Africa.
[2]Department of Chemistry, University of South Africa, P. O. Box 392, Muckleneuk Ridge, City of Tshwane, 0003, South Africa.

Lead-based stabilisers, such as lead sulphate and lead stearate, currently used in the processing of polyvinylchloride (PVC) products are being substituted with more environmentally safe, economically viable and higher quality materials. This study aimed to use hydrotalcites (HTs), which are environmentally non-hazardous and simple to synthesise, to effectively quench the HCl gas evolved during the degradation of PVC. From the results obtained in this study, the effects of varying the metal ion ratios in Mg/Zn/Al-HTs showed many similarities with some variations. The synthesised Mg/Al-HTs, viz. $Mg_4Al_2(OH)_{12}CO_3.3H_2O$ and $Mg_6Al_2(OH)_{16}CO_3.4H_2O$, displayed the most superior stability properties, with the latter being the best PVC stabiliser, over the other HTs synthesised. The most important test conducted in this study, demonstrated the synthesised HTs ability to quench the HCl gas evolved as a result of dehydrochlorination, which occurs during the processing of PVC to useful polymeric products. Results from this study also confirmed that approximately 0.2 g of synthesised Mg/Al-HTs adsorbed 0.2 μmol of the HCl gas evolved, thus most effectively delayed the onset of degradation of powdered PVC (~10 g) than other HTs synthesised, including the non-stabilised PVC. It is thus evident that Mg/Al-HTs can function as effective heat stabilisers during the processing of PVC, with the potential of inhibiting the degradation of PVC, especially during the evolution of HCl gas.

Key words: Hydrotalcites, polyvinylchloride, thermal gravimetric analysis (TGA), dehydrochlorination, Brunauer–Emmett–Teller (BET) surface area, crystalline, amorphous.

INTRODUCTION

Lead compounds continue to assume an important role among stabilisers, especially in PVC derived products. Heat stabilisers are applied to polyvinylchloride (PVC) processing in order to prevent or minimise thermal instability, which is autocatalytic and leads to polymers adhering to the equipment (Gupta et al., 2008), discolouration, brittleness and insolubility. Common heat stabilisers used include metal salts, metal soaps and organometallic compounds. The toxicity and affinity for environmental pollution, renders these types of stabilisers harmful, therefore alternatives are required.

During the processing of PVC, stabilisers and additives are mixed and heated to form a solid, which can be subsequently moulded into a range of useful household and industrial products, examples: raincoats, toys, food packaging and conduits. When PVC is heated, chlorine and hydrogen atoms in the molecules are eliminated, and the release of HCl gas (dehydrochlorination) becomes apparent. This is an undesirable by-product that can negatively affect the properties and quality of the end products. Therefore it is vital that the evolution of hydrogen chloride be prevented, by using metal

$$PVC(s) \xrightarrow{\text{dehydrochlorination}} PVC(s) + HCl(g) \longrightarrow$$

$$Pb(C_{17}H_{35}COO)_2\,(s)$$
$$+$$
$$C_{17}H_{35}COOH(s)$$

$$PbCl_2(s) + C_{17}H_{35}COOH(s) \xleftarrow[\displaystyle C_{17}H_{35}COOH(s)]{\displaystyle HCl(g)} [C_{17}H_{35}COO\text{-}Pb\text{-}Cl]$$

chloridostearate intermediate

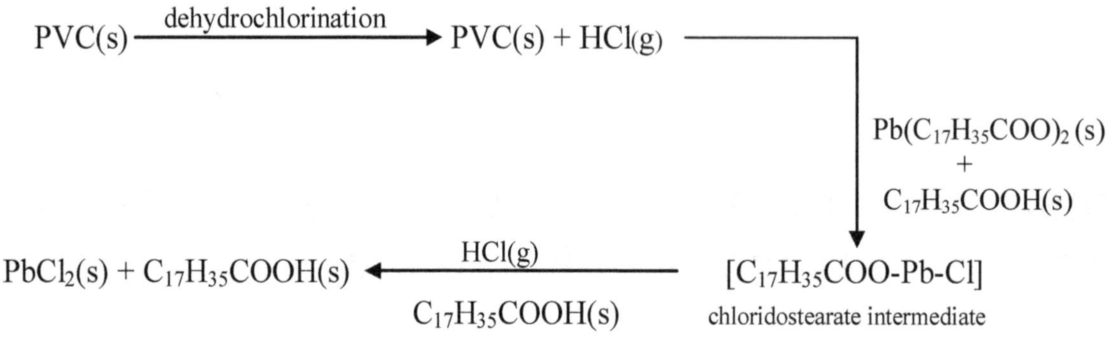

Scheme 1. Behaviour of lead stearate during the degradation of PVC.

Brucite-like sheet (cations)

Interlayer region (anions and water)

$[M^{z+}_{1-x}M^{3+}_x(OH)_2]^{b+}[A^{n-}_{b/n}] \cdot mH_2O$

Gallery height

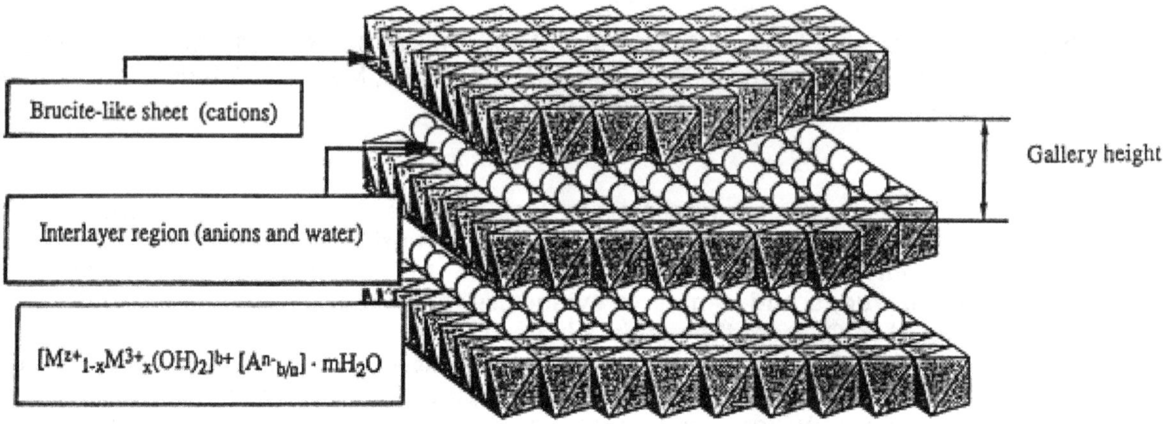

Figure 1. Layered structure of hydrotalcites with rhombohedral symmetry.

compound stabilisers.

A current additive, such as lead stearate $[Pb(C_{17}H_{35}COO)_2]$, is used to quench the evolved HCl gas, which results in the formation of lead (II) chloride, as illustrated in Scheme 1, where lead chloridostearate was formed as an intermediate product and stearic acid served as a catalyst (Kalouskova et al., 2004). Lead stearate readily reacts with the evolved HCl gas and favourably affects the stability of PVC. However, the main concern is that the $PbCl_2$ formed as a product can leach into the environment as the polymer degrades. This is extremely harmful to the environment and the public because exposure to $PbCl_2$ is a common source of lead poisoning. This is another reason why alternatives need to be investigated. In this study, the use of hydrotalcites (HTs) as a lead replacement stabiliser in PVC processing was investigated. The general formula for HTs is represented by Equation 1, where M^{2+} and M^{3+} represent divalent and trivalent cations respectively (Morioka et al., 1995; Borja and Dutta, 1992; Lin et al., 2006; Yong and Rodrigues, 2002; Tong et al., 2011).

$$[M^{2+}_{1-x}M^{3+}_x(OH)_2](A^{n-})_{x/n} \cdot mH_2O \qquad (1)$$

The synthesis of HTs involves the formation of positively charged (cations) layers and filling the interlayer region with anions. During this process, the initial solution contains carbonate and hydroxide ions. The metal ions are later added to this solution and bonds to the hydroxide ions to form the metal-hydroxide layers (indicated as 'brucite-like sheet' in Figure 1).

According to Gupta et al. (2008, 2009), the reaction between an Mg/Al-based HT and HCl gas is shown in Scheme 2. In comparison to lead stearate in Scheme 1, no harmful products are formed when an Mg/Al-HT reacts with the evolved HCl gas. This shows that HTs are more favourable (in terms of environmental health and safety) to use as stabilisers in the manufacturing of PVC polymeric products. In order to determine how HTs enhance the stability of PVC during degradation, synthesised HTs were used to quench HCl gas evolved as a result of thermal processing of PVC. Under conditions of elevated temperatures and UV radiation, PVC undergoes dehydrochlorination, where the evolved HCl gas is a polar covalently bonded gas. Signs of degradation during dehydrochlorination are apparent as severe discolouration of the PVC sample occurs. In this

$$PVC\ (s) \xrightarrow{\text{dehydrochlorination}} PVC(s) + HCl\ (g) \longrightarrow$$

$$Mg\text{-}A\ell\text{-}CO_3.mH_2O$$

$$Mg\text{-}A\ell_2Cl_2CO_3.mH_2O(s) + H_2O(\ell) + CO_2(g)$$

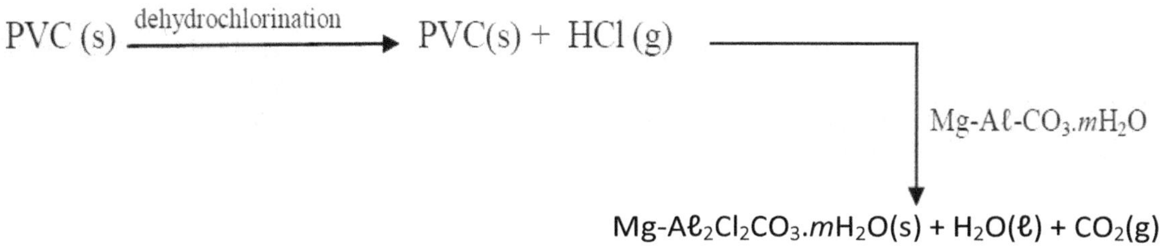

Scheme 2. Behaviour of a Mg/Al-Hydrotalcite during the degradation of PVC.

study, results from the dehydrochlorination and thermogravimetric analysis performed on the PVC and HTs, will show that the synthesised HTs (as heat stabilisers) can inhibit the degradation of PVC by quenching the evolved HCl gas.

Lin et al. (2005) in Gupta et al. (2008), van der Ven et al. (2000) and Kalouskova et al. (2004) have also used HTs as heat stabilisers in PVC processing and suggested that a possible reason for its stabilising properties is due to ion exchange, where CO_3^{2-} ions are replaced by the absorbed Cl^- ions. Contrary to Lin et al. (2005) and Kalouskova et al. (2004) findings, Gupta et al. (2008, 2009) reported that besides ion exchange, adsorption of the evolved HCl gas by HTs is a major factor in the stabilisation of PVC.

The phenomena of adsorption can only be explained, according to Gupta et al. (2009), if the chloride ions on the surface of the HT are adsorbed, instead of exchanged with the carbonate ions. A possible mechanism for scavenging the evolved HCl gas using HTs was suggested by Gupta et al. (2008, 2009). Ion exchange is likely to occur if the chlorine atoms from the HCl are converted into Cl^- ions, however in the gaseous state this is least likely to occur. As a result, it is the surface adsorption of HCl gas onto the HTs that plays a significant role in heat stabilisation.

The research undertaken aims to explore the stabilising effect on PVC products during thermal processing using synthetic hydrotalcites.

MATERIALS AND METHODS

Synthesis of hydrotalcites

The metal solution, containing 0.75 M magnesium nitrate, zinc nitrate and/or aluminium nitrate, was transferred to the carbonate solution, containing 0.5 M sodium carbonate, using a burette at room temperature. This was conducted at a flow rate of ~1.0 mL/min under constant stirring. The pH was maintained between 11 and 12, by adding 2 M NaOH. Once the transfer was complete, the resulting mixture was stirred for 1.5 h at room temperature. The resulting metal/carbonate mixture was refluxed at 80°C, in an oil bath for 18 h, cooled to room temperature and filtered under a vacuum. The precipitate obtained was rinsed several times with deionised water and dried overnight at 110°C.

Characterisation

FT-IR spectroscopy

All FT-IR spectra were recorded on a Perkin Elmer Spectrum 100 FT-IR Spectrometer with a universal ATR sampling accessory. The spectrum was recorded in the range of 4000-400 cm^{-1}.

Chemical analysis

To analyse the metal ions using ICP-OES, into a 50 mL centrifuge tube, 200 mg of each HT was accurately weighed and to it, 10 mL of aqua regia (3 parts HCl, 1 part HNO_3) was added. These tubes were sealed and placed on an orbital shaker for 2 h, centrifuged for 5 min then decanted into 100 ml volumetric flasks. The samples were filtered (by gravity) as it passed into the flasks and dilute to 100 mL with deionised water. These samples were prepared in triplicate and were analysed by a Perkin Elmer Optical Emission Spectrometer. The percentage of carbon, hydrogen and nitrogen present in the HTs was determined by accurately weighing approximately 2 mg of HT into a pre-shaped casing, closed and pressed into a pellet then analysed by a LECO CHNS-932 elemental analyser.

Physical analysis: PXRD, Surface area and pore volume

Powder X-Ray Diffraction (PXRD) studies was conducted on a Bruker AXS Diffraktometer D8, at room temperature, where a PXRD pattern with d-values was obtained using a scan range between 10 and 90° 2θ (theta) and a step width of 0.014°. The unit cell values were determined by the following equation:

$$\sin^2\theta = (\lambda^2/4a^2) \times N^{12} \tag{2}$$

Determination of surface area and pore volume of the HTs was obtained using a Perkin Elmer instrument with Gemini 2375 V5.00 software.

Thermogravitmetric analysis

This analysis was performed by a SDT Q600 Thermogravimetric analyser, under nitrogen (100 mL/min) from 50 to 700°C at a heating rate of 10°C/min, where approximately 20 mg of sample was required. The HT and PVC powder were analysed individually, followed by the PVC-HT samples prepared by mixing 10 g of PVC powder and 200 mg of the synthesised HTs. From this analysis, a thermogram was obtained illustrating the sample's weight loss, derivative weight and heat flow as a function of temperature.

Dehydrochlorination

This is a thermal stabilising efficiency test, where the efficiency of the HT's ability to quench the evolved HCl was determined. Samples were placed in a round bottom flask and heated at 180°C in an oil bath, under a steady flow of nitrogen (60 mL/min). The HCl gas evolved was collected in a 100 mL Erlenmeyer flask, which was placed at the end of the condenser containing 50 ml of 4 μM NaOH solution with phenolphthalein indictor. The time taken for 0.2 μmol of HCl gas to evolve was recorded.

RESULTS AND DISCUSSION

The method for synthesising HTs is a well-established technique, and in this study the co-precipitation method was employed, which yielded a white precipitate in all cases. Our laboratory yields for all HTs ranging from 17 to 24% are considered low in comparison to yields recorded by others ranging from 80 to 90% (Mohan et al., 2005). According to Kloprogge et al. (2004a), HTs synthesised at pH 12 are more highly crystalline than the HTs synthesised at pH values other than 12. Thus, pH 12 was used as a standard bench mark for the synthesis of HTs in this study. However, according Vaccari et al. (1998), a pH range of 8-10 is also considered suitable for the synthesis of HTs. It is therefore possible that should a pH range of 8-10 be used, a higher yield of HTs would result, consequently, with a possible variation in the degree of crystallinity.

The use of FT-IR spectroscopy (generally used to identify the functional groups present in a HT structure composed of two or three metals) showed a broad band at ~3400 cm^{-1}, which is assigned to the stretching of the O-H bonds in the brucite-like sheets (Figure 1) in all six samples. However, a shift of this OH band to a higher wavenumber with increasing M^{2+}/M^{3+} ratios (HT 2 > HT 1 and HT 4 > HT 3) was observed. This is possibly due to a decrease in ionic radius across a period (from left to right) and an increase down a group (Zn^{2+} = 0.74 Å, Mg^{2+} = 0.65 Å, Al^{3+} = 0.50 Å). As a result, the electrostatic forces between Al^{3+} and OH^- (1:3) are stronger than that between Mg^{2+} or Zn^{2+} and OH^- (1:2). As the metal/hydroxide ion ratios decrease, less electrostatic forces are present, thus resulting in the hydroxide ions vibrating at higher wavenumbers.

As the number of metal ions in synthesised HTs were increased, the IR bands were observed at higher wavenumbers (Mg/Zn/Al-HT > Mg/Al-HT > Zn/Al-HT). Although Zn has a greater ionic radius than Mg, it is a transition metal with a completely filled d-orbital. Thus the electrostatic forces between Zn^{2+} and OH^- would be stronger than Mg^{2+} and OH^-, resulting in the O-H stretch of Zn-containing HTs vibrating at a slightly lower wavenumber than the initially observed 3400 cm^{-1}. The band observed at 3459.19 cm^{-1} was due to HT 5, which contained more Mg ions than Zn. All synthesised HTs showed two weak bands occurring as shoulders at ~2900-3000 cm^{-1} and ~1620 cm^{-1} respectively (as a result of the low percentage yields). The first according to Kloprogge and co-workers et al. (2004b) is due to the CO_3-H_2O bridging vibrations, and the second is assigned to the bending modes of H_2O caused by the hydrogen bonding between the molecules.

Table 1. Ratios of metals ions present in HTs synthesised determined by ICP-OES.

HT	Calculated ratios (Mg^{2+}:Zn^{2+}:Al^{3+})	Experimental ratios (Mg^{2+}:Zn^{2+}:Al^{3+})
1	4 : 0 : 2	4 : 0 : 2
2	6 : 0 : 2	6 : 0 : 2
3	0 : 4 : 2	0 : 5 : 1
4	0 : 6 : 2	0 : 7 : 1
5	4 : 2 : 2	4 : 3 : 1
6	2 : 4 : 2	2 : 5 : 1

The carbonate ions found in the interlayer region of the HT structure shown in Figure 1 is indicative of a sharp intense band at ~1360 cm^{-1} due the antisymmetric stretching. For all synthesised HTs, three FT-IR bands were observed in the fingerprint region at ca. 550 cm^{-1}, ca. 630 cm^{-1} and ca. 770 m^{-1}. These bands are typically found in HTs and are due mainly to the lattice and translational vibrations of the metal ions, or the flexural vibrations of CO_3^{2-} (that is, 'rocking' of the molecule) (Seftel et al., 2005). A sharp band observed at approximately 440 cm^{-1}, is due to the lattice vibration of Al^{3+} which was observed in HT 1 and 6 only. Hence, it can be deduced that all relevant functional groups for example, CO_3^{2-} and OH^- are present in HTs synthesised. Also the addition of Zn to the Mg/Al-HT did not result in any major variation in the IR bands observed.

The concentration of metal ions in HTs was determined from the relative intensities of metal ions using ICP-OES. The calculated and experimental values for the metal ion ratios in Mg/Al-HTs shown in Table 1 were found to be the same. However, laboratory synthesised Zn-containing HTs are found to contain one more Zn atom than the calculated amount. This results in one less Al atom being present in the Zn-containing HTs. It was suggested by Rajamathi et al. (2001) that Zn^{2+} ions occupy the tetrahedral as well octahedral sites in the layered structure. This can lead to competition between the Al^{3+} and Zn^{2+} ions for the tetrahedral sites in the structure.

From the C, H, N instrumental analysis, nitrogen, as expected, was absent in all HTs synthesised. This indicates that all nitrates were successfully removed from the precipitate during the washing stages using deionised water. Applying the general formula for HTs (Equation 1) theoretically and calculating it experimentally from C, H, N analysis, all synthesised HTs were found to contain one carbon atom. The variation of the number of hydrogen atoms in each HT, based on predicted and calculated results indicate that the amount of water

Table 2. Chemical formulae of synthesised HTs determined theoretically and experimentally.

HT	Calculated chemical formula	Actual chemical formula
1	$Mg_4Al_2(OH)_{12}CO_3.3H_2O$	$Mg_4Al_2(OH)_{12}CO_3.4H_2O$
2	$Mg_6Al_2(OH)_{16}CO_3.4H_2O$	$Mg_6Al_2(OH)_{16}CO_3.4.5H_2O$
3	$Zn_4Al_2(OH)_{12}CO_3.3H_2O$	$Zn_5Al(OH)_{12}CO_3.3H_2O$
4	$Zn_6Al_2(OH)_{16}CO_3.4H_2O$	$Zn_7Al(OH)_{16}CO_3.3.5H_2O$
5	$Mg_4Zn_2Al_2(OH)_{12}CO_3.3H_2O$	$Mg_4Zn_3Al(OH)_{12}CO_3.8H_2O$
6	$Mg_2Zn_4Al_2(OH)_{12}CO_3.3H_2O$	$Mg_2Zn_5Al(OH_{12}CO_3.6.5H_2O$

molecules in the interlayer region does not correlate to the expected values. Therefore from chemical analyses, the synthesised HTs were found to have an experimentally determined chemical formula (Table 2) that varied from the theoretically calculated formula.

In order to establish which HT that has the potential to function as an effective heat stabiliser for PVC, its physical properties needs to be investigated and considered. The crystalline or amorphous structural identification of HTs were investigated by PXRD (Figure 2a-f). In addition, the surface area and pore volume, from BET surface area analysis, is a direct indication of the synthesised HTs ability to quench the HCl gas.

According to Bastiani et al. (2004), broad and narrow symmetrical peaks observed in PXRD for HTs is a characteristic feature of a layered crystal structure HT. The PXRD patterns (Figure 2) obtained for all synthesised HTs in this study showed a similar trend thus confirming the formation of a layered crystal structure. The unit cell values calculated (Equation 2) for HTs synthesised in this study correlated to the literature value of 3.054 Å (Cavani et al., 1991). This confirms the rhombohedral symmetry in the layered structures of Mg/Al-HTs, with some variations in the Zn-containing HTs. The variations in the Zn-containing HTs may be attributed to a large cationic distance between Zn^{2+} ions. Consequently, the Zn-containing HTs were found to have a smaller pore volume than the Mg/Al-HTs, which was confirmed by BET surface area analysis (Table 3). Kloprogge et al. (2004) suggested that the higher the degree of crystallinity of the HT structure, the higher its thermal stability. However, the degree of crystallinity of the synthesised HTs using data obtained from PXRD patterns was uncertain.

The Mg/Al-HTs were found to have a larger surface area and pore volume, in comparison to the other HTs synthesised. From the BET surface area results, it is observed that the surface area and pore volume of the Mg/Al-HTs are much larger than the Zn-containing HTs. This suggests that the Mg/Al-HTs can act as an effective adsorbent, which is consistent with its industrial applications, and would be an ideal heat stabiliser for PVC. Therefore from the physical properties investigated, the Mg/Al-HTs are most favourable to show an inhibition of the degradation of PVC during thermal gravimetric analysis (TGA) and dehydrochlorination.

From the TGA analysis, thermograms showed three distinct (temperature ranges 25-230°C, 230-450°C and 450-700°C) of weight loss. Since all synthesised HTs are highly crystalline (according to the PXRD results) and comprised of the same anions, any variation in the temperature range was therefore due to the different metal ion ratios. This was suggested by Rao et al. (2005) who indicated that HTs generally undergo two weight losses around 270 and 450°C, which depend qualitatively and quantitatively on many factors such as crystallinity, type of anions (Borja and Dutta, 1992), as well as nature and relative amounts of cations (Vaccari, 1998). Two sample weight losses are evidently present in the thermograms (Figure 3) for Mg/Al-HTs, verifying the formation of the layered structure illustrated in Figure 1. The thermograms for the Zn-containing HTs showed only one major loss in weight between 25-230°C, while PVC decomposed completely with two major weight losses between 230-500°C (Figure 4).

According to TGA-MS studies on HTs, the temperature at which the weight losses occur, suggest that the first weight loss corresponds to the loss of water molecules (Gupta et al., 2009; Lin et al., 2005; Mohan et al., 2005; Kloprogge et al., 2004a) present in the interlayer region. Upon further heating the second weight loss (subsequent to the first weight loss) involve dehydroxylation of the brucite-like sheets (Figure 1), as well as the removal of the carbonate anions as confirmed by Kloprogge et al. (2004a). The derivative weight loss curves for HTs 3, 4, 5 and 6, show only one major mass change, which resulted in a low total weight loss (reported in Table 4), as suggested by Cavani et al. (1991), who found that the carbonate ions can be removed from the interlayer without destroying the structure of the HTs.

Further results from TGA showed that HTs 1 and 2 were most stable compared to the rest of them. It was found that HTs 1 and 2 decomposed at 250 and 225°C respectively, while HTs 4 and 5 decomposed at lower temperature of 205°C. A higher thermal stability of HTs 1 and 2 is inherent arising out of their composition made up mainly of $Mg(OH)_2$ and $Al(OH)_3$ (which decompose at 350 and 300°C, respectively). As a result, the Mg/Al-HTs are mostly likely to be better heat stabilisers than Zn-containing HTs (comprising mainly of $Zn(OH)_2$ which decomposes at 125°C) (Cavani et al., 1991), as indicated by BET surface area analysis. Thermograms generated for HTs 1 and 2 were the only compounds which corresponded to data reported by Rao et al. (2005), Gupta et al. (2009), Lin et al. (2006) and Kloprogge et al. (2004) thus confirming high thermal stability over other HT's synthesised. Therefore based on derivative weight loss, HTs in order of increasing thermal stabilities are:

HT 4 ≈ HT 5 < HT 3 < HT 6 < HT 2 < HT 1

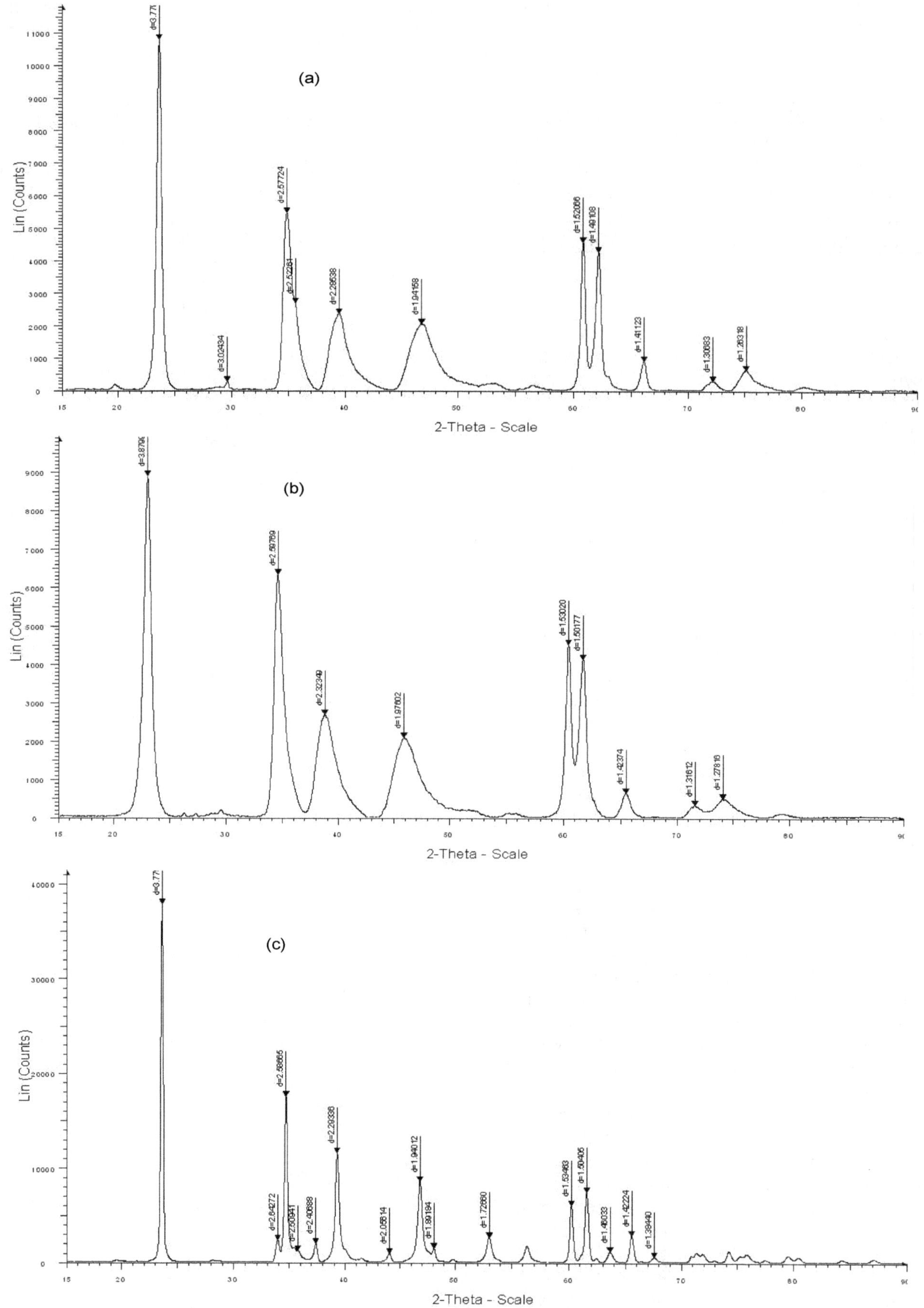
(a)
Lin (Counts)
2-Theta - Scale
(b)
Lin (Counts)
2-Theta - Scale
(c)
Lin (Counts)
2-Theta - Scale

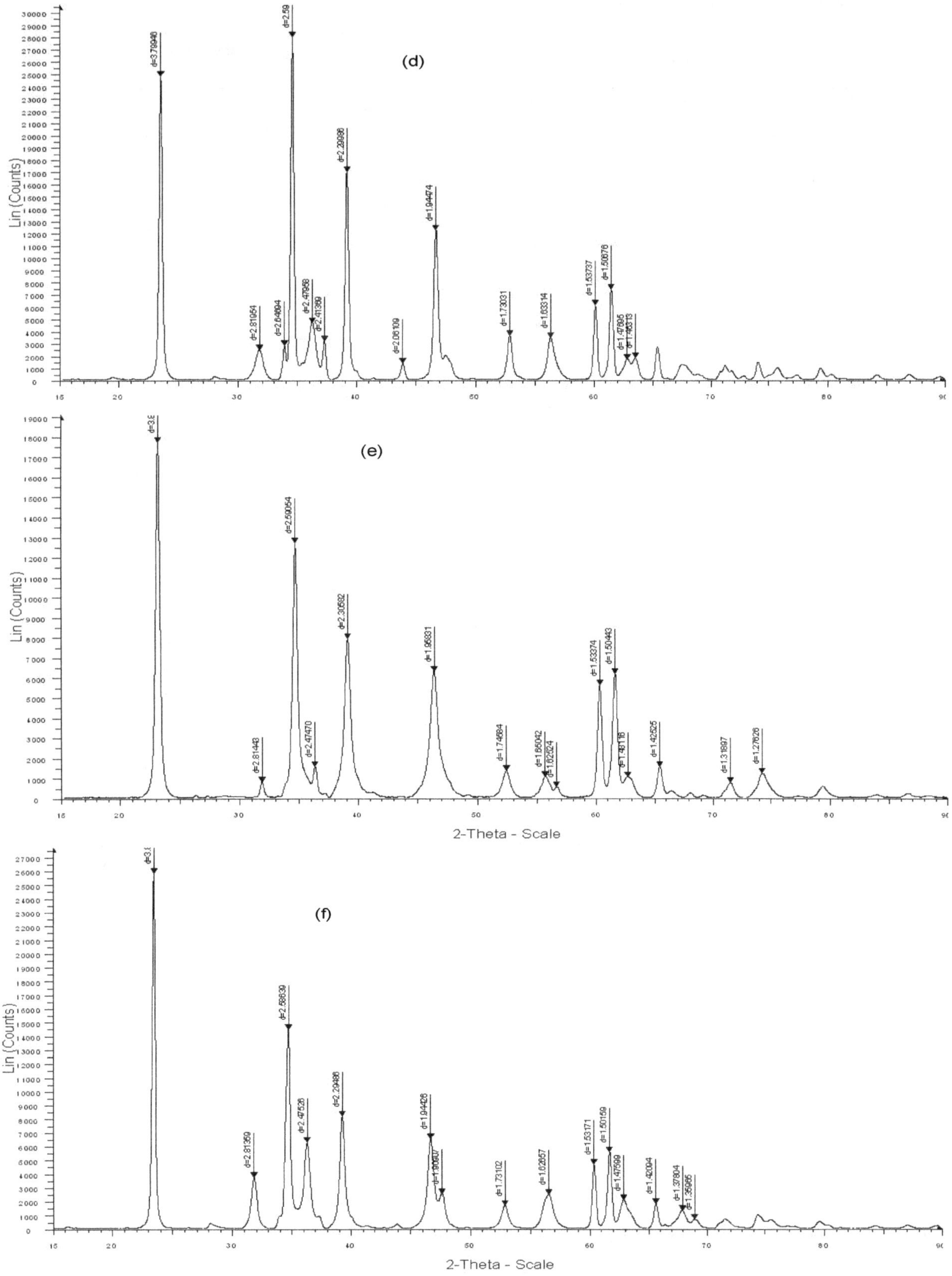

Figure 2. PXRD patterns obtained for HT's. (a) PXRD pattern obtained for $Mg_4Al_2(OH)_{12}CO_3.3H_2O$, (b) PXRD pattern obtained for $Mg_6Al_2(OH)_{16}CO_3.4H_2O$, (c) PXRD pattern obtained for $Zn_4Al_2(OH)_{12}CO_3.3H_2O$, (d) PXRD pattern obtained for $Zn_6Al_2(OH)_{16}CO_3.4H_2O$, (e) PXRD pattern obtained for $Mg_4Zn_2Al_2(OH)_{12}CO_3.3H_2O$, (f) PXRD pattern obtained for $Mg_2Zn_4Al_2(OH)_{12}CO_3.3H_2O$.

Table 3. Surface area and pore volume obtained for HTs synthesised.

HT	Sample mass	Surface area		Pore volume	
	/ mg	/ m^2 g^{-1}	/ m^2	/ cm^3 g^{-1}	/ dm^3
1	15.5	88.7000	1.375	0.5476	8.49
2	19.6	103.7092	2.033	0.5205	10.2
3	15.2	69.1270	1.051	0.0639	0.97
4	17.5	42.4833	0.743	0.1058	1.85
5	18.7	48.6898	0.910	0.2112	3.95
6	19.8	24.4649	0.484	0.0593	1.17

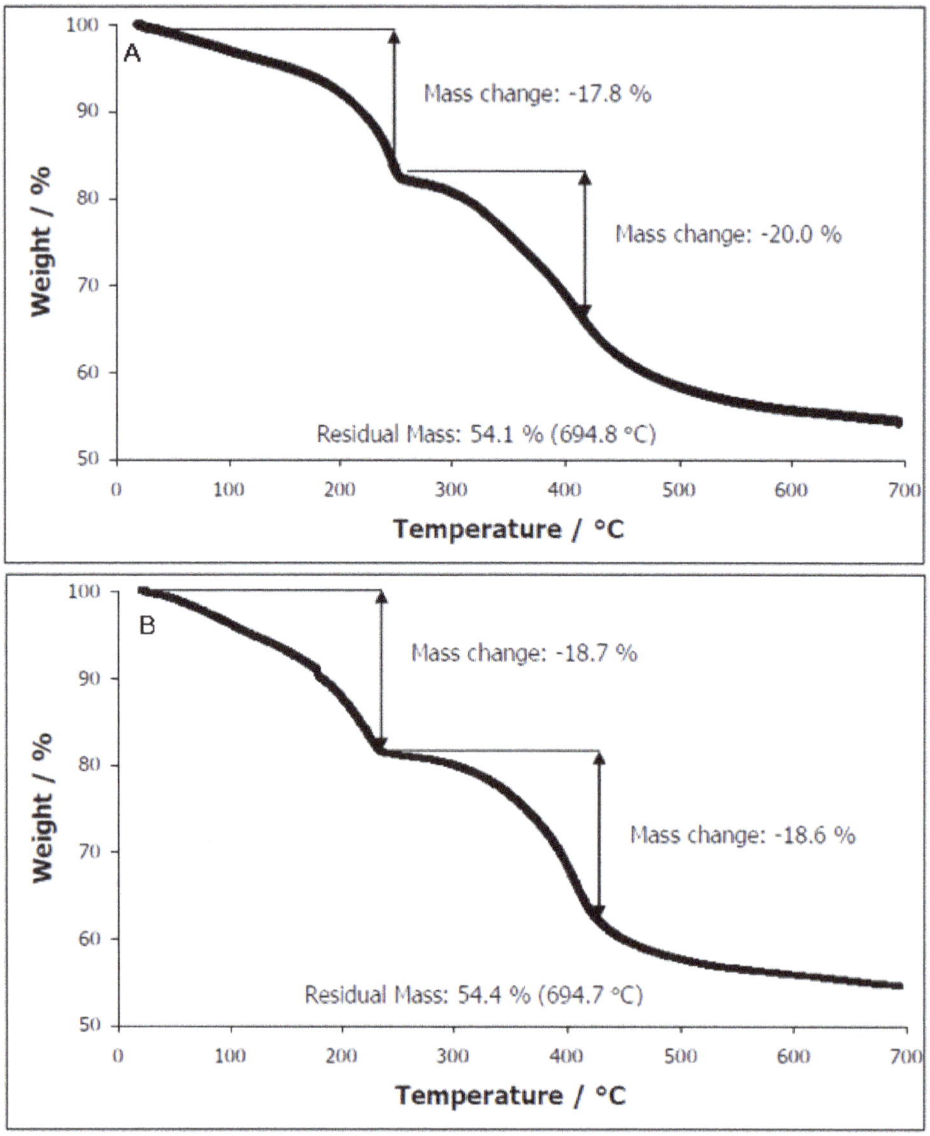

Figure 3. Thermograms for (A) Mg_4Al_2-HT and (B) Mg_6Al_2-HT.

Heating of the PVC powder, resulted in the evolution of HCl gas at temperatures greater than 180°C. From the thermogram for PVC (Figure 4), it was evident that the structure of PVC becomes unstable due to its facile

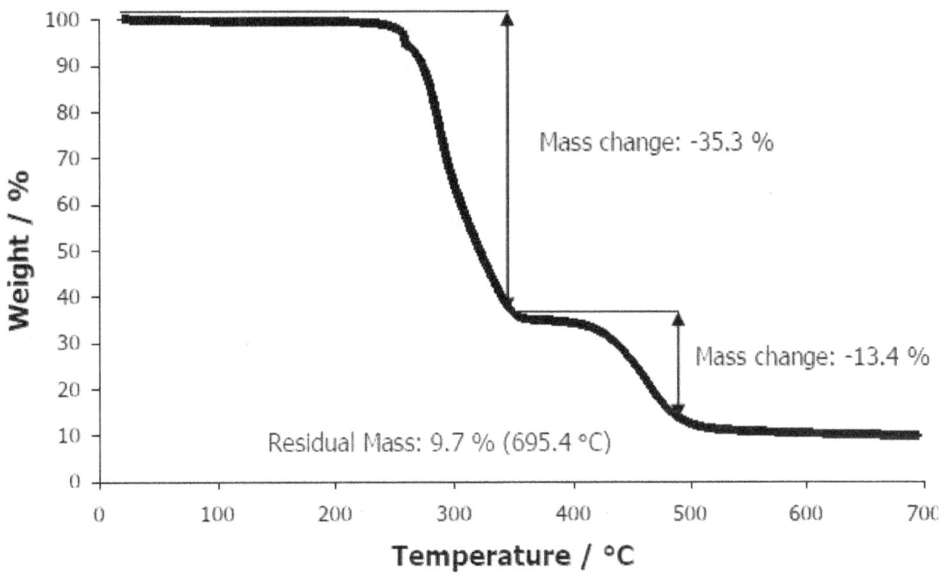

Figure 4. Thermogram for PVC.

Table 4. TGA data obtained for HTs and PVC.

Samples	Temperature / °C	Weight loss / %	Total weight loss / %
HT 1	250; 415	17.8; 20.0	45.9
HT 2	225; 415	18.7; 18.6	45.6
HT 3	215	18.0	33.0
HT 4	205	15.1	27.3
HT 5	205	15.0	37.6
HT 6	220	13.5	29.8
PVC	300; 460	35.3; 13.4	90.3

decomposition observed between 250-350°C. The process of dehydrochlorination (release of HCl gas from PVC) accelerates the degradation of PVC as observed from the continuous weight loss up to approximately 500°C. For this reason, heat stabilisers are an essential part of the PVC industry to prevent subsequent decomposition of the polymer upon heating. Potential inhibition of degradation of PVC by synthesised HTs is clearly evident from their characterisation using TGA, PXRD and BET surface area analysis. This can also be confirmed by the fact that at 700°C, all HTs maintained at least 50% of residual mass, indicating favourable thermal stability properties of the structure.

Since thermal stability is temperature dependant, the most stable synthesised HTs appear to be HT 4 and 6 showing lowest total weight loss (Table 4). However, the BET surface area results suggest that HTs 4 and 5 are most unlikely to quench the evolved HCl gas, due to its small pore volumes and surface areas. Although HT 4 and 6 show high thermal stability from the TGA results, it is the HTs ability to quench the HCl gas that is of importance to this study. Therefore from this analysis, it was observed that the HT thermal stability in increasing order is as follows:

HT 1 ≈ HT 2 < HT 5 < HT 3 < HT 6 < HT 4

The results of dehydrochlorination are reported in Table 4, where all samples were analysed under the same conditions. A blank analysis, containing PVC only, showed signs of degradation (decolourised to light brown) after 5 min. The Mg/Al-HTs (0.2 g) showed the highest thermal stabilisation by delaying the degradation of 10 g of PVC powder by 14 to 15 min in comparison to the blank analysis, where the PVC completely degraded after 12 min. Such results are consistent with the BET surface area results, where the Mg/Al-HTs had the largest surface areas and pore volumes. Therefore, the Mg/Al-HTs would be able to quench the most HCl gas in comparison to the other HTs synthesised.

PVC samples containing HTs 3 and 4 were the only samples to turn black after a short time, while all the other samples decolourised to light brown. This shows that Zn did not offer any stability towards the PVC, as the PVC-HT 6 sample degraded the fastest at 6.45 min. Therefore, the PVC samples containing HTs 3, 4 or 6 instability is consistent with the small surface area and pore volume observed. The time taken for the PVC samples containing HTs 3, 4 or 6 to degrade was less than the PVC containing no HTs. It can be suggested that the reason for the PVC samples containing HTs 3, 4 or 6 not being stable for at least 12 min was due to the extra Zn atom present in the HT, reported in Table 1.

Table 5. Adsorption rate of 0.2 μmol evolved HCl gas (dehydrochlorination) by HTs at 180°C.

Samples	Degradation Time/ mins	Observations
PVC (Blank)	12.00	Sample decolourised to light brown after 5 minutes.
PVC-HT 1	26.45	Sample decolourised partially from white to light brown.
PVC-HT 2	28.20	
PVC-HT 3	12.51	Sample turned black after 11 minutes.
PVC-HT 4	8.30	Sample turned black with fumes of HCl rapidly being given off after 8 minutes.
PVC-HT 5	12.42	Sample decolourised partially from white to light brown.
PVC-HT 6	6.45	

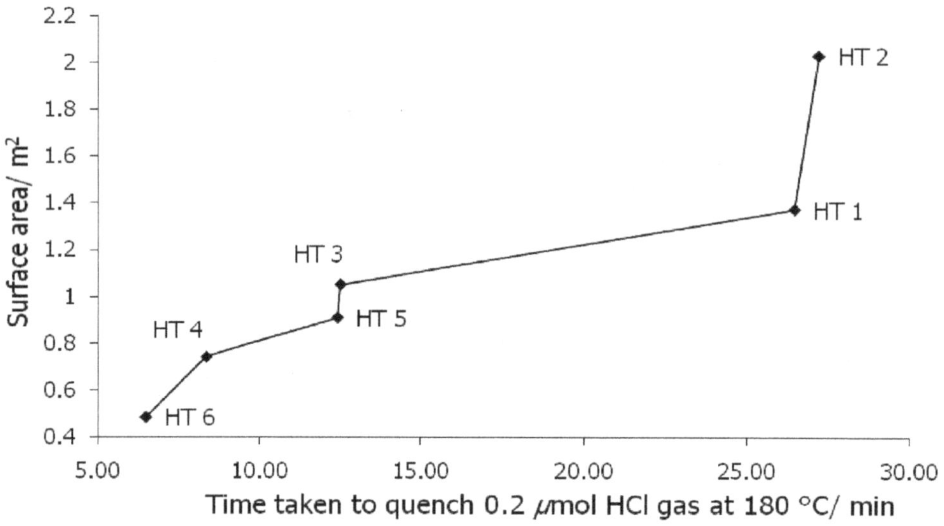

Figure 5. Correlation between the BET surface area and dehydrochlorination results obtained.

From the BET surface area and TGA results obtained, it is possible to assume that HT 3, 4, 5 and 6 did not form the expected layered structure and therefore a second weight loss was not observed on each thermogram recorded. Further investigation is deemed necessary to elucidate the structure of HTs 3, 4, 5 and 6. Although PVC-HT 5 has a small surface area and pore volume relative HTs 1 and 2, but larger than HTs 3, 4 or 6, showed signs of stabilising the PVC for approximately 42 seconds due to the presence of Mg in the HT. The observed thermal stability of synthesised HT from this analysis in order of increasing ability to stabilise the PVC is: HT 6 < HT 4 < HT 5 < HT 3 < HT 1 < HT 2.

It is also envisaged from the results obtained that the type of metal ions present in a synthesised HT has a significant effect on the thermal stability of PVC during dehydrochlorination as observed from the adsorption rate of evolved HCl gas shown in Table 5. This is confirmed by the relatively linear relationship between the HTs surface area and its ability to inhibit the degradation of PVC (Figure 5). Surface adsorption and not absorption, according to Figure 5, is the mechanism in which Mg/Al-HTs quench evolved HCl gas. Stabilisation of PVC by HT 3 was more effective than HT 4 due to its surface area rather than to its pore volume. HT 2 was found to have the largest surface area over all synthesised HTs and consequently was the most effective in stabilising the PVC powder. HT 6 was found to have the smallest surface area and as a result, did not stabilise the PVC at all. Hence from the HTs synthesised in this study, HT 2 ($Mg_6Al_2(OH)_{16}CO_3.4H_2O$) has proven to be the most effective heat stabiliser in processing of PVC to useful polymeric products.

Conclusion

The results from this study show that Mg/Al-HTs have proven to potentially function as ideal heat stabilisers in processed PVC products. From the PXRD studies, it can be confirmed that the Mg/Al-HTs synthesised in this study were the only HTs which formed layered structures with rhombohedral symmetry. A combination of favourable properties, such as a large surface area confirmed by the BET surface area analysis, thermal stability and quenching ability of a PVC-HT blend determines the inhibition potential of dehydrochlorination. The addition of Zn to the Mg/Al-HT did not benefit the HT and in fact accelerated the degradation of PVC during dehydrochlorination. Mg/Al-HTs were able to adsorb 0.2 μmol of HCl gas evolved during the degradation of PVC and significantly delayed the onset of degradation of PVC relative to the other HTs synthesised, including the non-stabilised PVC sample. Therefore scavenging of HCl gas by synthesised Mg/Al-HTs in this study has shown an increase toward the inhibition of degradation and thermal stability of PVC during its processing.

REFERENCES

Bastiani R, Zonno IV, Santos IAV, Henriques CA, Monteiro JLF (2004). Influence of thermal treatments on the basic and catalytic properties of Mg, Al-mixed oxides derived from hydrotalcites. Brazilian J. Chem. Eng. 21(2):193-202.

Borja M, Dutta PK (1992). Fatty acids in layered metal hydroxides; membrane-like structure and dynamics. J. Phys. Chem. 96(13): 5434-5444.

Cavani F, Trifiro F, Vaccari A (1991). Hydrotalcite-type anionic clays: preparation, properties and applications. Catalysis today. 11(2):176-190.

Gupta S, Agarwal DD, Banerjee S (2008). Synthesis and characterization of hydrotalcites: Potential thermal stabilizers for PVC. Indian J. Chem. 47A:1004-1008.

Gupta S, Agarwal DD, Banerjee S (2009). Thermal stabilization of poly (vinyl chloride) in presence of hydrotalcites, Zeolites and conventional stabilizer system. J. Vinyl Additive Technol. 15:164-170.

Haynes D (2007). Materials Practical Manual - APCH 312. 2007, P. 19.

Kalouskova R, Novotna M, Vymazal Z (2004). Investigation of thermal stabilization of poly(vinyl chloride) by lead stearate and its combination with synthetic hydrotalcite, Polymer Degradation and Stability. Polymer Degradation and Stability, 85:903-909.

Kloprogge JT, Hickey L, Frost RL (2004a). The effects of synthesis pH and hydrothermal treatment on the formation of zinc aluminum hydrotalcites. J. Solid State Chem. 177:4047-4057.

Kloprogge JT, Hickey L, Frost RL, Raman J (2004b). FT-Raman and FT-IR spectroscopic study of the local structure of synthetic Mg/Zn/Al-hydrotalcites. Spectroscopy, 35:967-974.

Lin Y, Wang J, Evans DG, Li D (2006). Layered and intercalated hydrotalcite-like materials as thermal stabilizers in PVC resin. J. Phys. Chem. Solids, 67:998-1001.

Lin YJ, Li DQ, Evans DG, Duan X (2005). Modulating effect of Mg–Al–CO_3 layered double hydroxides on the thermal stability of PVC resin. Polymer Degradation and Stability, 88:286-293.

Mohan RM, Ramachandra RB, Jayalakshmi M, Swarna JV, Sridhar B (2005). Hydrothermal synthesis of Mg–Al hydrotalcites by urea hydrolysis. Mater. Res. Bulletin, 40:347-359.

Morioka H, Tagaya H, Karasu M, Kadokawa J, Chiba K (1995). Preparation of new useful materials by surface modification of inorganic layered compound. J. Solid State Chem.117:337-342.

Rajamathi M, Thomas GS, Kamath PV (2001). The many ways of making anionic clays. Proceedings of the Indian national Academy of Sciences (Chemistry Sciences). 113(5-6):671-680.

Seftel EM, Dvininov E, Lutic D, Popovici E, Ciocoiu C (2005). Synthesis of hydrotalcite-type anionic clays containing biomolecules. Optoelectronics Adv. Mater. 7(6):2869-2874.

Tong M, Chen H, Yang Z, Wen R (2011). The Effect of Zn-Al-Hydrotalcites Composited with Calcium Stearate and β-Diketone on the Thermal Stability of PVC. Int. J. Molecular Sci. 12:1756-1766.

Vaccari A (1998). Preparation and Catalytic properties of cationic and anionic clays. Catalysis Today. 41:53-71.

van der Ven L, van Germert MLM, Batenburg LF, Keern JJ, Gielgens LH, Koster TPM, Fischer HR (2000). On the action of hydrotalcite-like clay materials as stabilizers in poly(vinyl chloride. Appl. Clay Sci. 17:25-34.

Yong Z, Rodrigues AE (2002). Hydrotalcite-like compounds as adsorbents for carbon dioxide. Energy Conver. Manage. 43:1865.

Numerical investigation on the effect of different parameters on the performance and the emission of a spark ignition (SI) engine

H. Mirgolbabaei[1] , M. Bozorgi[2] and M. M. Etghani[2]

[1]Young Researchers Club, Islamic Azad University, Jouybar Branch, Jouybar, Iran.
[2]Automotive Engineering Department, Iran University of Science and Technology, Tehran, Iran.
[3]Department of mechanical engineering, Babol Noshirvani University of Technology, Babol, Iran.

In this work, NISSAN Z24 engine was simulated numerically and valves and ignition timings were modified in order to improve the engine performance and emissions. The engine was simulated using a version of the Los Alamos code KIVA-3V2 which is a computer program for the numerical calculation of transient, two and three dimensional chemically reactive flows with sprays. The computational grid was generated by the commercial software, ANSYS ICEM CFD. To validate the code, a comparison between experimental and theoretical results has been made, which confirm good qualitative agreement between these results. A series of parametric studied is performed to gain a better understanding of the effects of valves and ignition timings on the engine performance and emission. The results show that the engine volumetric efficiency, output power, emissions and performance are well improved by applying the optimal values.

Key words: Spark ignition (SI) engine, valves and ignition timings, engine performance and emissions.

INTRODUCTION

Nowadays, due to government server rules on environmental standards and fuel consumption, car manufacturers are under pressure to reduce regulated emissions and fuel consumption. Valves and ignition timings are known as a very effective process on output power, emissions and engine performance. The improvements can be achieved with the lowest change in the engine and hence less cost. Shayler (2007) investigated the effect of valve timing on the amount of burned gas that remains from pervious cycle in the combustion chamber. He concluded that valve timing has a noticeable effect on the entrance of a fresh mixture. This effect is because of valves overlap. At this moment, some burned gas that remains from pervious cycle enters into the intake manifold, pushes back the fresh mixture and reduces the volumetric efficiency. This event is more important in low speed of the engine, since at engine low speed, the engine has a longer overlap. Siewert (1971) studied the effect of valve timing on the emissions. He found that these effects are more noticeable in NOx and HC emission. Also he found that by controlling the temperature of the combustion chamber according to the amount of burned gas that remains from pervious cycle, the amount of NOx can be reduced. This event can take place by adjustment of valve timing and valve overlap. Furthermore, the amount of vapor of gasoline that leaves the combustion chamber through the exhaust valve at valve overlap can be reduced by exact adjustment of valve timing. Tuttle (1980, 1982) considered the effect of early or late intake valve opening on the output power of the engine. Smith et al. (1984) performed some experimental research about knock and the factors affect on it. They found that compression ratio and ignition timing are tow the most important factors that have a great effect on knock. Zhu et al. (2004) investigated the ignition timing and some limitation for choosing the ignition timing. They concluded that if we advance the ignition timing more than a suitable point it leads to knock and if we retard the ignition timing it may lead to misfiring. Boyce et al. (1999) did some research in the effect of an ignition timing and air/fuel ratio on the release of energy in combustion chamber and temperature of a combustion chamber. They found advancing the ignition timing increase the rate of energy release and temperature of the combustion chamber. Duclos et al. (1996) simulated combustion and emissions in an SI engine using KIVA and found the effect of a residual gas and spark plug situation on the engine performance. Eckert et al. (2003) simulated knock by KIVA-3V. They validated their results with an experimental test. In the present study, the improvement of the engine volumetric efficiency is performed numerically, using the method "Improving" considering the valve timing of the engine.

VOLUMETRIC EFFICIENCY

The intake system (the air filter, throttle plate, intake manifold, intake port, intake valve) restricts the amount of air which an engine of given displacement can induct. The parameter used to measure the effectiveness of an engine's induction process is the volumetric efficiency. It is defined as the volume flow rate of air into the intake system divided by the rate at which volume is displaced by piston:

$$\eta_v = \frac{2\dot{m}_a}{\rho_{a,i} V_d N} \tag{1}$$

Volumetric efficiency is mainly affected by the fuel, engine design and engine operating variables in terms of Heywood (1988): Fuel type, fuel/air ratio, fraction of fuel vaporization in the intake system and fuel heat of vaporization, mixture temperature as influenced by heat transfer, ratio of exhaust and intake manifold pressure, compression ratio, engine speed, intake and exhaust manifold and port design, intake and exhaust valve geometry, size, lift and timing. There are lots of methods can be used for the better operating condition of engine (more output power and less emission). For example we can consider these methods (Pulkrabeck, 2003): Increasing the intake and exhaust valves diameter, increasing the number of intake valve, using of some new methods like Gasoline Direct Injection (GDI), variable valve timing (VVT), improving the valve timing of the engine. In this paper, the last method (improving the valve timing of the engine) is used for improving the engine volumetric efficiency. Intake and exhaust valve timing has great effect on volumetric efficiency. Reverse flow into the intake manifold (it takes places at the valve overlapping) and the temperature of fresh air and fuel mixture when the intake valve opens are two important items depending on valve timing and have great effect on volumetric efficiency (Siewert, 1971). Furthermore, valve timing has a great effect on HC and NOx. We can reduce the amount of HC by adjusting the valve overlap. Also we can reduce the amount of NOx by adjusting the amount of burned gas that remains from pervious cycle in a combustion chamber. Since this burned gas causes the leaner mixture in the combustion chamber, it leads to decrease in the maximum temperature of a combustion chamber. NOx has a direct relationship with the maximum temperature of a combustion chamber and it can be adjusted by exact adjusting the valve timing (Shayler, 2007).

Combustion in a spark ignition (SI) engine and the factors which control it

In a conventional spark ignition engine, the fuel and air are mixed together in the intake system, inducted through the intake valve into the cylinder, where mixing with residual gas takes place and then compressed. Under normal operating conditions, combustion is initiated towards the end of the compression stroke at the spark plug by an electric discharge. Following inflammation, a turbulent flame develops, propagates through this essentially premixed fuel, air, burned gas mixture until it reaches the combustion chamber walls, and then extinguishes. According to our knowledge about a SI engine's combustion progress (ignition and development of flame, propagation and quenching), we can find the relation between physical and chemical factors which control the combustion and the design of engine. Numbers of factors affecting combustion and propagation of the flame are (Ferguson and Kirkpatrick, 2001): Geometry of combustion chamber and situation of spark plug in the cylinder head; fluid flow's characteristic which include velocity of the fluid; turbulence intensity and unburned gas turbulence characteristic, and unburned mixture state and combination like fuel equivalence ratio, pressure and temperature of mixture and ignition timing. In this paper, improving the ignition timing is used for improving the combustion characteristics. Ignition timing in SI engine has a great effect on air and fuel mixture combustion quality, since by exact adjustment of ignition timing we can adjust the maximum pressure angle of the combustion chamber (around 10 to 15° ATDC) in the best situation. Therefore, the output power of the engine

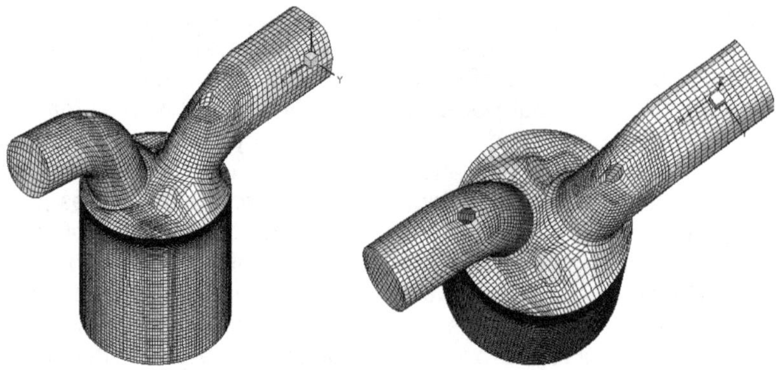

Figure 1. Computational grid generated in ICEM CFD.

Table 1. Specification of NISSAN Z24.

Bore × stroke (mm)	89 × 96
Compression ratio	8.4
Power (kW/cyl)	137
Air/fuel ratio	14
Number of holes/hole diameter (mm)	8/0.4
Fuel amount (g/cycle)	0.0453
Rpm	2800
Cylinder pressure at IVC (bar)	3.6
Air temperature at IVC (K)	400
Wall temperature (K)	Wall: 45 Head: 500 Piston: 550
Valve timing	IVO: 9 BTDC IVC: 37 ABDC EVO: 41 BBDC EVC: 5 ATDC
Type	4 cylinder,4 cycle
Max. motoring / firing pressure at full load and 2800 rpm (bar)	14.5/52

increases and the HC and CO emission decreases (Heywood, 1988). Of course, it should be noticed that a little change in ignition timing can lead to misfire or knock (Ferguson and Kirkpatrick, 2001).

Abnormal combustion reveals itself in many ways. One of the most important abnormal combustion processes important in practice is knock. Knock originates in the extremely rapid release of much of the energy contained in the end-gas ahead of the propagation turbulent flame, resulting in high local pressures. The non uniform nature of this pressure distribution causes pressure waves or shock waves to propagate across the chamber, which may cause the chamber to resonate at its natural frequency (Heywood, 1988). The two important factors, which affect knock are air/fuel ratio and ignition timing. We can prevent it by exact adjustment of these two factors (Pulkrabeck, 2003).

Numerical simulation

Figure 1 shows the computational mesh of the engine combustion chamber and ports, which is created in ICEM-CFD. There are about 67000 cells in the entire computational domain. The reference point of z = 0 cm is defined as the lowest axial point of the piston bowl at BDC. The engine is the NISSAN Z24 engine. The specifications of the engine and calculating conditions are given in Table 1. The calculation starts 251 crank angles before IVO and continues 720 crank angles for example, a full cycle calculation including intake and exhaust processes. For initial conditions, the gas temperature,

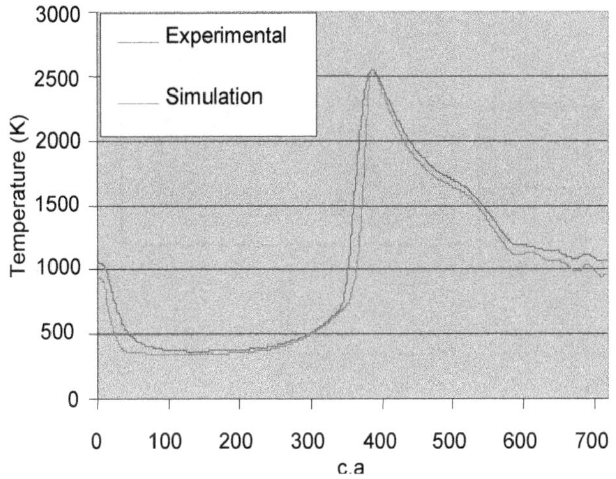

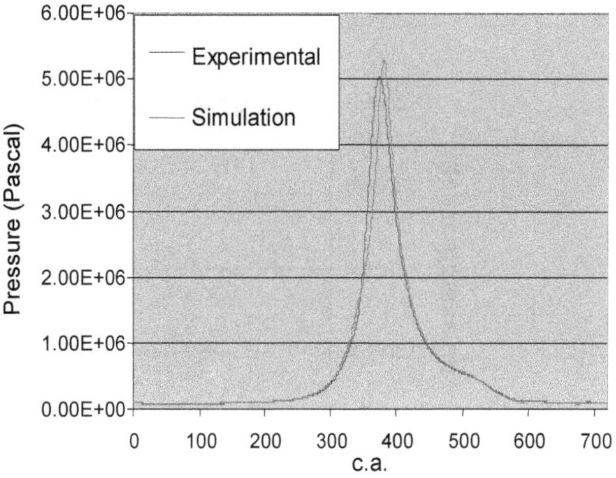

Figure 2. Validation of numerical results against measured data.

Table 2. NISSAN Z24 engine valve timing.

IVO	9 c.a BTDC
IVC	37 c.a ABDC
EVO	41 c.a BBDC
EVC	5 c.a ATDC

Table 3. Different valve timing.

State	IVO at 5 c.a BTDC	IVO at 12 c.a BTDC
IVO	5 c.a BTDC	12 c.a BTDC
IVC	43 c.a ABDC	35 c.a ABDC
EVO	36 c.a BBDC	43 c.a BBDC
EVC	10 c.a ATDC	3 c.a ATDC

pressure and species densities were assumed to be uniform in the entire combustion chamber.

Validation

For validation, the result that is driven from KIVA 3 V, the experimental data like the pressure and the temperature of the cylinder is used as shown in Figure 2. The curves in intake, compression and expansion phases are close together but there is around 5% difference between the experimental and simulation curves in ignition phase. There is an important reason for this difference; the fuel that we use in the experimental test is not standard. Consequently, the factors of chemical reaction rate of KIVA 3 V are calibrated for this case by exact adjustment. It should be noticed that all the calculations are performed in this situation. The engine speed is 2800 rpm and Ø=1, since the engine maximum torque happens in these situations.

Improving the volumetric efficiency of the engine by changing the valve timing

The NISSAN Z24 engine valve timing is mentioned in Table 2. The volumetric efficiency of NISSAN Z24 is 87% and the amount of volumetric efficiency that is calculated with KIVA 3 V is 92%. This difference is because of ignoring the ram effect in the intake system. To improve the volumetric efficiency of the engine by changing the valve timing, without changing the cam profile, we can only replace the cam angle according to its last position and the TDC position. Therefore, we considered the various situations of this concept. Table 3 presents these various situations. Figure 3 shows the effect of these changing on cylinder pressure. As shown in Figure 3, the IVO at 12 C.A BTDC is the best because between all situations; the maximum cylinder pressure and consequently, the maximum output power of the engine are obtained. Table 4 presents the exact changing amount; also Figure 4 shows the difference of cylinder pressure between Z24 and improved Z24. As shown in Table 4, the volumetric efficiency of improved Z24 is 94%, which has 2% improvement in comparison with Z24; also the engine power increased almost 1%. On the other hand, as shown in Table 5, the efficiency of exhaust process is calculated and it shows the new valve timing has the better efficiency of exhaust process almost 0.1%.

Improving the ignition timing

Ignition timing was optimized in order to increase the engine power and decrease the emission. There are some limitations in changing the ignition timing. For

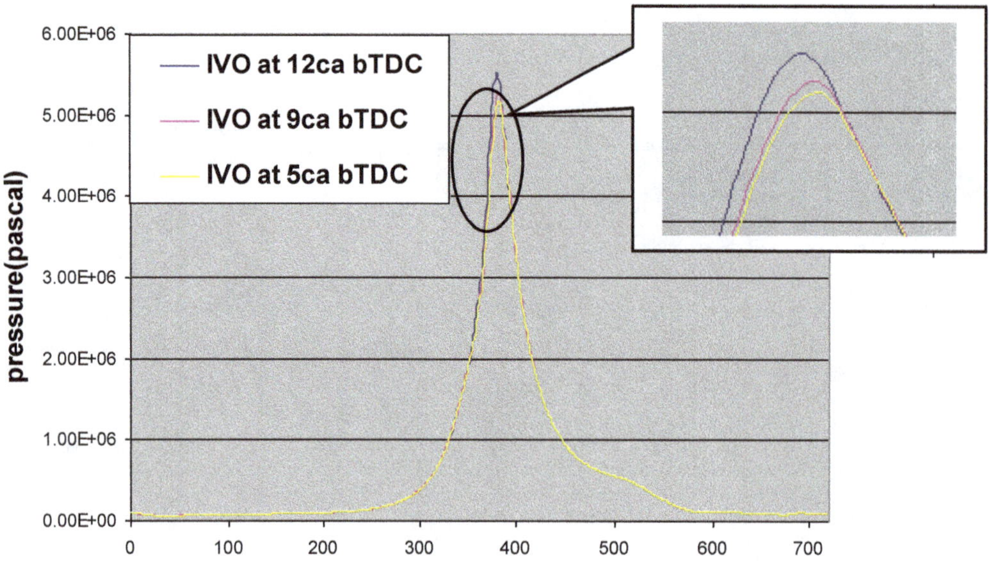

Figure 3. Combustion chamber pressure in 3 different valve timing.

Table 4. Exact changing amount of Z24 in new situation.

Parameter	Unit	Z24	Improved Z24
Mass of mixture	g	0.62313	0.64762
Mass of fuel	g	0.0453	0.0453
Displacement volume	cm^3	600	600
Air temperature before entrance of cylinder	K	332	325
Air density in air temperature before entrance of cylinder	g/cm^3	0.00106	0.001079
Volumetric efficiency	%	0.92	0.94

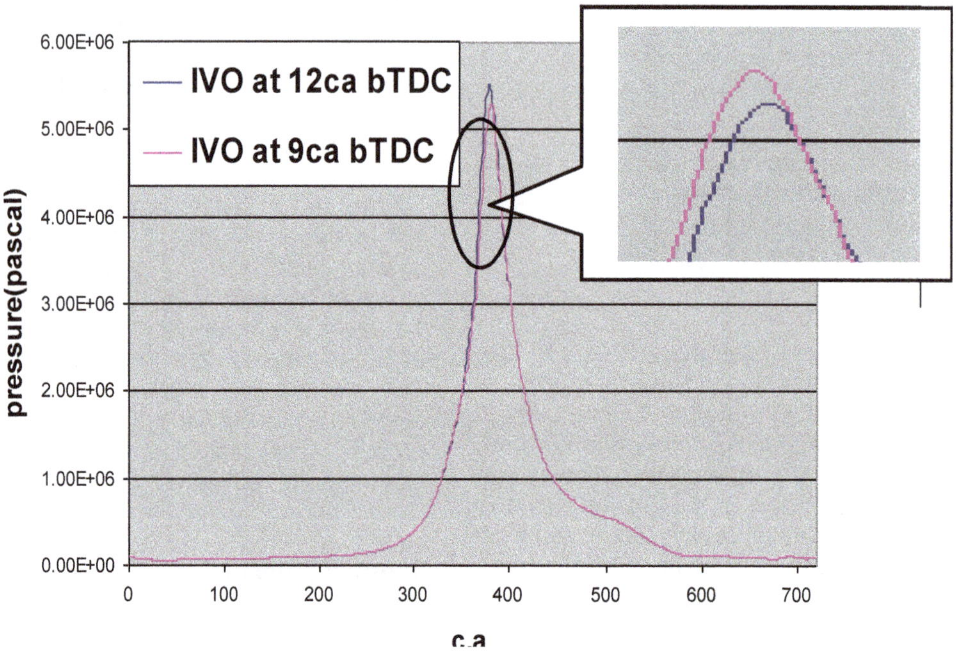

Figure 4. Combustion chamber pressure in Z24 and improved Z24.

Table 5. The efficiency of exhaust process.

State	Unit	IVO at 9 c.a BTDC	IVO at 12c.a BTDC (improved η_v)
Mass of mixture	g	0.62312	0.64762
Mass of burned gas remained in combustion chamber from pervious cycle at IVO	g	0.0247	0.02066
Efficiency of exhaust process 3 c.a ATDC	% EVC	96.04	96.16

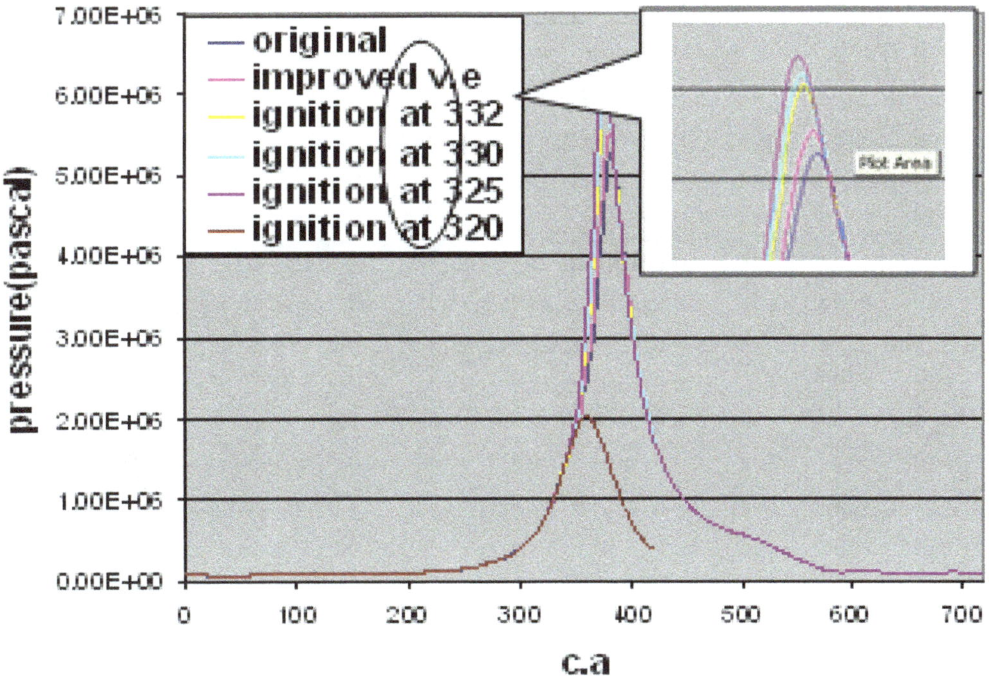

Figure 5. Pressure of combustion chamber in different ignition timing.

example, if we want to retard the ignition timing, although it leads to decrease in the NOx emission and prevention of knock, it may leads to increase in the HC emission. And advancing the ignition timing leads to knock, increasing the NOx emission and decreasing the engine power, so we should choose the best ignition timing that leads to the more power and less emission. Furthermore, in advancing the ignition timing, we have to consider the endurance of the cylinder head against the cylinder pressure. Another important item that we must notice to it is the probability of knock formation. Furthermore, after each change, we have to check it. The ignition timing of Z24 engine in 2800 rpm is 23° BTDC. So if we assume TDC at intake process, as our basis, the time of ignition is 336°. To reach the best ignition timing, we did simulation in some different ignition timing. Simulation was done in 332, 330, 325 and 320°. Outputs of these calculations are shown in Figure 5. You can see the cylinder pressure of these states in this figure. Also the amount of NOx, HC, CO and CO_2 in these states is shown in Figures 7 to 10. It should be noticed that all the calculations in this part were done on the base of new valve timing that was done in the perviously.

As shown in Figure 5, advancing the ignition timing 35° is the best because the maximum output power is released. If we increase the advancing angle of the ignition, it leads to decreasing the output power and finally misfiring. We can confirm this claim with ignition timing at 320° as shown in Figure 5. So we choose the ignition timing at 325° and for stabilizing this angle, we should check these points:

1. Calculation of the output power in this case and compare it with original engine.

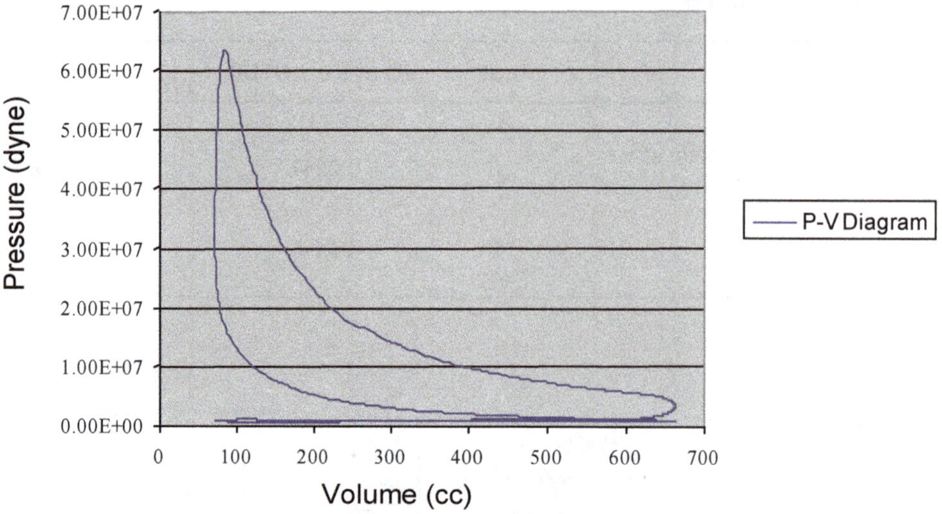

Figure 6. (P-V) diagram of improved Z24.

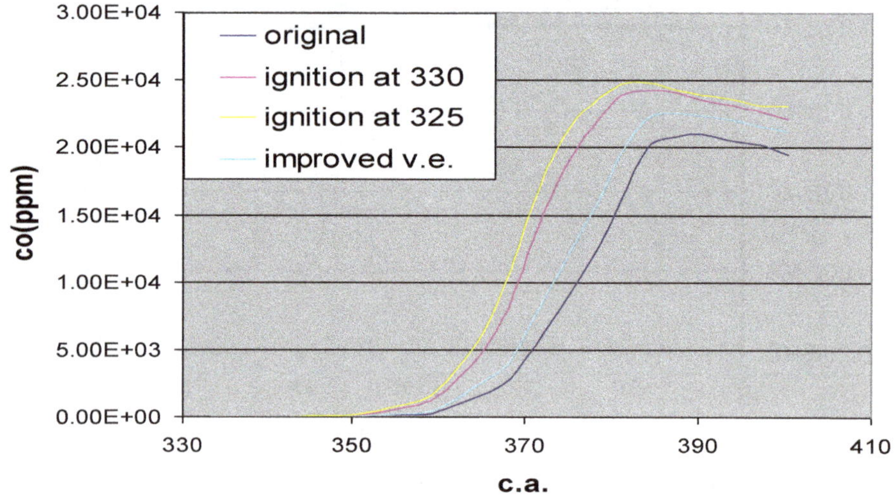

Figure 7. CO emission.

2. Consideration of the amount of emissions and compare it with original engine.
3. Consideration of probability of knock in this ignition timing.

Output power

Knowing the cylinder pressure in each crank angle and using the integral $\int_0^{720} PdV$, output power is driven. For this calculation, we use MATLAB by using traps () function. This function calculates the integral, numerically. So it calculates the area under the curve in (P-V) diagram. It is shown in Figure 6. Calculation shows us the output power increase almost 10.9%. It is shown in Table 6.

Emission

Here, we consider the emission's diagram. As shown in emission's figure, in state ignition timing, the emission CO and NOx have the maximum amount and in these tow cases, the amount of each emission increases according to increasing the advanced angle of ignition. The HC emission's behavior is completely reverse. According to increase the advanced angle of ignition, it decreases. So it has a good condition. As shown in Figure 8, the behavior of CO_2 is different from the other. The amount of this emission increases at first up to around 385° and then decreases according to increasing the advanced angle of ignition. As the advanced angle of ignition increases, the ignition process progresses better. We can confirm this claim with considering Figure 10, the

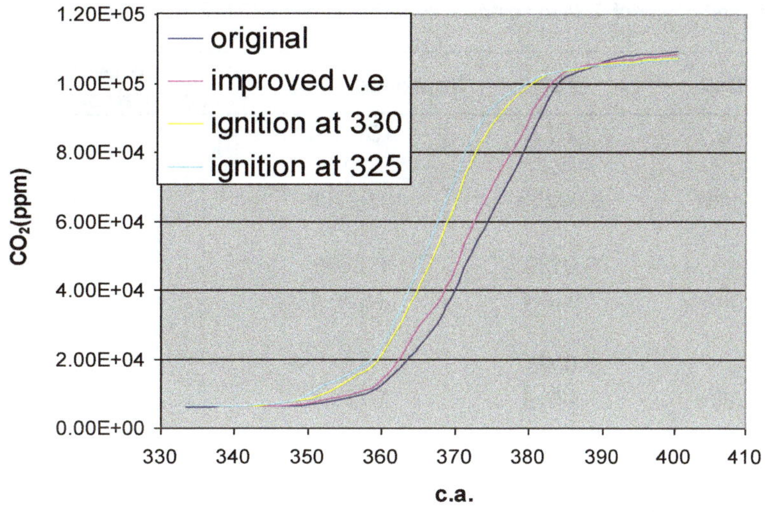

Figure 8. CO_2 emission.

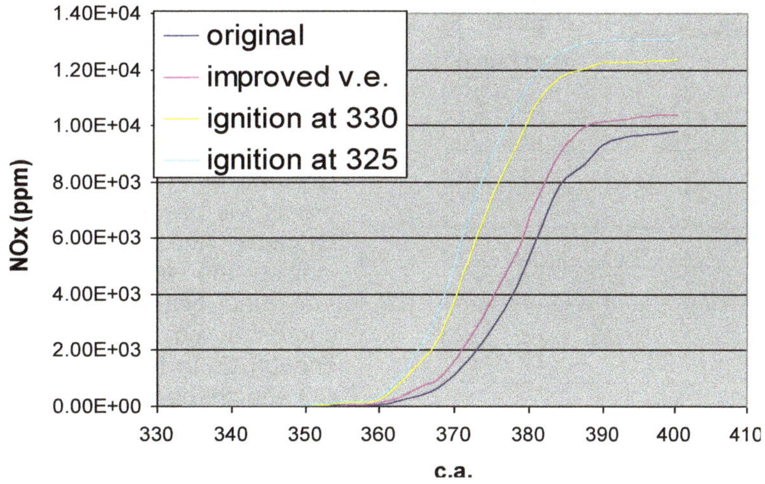

Figure 9. NOx emission.

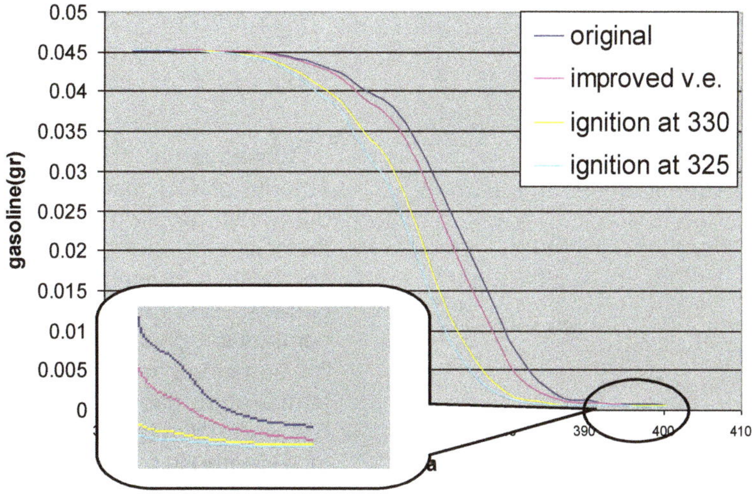

Figure 10. HC emission.

Table 6. Emission and power of Z24 in comparison with improved Z24 (Ø = 1).

Variables	Unit	Z24	Improved Z24(Ø =1)	Percentage of changes in comparison with original state (%)
Power	kw	52.77	57.52	10.90
CO_2	g	0.123	0.12	-2.50
	g/kw	0.00233	0.00208	-11
CO	g	0.0139	0.0169	12.10
	g/kw	2.63e-4	2.93e-4	11.10
NO	g	0.0075	0.01	13.30
	g/kw	1.42e-4	1.73e-4	12.10
Gasoline	g	7.07e-4	5.58e-4	-22.10
	g/kw	1.33e-5	9.76e-6	-28.10

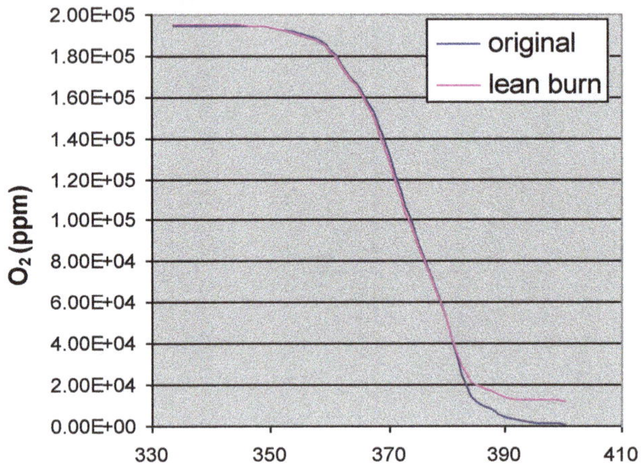

Figure 11. Consumption of O_2 in idle and lean mixture.

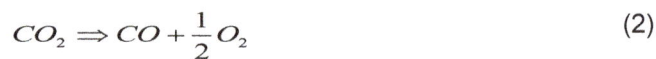

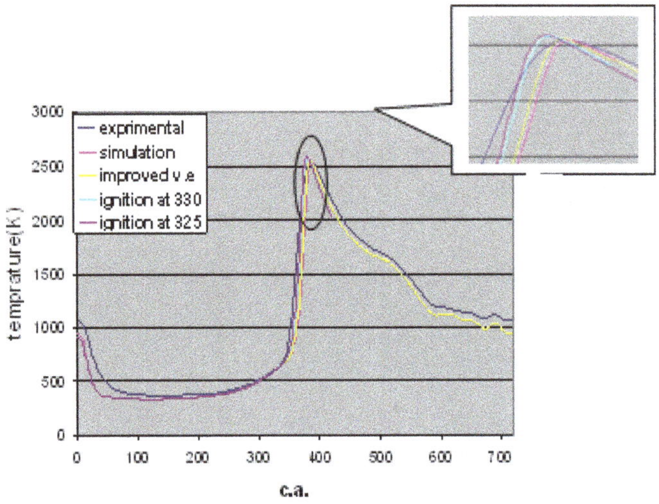

Figure 12. Temperature of combustion chamber in different ignition timing.

HC emission's behavior. As shown, it decreases according to increase in the advanced angle of ignition, so we can conclude that more fuel ignites more carbon molecules release in comparison with the other conditions. Consequently, the amount of CO increases. About CO_2, this theory is true until it is increasing. Decreasing the amount of CO_2 at the end of ignition process is because of shortage of O_2 in the combustion chamber. According to increasing the advance angle of ignition, the rate of ignition process and the amount of fuel that ignites, increases consequently the ignition confronts with more shortage of O_2, so there is not enough O_2 in the combustion chamber for oxidizing the CO and change it to CO_2. We can solve this problem with decreasing the amount of injected fuel. So our mixture becomes lean. This claim is confirmed with a new simulation. In this case, we simulated the ignition process in tow different air/fuel ratio Ø = 1 and Ø = 0.9. As shown in Figure 11, in the original state (Ø = 1). The ignition process encounters shortage of O_2 after around 385° and compensates this lack with this process:

$$CO_2 \Rightarrow CO + \frac{1}{2}O_2 \tag{2}$$

Consequently, the amount of CO_2 decreases and the amount of CO increases. But in the second state (Ø = 0.9), this problem is solved because there is a little excessive O_2 in the combustion chamber at the end of a combustion process. Increasing the amount of NOx is because of increasing the maximum temperature of the combustion chamber. As shown in Figure 12, increasing the advanced angle of ignition, the combustion chamber's temperature rises. Consequently, the amount of NOx that has a direct relationship with the maximum temperature of a combustion chamber rises. Figures 13 to 15 show the trend of amount of emissions in term of g/kW. The amount of HC emission decreases by almost 28%.

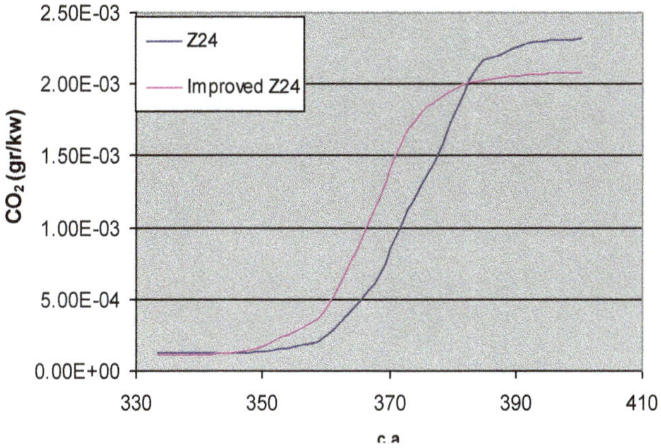

Figure 13. CO_2 emission.

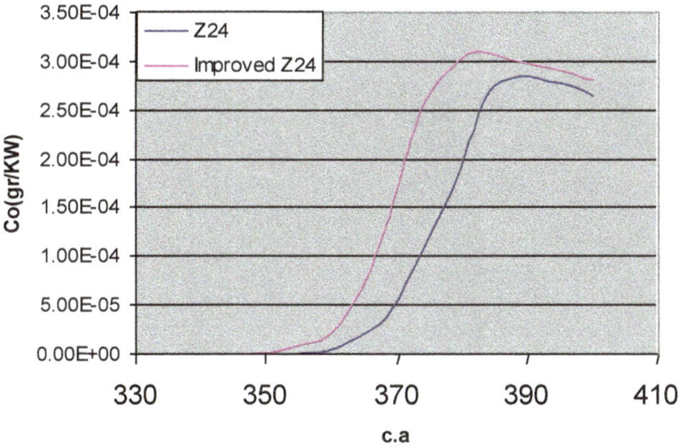

Figure 14. CO emission.

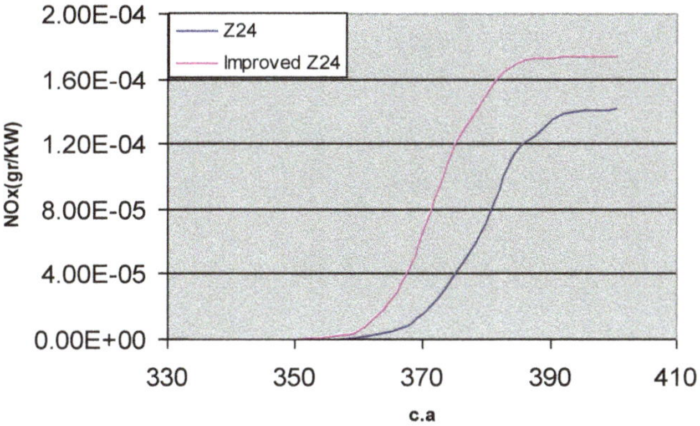

Figure 15. NOx emission.

In addition, the amount of emission decreases by almost 11%. However, the amount of CO and NOx decrease by 21 and 11%, respectively. Table 6 shows the results. As can be seen, our improved engine does not satisfy our aim in the amount of CO and NOx emission. So we decided to increase the air/fuel ratio to solve this problem.

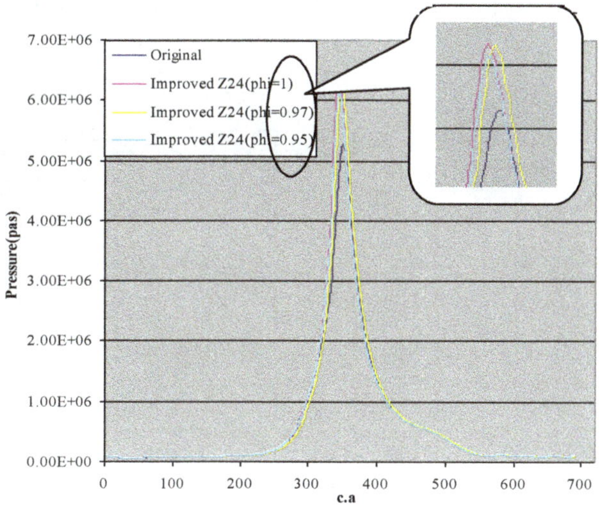

Figure 16. Pressure of combustion chamber in different air/fuel ratio.

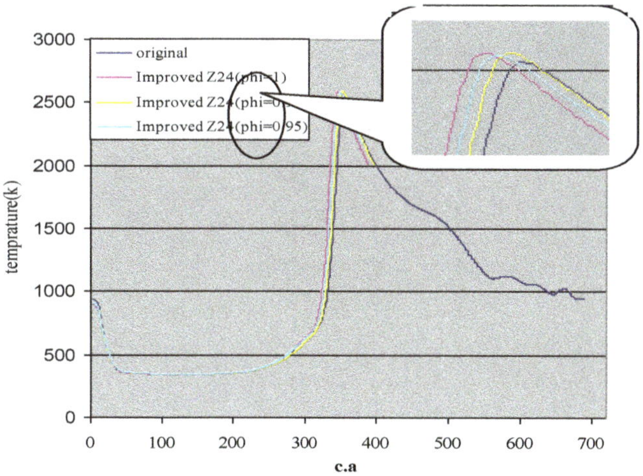

Figure 17. Temperature of combustion chamber in different air/fuel ratio.

As shown previously, by decreasing the amount of injected fuel, the ignition in the combustion chamber can compensate the shortage of O_2. Moreover, it should be noticed that decreasing the amount of injected fuel may lead to increase in the amount of NOx emission, so we should consider the entire situation, do this new simulation and then choose the best specification.

In this step, we did the simulation in tow air/fuel ratio, Ø = 0.97 and Ø = 0.95. When we decrease the amount of injected fuel poured into the combustion chamber, output power of engine decreases but in these tow conditions it is negligible, also the emission is very important. Consequently, this decrease can be ignored. The results are shown in Figures 16 to 20. According to these figures, we conclude the state is the best because the output power is almost constant and the CO, CO_2 and HC

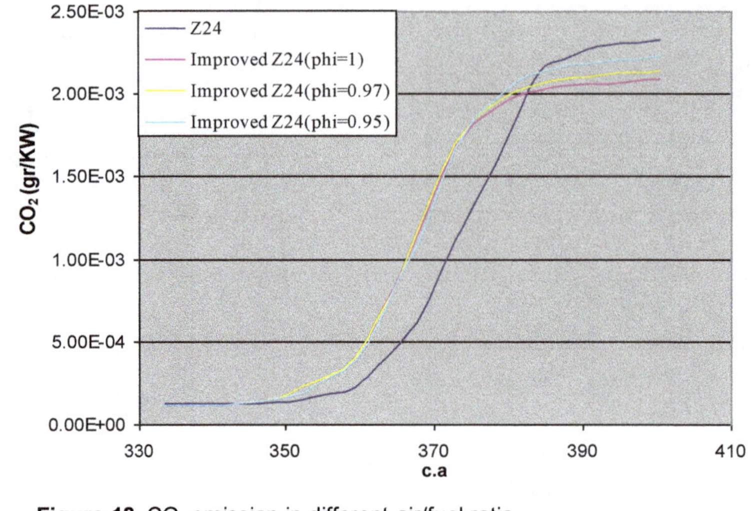

Figure 18. CO_2 emission in different air/fuel ratio.

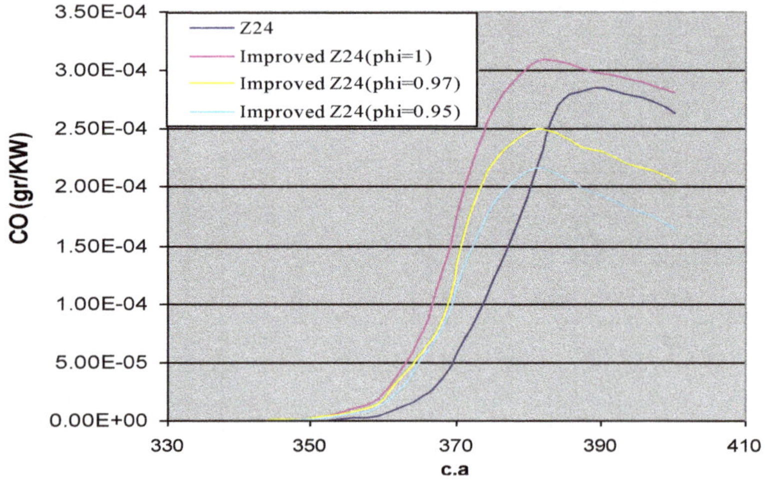

Figure 19. CO emission in different air/fuel ratio.

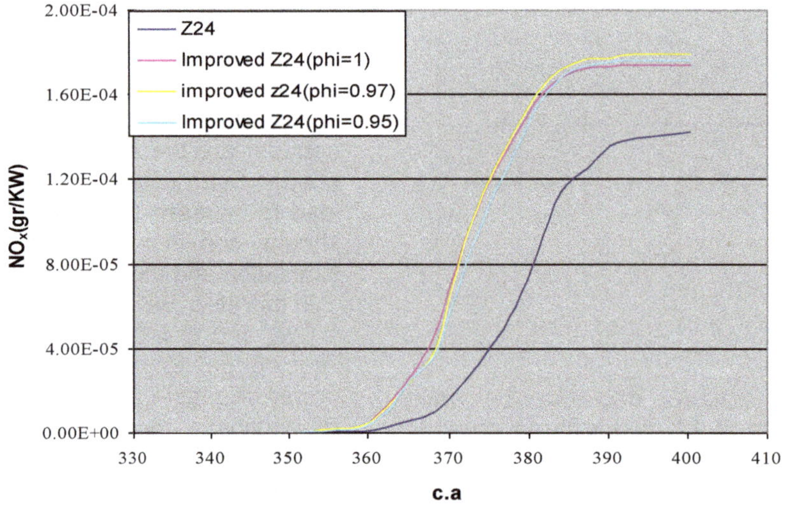

Figure 20. NOx emission in different air/fuel ratio.

Table 7. Emission and power of Z24 in comparison with improved Z24 (Ø = 1)

Parameter	Unit	Z24	Improved Z24 (Ø = 0.97)	Percentage of changes in comparison with original state (%)
Power	kw	52.77	57.12	10.82
CO_2	g	0.123	0.123	0
	g/kw	0.00233	0.00215	-8
CO	g	0.0139	0.0119	-15.10
	g/kw	2.63e-4	2.06e-4	-22.70
NO	g	0.0075	0.0103	13.70
	g/kw	1.42 e-4	1.80e-4	12.60
Gasoline	g	7.07e-4	4.48e-4	-36.70

Table 8. Specification of improved Z24.

Quantity	Specification
35°	Ignition timing
250°	Injection timing
238°	During injection
44.135 mg	Mass of injected fuel (g/cycle)

Table 9. Improved Z24 engine valve timing.

IVO	12 c.a BTDC
IVC	35 c.a ABDC
EVO	43 c.a BBDC
EVC	3 c.a ATDC

emissions decrease. Of course, the trend of NOx emission is diverse but its amount decreases negligibly in comparison with previous state. All the results are shown in Table 7. Finally improved Z24 with these specifications is derived Table 8. To fix these specifications, we must examine the probability of knock.

Conclusion

In this work, NISSAN Z24 engine was simulated numerically to investigate the effects of valves and ignition timings and also modifying them to improve the engine performance and reducing the emissions Table 9. It has been concluded that: (a) By changing and improving the intake and exhaust valve timing and ignition timing, the engine performance (output power and emission) gets better; (b) The HC emission decreases by increasing the advanced angle of ignition; (c) The CO and CO_2 emission decreases by decreasing the air/fuel ratio, and (d) The NOx emission increases by increasing the advanced angle of ignition as increasing the advanced angle of ignition leads to increasing the combustion chamber temperature.

REFERENCES

Boyce BP, JK Martin, Poehlman AG, Shears D (1999). Effects of Ignition Timing and Air-Fuel Ratio on In-Cylinder Heat Flux and Temperatures in a Four-Stroke, Homogeneous Charge Engine, International Congress and Exposition, Detroit, MI, USA.

Duclos JM, Bruneaux G, Baritaud TA (1996). 3D Modeling of Combustion and Pollutants in a 4-Valve SI Engine; Effect of Fuel and Residuals Distribution and Spark location, SAE Technical Paper #961964.

Eckert P, Kong SC, Reitz RD (2003). Modelling Autoignition and Engine Knock Under Spark Ignition Conditions", SAE World Congress and Exhibition, Detroit, MI, USA.

Ferguson CR, Kirkpatrick AT (2001). Internal Combustion Engines, Applied Thermodynamics (2nd ed) Wiley.

Heywood JB (1988). Internal Combustion Engine Fundamentals", McGraw-Hill Inc.

Pulkrabeck WW (2003). Engineering Fundamental of Internal Combustion Engine (2nd ed), Prentice Hall.

Shayler PJ (2007). Experimental Investigations of Intake and Exhaust Valve Timing Effects on Charge Dilution by Residuals, Fuel Consumption and Emissions, SAE World Congress and Exhibition, April.

Siewert RM (1971). How Valve Timing Events Affect Exhaust Emissions, General Motors Corp, SAE Technical Papers #710609.

Smith JR, Green RM, CK WestBrook, Pitz WJ (1984). An Exprimental and Modeling Study of Engin Knock," Twentieth Symposium on Combustion Institute, Pittsburgh, PA.

Tuttle JH (1980). Controlling Engine Load by Means of Late Intake Valve Closing, SAE paper #800794.

Tuttle JH (1982). Controlling Engine Load by means of Early Intake-Valve Closing, SAE paper # 820408.

Zhu GG, Haskara I, Winkelman J, Corp V (2004). IC Engine Retard Ignition Timing Limit Detection and Control Using In-Cylinder Ionization Signal, Powertrain and Fluid Systems Conference and Exhibition, Tampa, FL, USA.

4

Study of the effect of particles size and volume fraction concentration on the thermal conductivity and thermal diffusivity of Al_2O_3 nanofluids

Faris Mohammed Ali, W. Mahmood Mat Yunus and Zainal Abidin Talib

Department of Physics, University Putra Malaysia, 43400 UPM, Serdang, Selangor, Malaysia.

In this study we present new data for the thermal conductivity enhancement in four nanofluids containing 11, 25, 50, and 63 nm diameter Aluminum oxide (Al_2O_3) nanoparticles in distilled water. The nanofluids were prepared using single step method (that is, by dispersing nanoparticle directly in base fluid) which was gathered in ultrasonic device for approximately 7 h. The transient hot-wire laser beam displacement technique was used to measure the thermal conductivity and thermal diffusivity of the prepared nanofluids. The thermal conductivity and thermal diffusivity were obtained by fitting the experimental data to the numerical data simulated for Al_2O_3 in distilled water. The results show that, the thermal conductivity and thermal diffusivity enhancement of nanofluids increases as the particle size increases. Thermal conductivity and thermal diffusivity enhancement of Al_2O_3 nanofluids was increase as the volume fraction concentration increases. This enhancement attributed to the many factors such as, ballistic energy, nature of heat transport in nanoparticle, and interfacial layer between solid/fluids.

Key words: Thermal conductivity, thermal diffusivity, effect of particle size, effect of volume fraction.

INTRODUCTION

Nanofluids are defined as the dispersed of solid nanoparticles (metal or metal oxide) in the fluid. Nanofluids have attracted much attention recently because of their potential as high performance heat transfer fluids in electronic cooling and automotive (Wen and Ding, 2006; Maiga et al., 2006). This interest begins with the work of Eastman et al. (1997); Lee et al. (1999); and Wang et al. (1999). They have reported the thermal conductivity enhancements of common heat transfer fluids such as ethylene glycol, water, oil, and small addition of solid nanoparticles. For example, Eastman has showed that, the enhancement of the thermal conductivity of Copper (Cu) nanoparticle suspension in ethylene glycol increases by 40% at volume fraction concentration of 0.3% (Eastman et al., 1997). Similarly, Choi and his co-researchers has illustrated that, the enhancement in the thermal conductivity of carbon nanotubes (NCT) suspension in oil was 160% measured at volume fraction concentration of 1% (Choi et al., 2001).

The effects of particle size, volume fraction, and temperature on thermal conductivity have been investigated by Michael et al. (2009a). They used a liquid metal transient hot wire device to measure the thermal conductivity of alumina dispersed in water, ethylene glycol, and ethylene glycol-water mixtures base fluids. The measurements have been carried out at a specific concentration and temperature, with an error of ± 2%. The results showed that, the thermal conductivity as a function of temperature was close to the value of base fluids. However, the maximum enhancement of thermal conductivity appeared at 400K and 380K, respectively. In general their results showed that, the thermal conductivity enhancement of alumina nanofluids at room temperature decreased as the particle size decreases, but increased with volume fraction concentration. The effect of particle size on the effective thermal conductivity of Al_2O_3 suspension in water nanofluids have been investigated using steady-state method (Calvin and Peterson, 2007). The nanoparticle with 36 and 47 nm were dispersed in water base fluid at different volume fraction ranging from 0.5 to 6%. The measurements were carried out for a temperature range of 27 to 37°C. They found out that, the enhancement in thermal conductivity of two types of nanofluids (36 and 47 nm) increased nonlinearly with volume fraction and temperature. The most significant finding was that, the largest enhancement was occurred at a temperature of approximately 32°C for volume fraction ranging from 2 to 4%.

The effect of particle size on some thermophysical properties such as thermal conductivity and viscosity of the nanofluids have been studied experimentally using a transient hot wire method and numerically studied using a simplified molecular dynamics method (Wen and Qing, 2008). In this case, the Aluminum oxide (Al_2O_3) nanoparticle with particle sizes of 35, 45, and 90 nm were suspended in water and ethylene glycol at different volume fraction concentrations were studied. The results showed a good agreement between the experimental data and numerical results. According to the numerical results, increasing the volume fraction concentration or decreasing the particle size of nanoparticle could increase the thermal conductivity of nanofluids. This is due to the increasing of viscosity of the nanofluid. For suitable particle size and volume fraction, the increasing of viscosity gives an enhancement of heat transfer capability.

The effect of particle size on the thermal conductivity of alumina Al_2O_3 nanoparticle suspension in water and ethylene glycol base fluids for particle size ranging from 8 to 282 nm have been conducted by Michael et al. (2009b). The thermal conductivity of the nanofluid was measured using a transient hot wire technique. The results show that, the enhancement in thermal conductivity of nanofluids increases with the increasing of particle size. They also developed the correlation between thermal conductivity enhancements with particle size.

The thermal conductivity of titanium dioxide, zinc-oxide, and alumina dispersed in water and ethylene glycol as base fluids have been measured using the transient hot-wire method (Kim et al., 2007). In this study, the effect of particle size and laser irradiation has been investigated. Emphasis was placed on the effect of particle size suspension in base fluid on the enhancement thermal conductivity. Also, the effect of laser pulse irradiation, which means, the changing of the particle size by laser ablation, was tested only for ZnO nanofluids. The results show that, the enhancement thermal conductivity of nanofluids increases linearly with decreasing the particle size, but no existing empirical or theoretical correlation can explain this behavior. In addition, the results illustrated that, high power laser irradiation can lead to substantial enhancement in the effective thermal conductivity although only a small volume fraction concentration of the nanoparticles are fragmented. However, the results show that, the thermal conductivity of the nanofluids increases linearly with the volume fraction concentration for Al_2O_3, ZnO, and TiO_2 nanofluids. Generally, the enhancement in the thermal conductivity that were observed in the approximately range of 15 to 25% were relative to the base fluid. The maximum enhancement was observed when the smallest nanoparticle size (10 nm) was suspended in ethylene glycol. The measurements of thermal conductivity of nanofluids using Transient Hot-Wire method was recently reported by Walvekar et al. (2012) and Pang and Kang (2012). They have carried out the investigation of the effect of particle volume fraction and the concenration of stabilizing agent as well as the fluid temperature on the thermal conductivity of NCT nanofluid. Their results show that, the thermal conductivity of the sample increases with the increasing of concentration of CNT, stabilizing agent, and temperature.

In the present study, we report measurements of the thermal conductivity and thermal diffusivity of Al_2O_3 nanofluids using transient hot-wire laser beam displacement technique. Four different particle sizes of nanofluid were studied in order to investigate the effect of particle size on the thermal conductivity and thermal diffusivity enhancement. The transient hot-wire laser beam displacement technique was chosen with two reasons; first we can obtained the thermal conductivity and thermal diffusivity values from a single measurement; secondly, we can simply eliminate the experiment error due to the convection process in liquid.

MATERIALS AND METHODS

Preparation of nanofluids

The Al_2O_3 nanoparticles with average particle sizes of 11, 25, 50, and 63 nm were purchased from Nanostructured and Amorphous Materials, Inc. USA. The nanofluid samples were prepared using the one-step method, similar to the one reported by Choi (1998). The Al_2O_3 nanoparticles were suspended in the distilled water base fluid to produce 5 different volume fraction concentrations (0.225, 0.55, 0.845, 1.124, and 1.4%) samples. The nanoparticles and base fluid were mixed and kept in an ultrasonic bath for more than 7 h. Acetyl trimethyl ammonium bromide (CTAB) was added as a surfactant for the mixtures. The size and the particle distribution of the nanoparticles in base fluids were verified and measured using TEM (Hitachi 7100 TEM).

Experimental setup

The experimental setup is based on the Transient Hot-Wire-laser beam displacement technique. It consists of a heating a wire, a CW He-Ne laser, as the probe beam, a lens with a focal length of 50

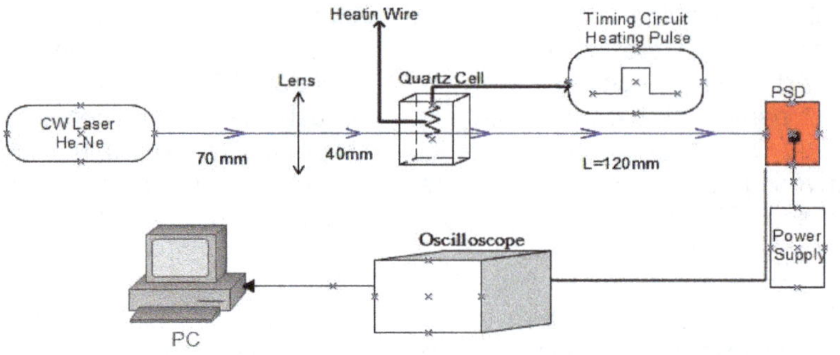

Figure 1. Schematic diagram of the hot wire-laser probe beam displacement technique

mm and a quartz cell as the nanofluids container. A position sensitive detector (PSD) was used to collect the signal of the deflected laser beam. The beam deflection signal was monitored using a Lecroy 9310A Digital Oscilloscope. Figure 1 shows the schematic diagram of the set up. A CW He-Ne laser operated at 632.8 nm with an output power of 2 mW serves as the probe beam. The measurements were carried out using the nanofluid quartz cell with the light path of 10.5 mm. The heating wire diameter and resistance was measured to be 254 μm and 2.38 Ω/cm, respectively. The wire was positioned vertically and perpendicular to the probe beam. A home-made timing circuit was used to control the heating current through the resistance wire to provide a pulse heating source. The heating pulse was 0.35 s with a heating voltage of 5 volts.

Thermal and laser beam displacement models

The purpose of a thermal model is to derive the fundamental correlations for conduction heat transfer performance analysis of the nanofluid, which is considered as a single phase fluid in the conventional approach (Yimin and Wilfried, 2000). A thermal model of heat conduction with guess values of k_{nf} and α_{nf} needs to be established first to solve the temperature distribution within the test liquid. Then the temperature profile of the laser probe beam deflection can be determined by relating the deflection angle with the solved temperature gradient through a beam deflection model (Jonathan et al., 1990). The heat conduction equations in the wire, fluid and nanoparticles are established using the cylindrical coordinate system in dimensionless forms as shown in Equations 1, 2, and 3, respectively.

$$\gamma_1 \frac{\partial u_w(r,\tau)}{\partial \tau} = \frac{\partial^2 u_w(r,\tau)}{\partial r^2} + \frac{1}{r}\frac{\partial u_w(r,\tau)}{\partial r} + g(r,\tau) \tag{1}$$

$$\gamma_2 \frac{\partial u_f(r,\tau)}{\partial \tau} = \frac{\partial^2 u_f(r,\tau)}{\partial r^2} + \frac{1}{r}\frac{\partial u_f(r,\tau)}{\partial r} \tag{2}$$

$$\gamma_3 \frac{\partial u_{np}(r,\tau)}{\partial \tau} = \frac{\partial^2 u_{np}(r,\tau)}{\partial r^2} + \frac{1}{r}\frac{\partial u_{np}(r,\tau)}{\partial r} \tag{3}$$

Where τ is the dimensionless time constant, *u* is the dimensionless temperature and **γ** is the dimensionless variable.

In a conventional approach, it is assumed that, both the liquid and particle phases are in thermal equilibrium and flow at the same velocity, therefore, the nanofluids behaved like a common pure fluid. This implies that, the whole equation of a single-phase fluid has been applied directly to the nanofluids (Wongwises and Weerapun, 2007). As the thermal and flow equilibrium between the fluid phase and nanoparticles, a uniform shape and size for nanoparticles becomes the most important parameter to be considered for enhancement in the thermal conductivity and thermal diffusivity of the nanofluid (Yimin and Wilfried, 2000; Calvin and Peterson, 2010; Omid and Ahmad, 2009). Therefore, we get:

$$\frac{\partial u_f(r,\tau)}{\partial \tau} \cong \frac{\partial u_{np}(r,\tau)}{\partial \tau} \tag{4}$$

$$\frac{\partial u_f(r,\tau)}{\partial r} \cong \frac{\partial u_{np}(r,\tau)}{\partial r} \quad and \quad \frac{\partial^2 u_f(r,\tau)}{\partial r^2} \cong \frac{\partial^2 u_{np}(r,\tau)}{\partial r^2} \tag{5}$$

For this reason, we can say that Equation 2 is approximately equal to Equation 3 so the dimensionless heat conduction equations of a single-phase fluid and nanoparticles becomes:

$$\gamma_{nf} \frac{\partial u_{nf}(r,\tau)}{\partial \tau} = \frac{\partial^2 u_{nf}(r,\tau)}{\partial r^2} + \frac{1}{r}\frac{\partial u_{nf}(r,\tau)}{\partial r} \tag{6}$$

The dimensionless boundary conditions are written as:

$$\left.\frac{\partial u(r,t)}{\partial r}\right|_{r=0} = 0 \tag{7}$$

$$k_{12} \left.\frac{\partial u(r,t)}{\partial r}\right|_{\substack{r\to 1\\ r\langle 1}} = \left.\frac{\partial u(r,t)}{\partial r}\right|_{\substack{r\to 1\\ r\rangle 1}} \tag{8}$$

$$u(r,t)\big|_{\substack{r=1\\ r\langle 1}} = u(r,t)\big|_{\substack{r=1\\ r\rangle 1}} \tag{9}$$

$$u(r,t)\big|_{r\to\infty} \to 1 \tag{10}$$

With the dimensionless initial condition given as;

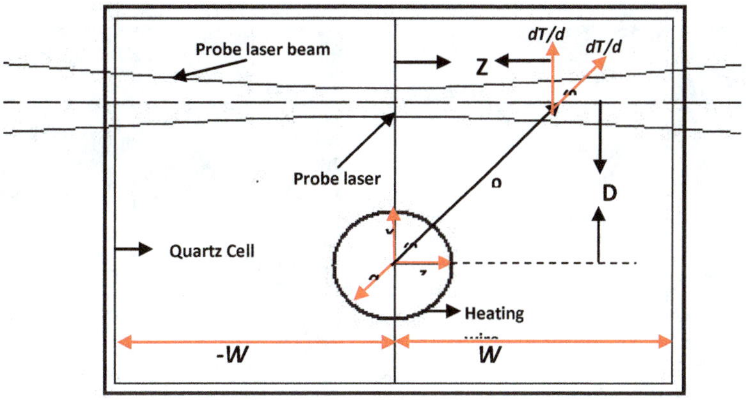

Figure 2. Geometry of heating wire and probe laser beam.

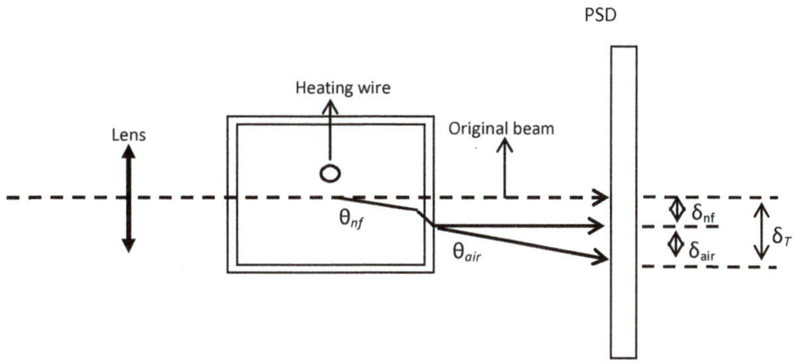

Figure 3. Optical path of probe laser beam.

$$u(r,0)=1 \tag{11}$$

and

$$\gamma_1=\frac{a^2}{\alpha_1 t_p},\qquad \gamma_{nf}=\frac{a^2}{\alpha_2 t_p},\qquad k_{12}=\frac{k_1}{k_{nf}} \tag{12}$$

Where t_p is the heating pulse duration; a is the radius of the wire, and k and α are thermal conductivity and thermal diffusivity respectively.

The volumetric heat generation $g(r,\tau)$ rate equation in dimensionless form is:

$$g(r,\tau)=\begin{cases}\dfrac{VI}{\pi T_0 k_1 l} & r\leq l,\ \tau\leq\tau_0\\ 0 & r>1,\ \tau>\tau_0\end{cases} \tag{13}$$

Where V and I are the voltage and current of heating wire respectively.

The beam deflection angle equation in the nanofluid caused by gradient of temperature (θ) can be expressed as:

$$\theta_{nf}=\int_0^{L_f}\frac{\nabla n\cdot\vec{a}_x}{n}dz=\frac{1}{n_0}\int_0^{L_f}\left(\nabla n\cdot\vec{a}_x\right)dz \tag{14}$$

Where n is the nanofluid refractive index , L_{nf} is the probe beam path length in nanofluid , $\vec{a}_x$ is the spatial unit vector perpendicular to the original probe beam, z is the spatial direction parallel to the original beam path and n_o is the normal value and can be ignored from the integral in Equation 14. The refractive index is a function of temperature; therefore the refractive index gradient and the temperature gradient can be related to each other by the equation:

$$\nabla n=\frac{dn}{dT}\nabla T \tag{15}$$

In order to derive the beam deflection position on the detector, we consider the sample cell geometry and the beam deflection as shown in Figures 2 and 3 respectively. From Figure 2 we can write Equation 14 in term of cell geometry as:

$$\theta_{nf}=\frac{1}{n}\frac{dn}{dT}\int_{-W}^{W}\frac{D}{\sqrt{D^2+Z^2}}\cdot\frac{dT}{d\rho}dz \tag{16}$$

Where D is the distance between the heating wire and probe beam,

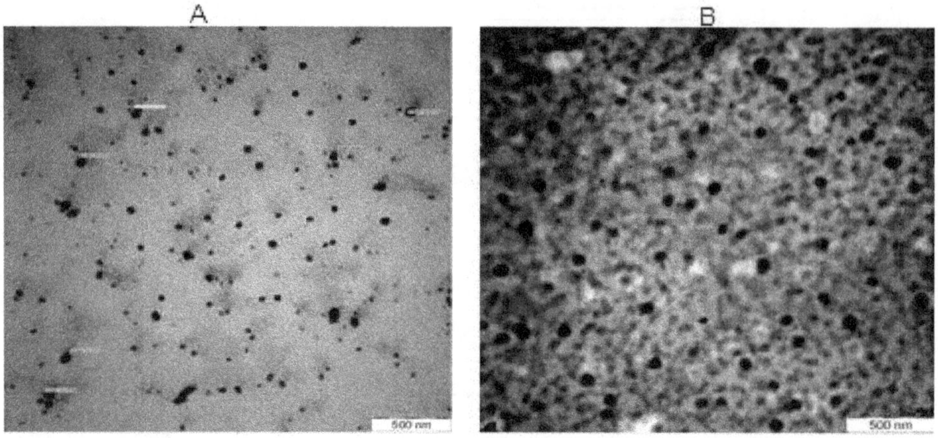

Figure 4. TEM image of Al2O3 (50 nm) nanoparticles dispersed in water after 7 h sonication, (a) volume fraction concentration 1.4% (b) volume fraction concentration 1.124%.

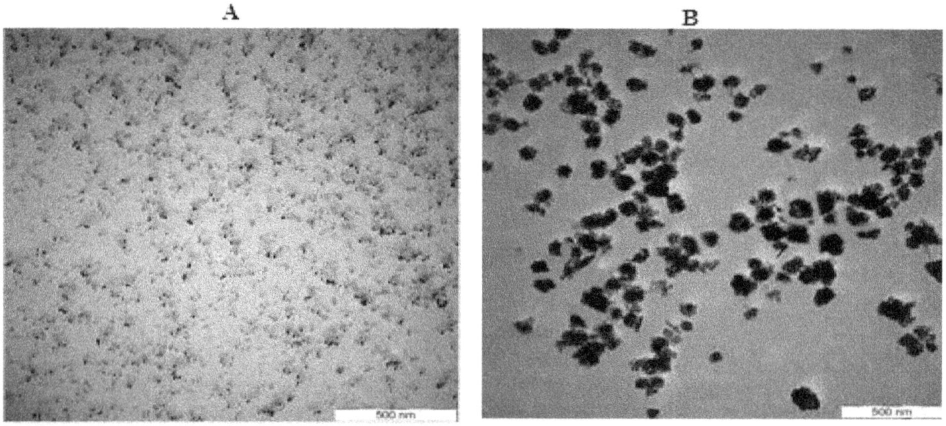

Figure 5. TEM image of Al_2O_3 (11 nm) nanoparticles dispersed in water after 7 h sonication, (a) volume fraction concentration 1.4% (b) volume fraction concentration 1.124%.

W the inner distant of the cell region within which the probe beam deflected as shown in Figure 2. Thus, the probe beam deflection angle in the nanofluid depends on the temperature gradient *dT/dρ* which can be obtained by solving the transient heat conduction equation in the heating wire and nanofluid numerically as discussed above. Therefore, the total beam deflection (δ_T , in Figure 3) on position sensitive detector (PSD) is the sum of the beam deflection in the nanofluid (δ_{nf}) and in air (δ_{air}) as shown in Figure 3, which can be expressed as:

$$\delta_T^n = \frac{w - W + n_o L_{air}}{n_o} \frac{dn}{dT} \frac{T_o}{a} \frac{\Delta z}{2} \sum_{i=0}^{N_z} \frac{D}{\sqrt{D^2 + Z_{i+1}^2}} \frac{du_{nf,i+1}^n}{dr} + \frac{D}{\sqrt{D^2 + Z_i^2}} \frac{du_{nf,i}^n}{dr} \tag{17}$$

Where *W* is the half width of the quartz cell, T_o Initial temperature, Δz is the grid spacing along z-axis and L_{air} is the probe beam path length in the air. The non-dimensional term, *du/dr* is related to the temperature distribution as:

$$\frac{du}{dr} = \frac{a}{T_o} \frac{dT}{d\rho} \tag{18}$$

The numerical data were obtained by solving the transient heat conduction equations of the heating wire and the nanofluid medium using the finite difference Method based on Crank-Nicolson algorithm. Thermal conductivity and thermal diffusivity of Al_2O_3 nanofluids were simultaneously obtained by fitting the experimental data to the generated numerical data for different particle size of nanofluids (11, 25, and 50 nm) samples.

RESULTS AND DISCUSSION

Figures 4 and 5 show the TEM images of nanoparticle distribution of Al_2O_3 nanoparticles in the base distilled water fluids after 6 h in the sonication process. The images show that, the Al_2O_3 nanoparticles were

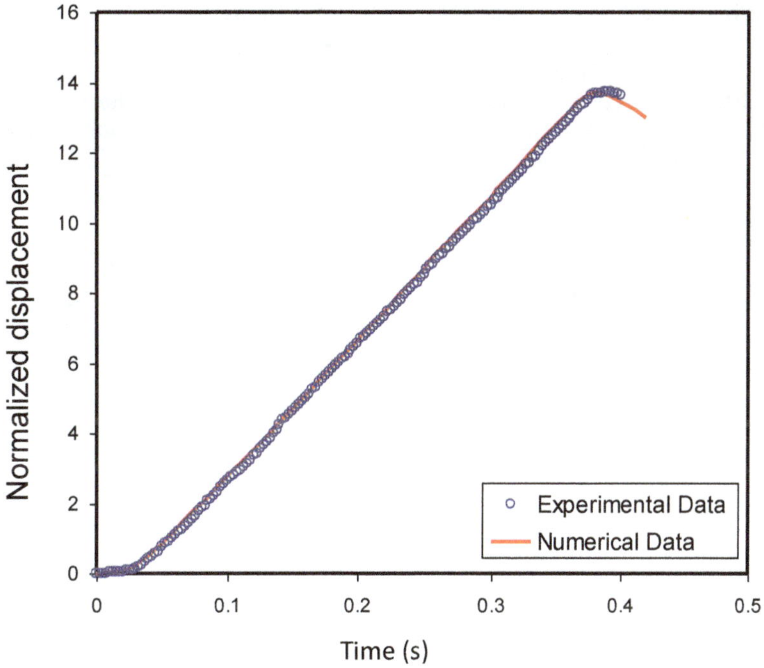

Figure 6. Normalized probe laser beam displacement for Al_2O_3 (50 nm) in water at D = 300 µm.

aggregated to form nanoparticle clusters and evenly distributed in the base fluids. These showed that, the nanoparticles are in nano-metric scale around (14 to 70 nm) and are semi-spherical in shape. The measurement of thermal diffusivity and the thermal conductivity is based on the energy equation for heat conduction and the deflection of the laser probe beam.

Figure 6 shows a typical plot of normalized displacement signal as a function of time for Al_2O_3 (50 nm) in water measured at D = 300 µm. The thermal conductivity and thermal diffusivity of were simultaneously obtained by fitting the experimental data to the numerical values calculated using Equation 17. The best value was recorded when the smallest total error between the numerical data and the experimental data was obtained that is:

$$\varepsilon = \sum_{j=0}^{N} (v_{exp} - v_{model})^2 \qquad (19)$$

Where v_{exp} and v_{model}, are the experimental and numerical model values of the normalized position, sensitive detector output voltage, respectively.

The results of the thermal conductivity and thermal diffusivity of Al_2O_3 (11 nm) nanoparticles suspension in distilled water were 0.646, 0.655, 0.661, 0.669, 0.676 W/m.K, and (1.5758, 1.6219, 1.6457, 1.6738, 1.7270) × 10^{-7} m^2/s at volume fraction concentrations of 0.225, 0.55, 0.845, 1.124 and 1.4%, respectively. Obviously, the thermal conductivity and thermal diffusivity of Al_2O_3 (63 nm) were higher than thermal conductivities and thermal diffusivities of Al_2O_3 with particle sizes of 11, 25, and 50 nm. Table 1 shows the thermal conductivity and thermal diffusivities of Al_2O_3 nanoparticles suspension in distilled water at different particle size and different volume fraction concentration. These results showed that, the thermal conductivity and thermal diffusivity of Al_2O_3 increased with the increasing of particle size and volume fraction.

Effect of particle size on thermal properties of nanofluids

The thermal conductivity and diffusivity of nanofluids measured at different volume fractions from 1.4 to 0.225% and particle sizes ranging from 11 to 63 nm are presented in Table 1 and Figures 7 to 16. All measurements were made at room temperature with the hot wire-laser probe beam displacement technique. The samples consisted of Al_2O_3 nanoparticles dispersed in distilled water. Figures 7 to 16 displayed the thermal conductivities and thermal diffusivities for Al_2O_3 nanofluids as a function of particle size. Moreover, it displayed the dependence of the thermal conductivity and thermal diffusivity on the particle size, as well as, note that the slope increases as the particle size increase. The thermal conductivity and thermal diffusivity values of these nanofluids generally increase nonlinearly with increases particle size.

Table 1. Thermal conductivity and diffusivity of Al_2O_3 (nanofluids in distilled water)

Particles size (nm)	Volume fraction (%)	Thermal conductivity (W/m.K)	Thermal diffusivity $(m^2/s)x10^{-7}$	Enhancement in thermal conductivity (%)	Enhancement in thermal diffusivity (%)
11	1.40	0.676	1.7270	9.56	16.84
11	1.124	0.669	1.6738	8.42	13.24
11	0.845	0.661	1.6457	7.13	11.34
11	0.55	0.655	1.6219	6.32	9.73
11	0.225	0.646	1.5758	4.70	6.61
25	1.40	0.681	1.7344	10.37	17.34
25	1.124	0.674	1.6954	9.23	14.7
25	0.845	0.666	1.6549	7.94	11.96
25	0.55	0.658	1.6246	6.64	9.91
25	0.225	0.647	1.5790	4.86	6.83
50	1.4	0.689	1.7477	11.6	18.24
50	1.124	0.676	1.7050	9.56	15.35
50	0.845	0.668	1.6683	8.26	12.87
50	0.55	0.660	1.6272	6.96	10.09
50	0.225	0.648	1.5813	4.93	6.98
63	1.4	0.705	1.7930	14.26	21.31
63	1.124	0.695	1.7407	12.64	17.77
63	0.845	0.680	1.6880	10.21	14.2
63	0.55	0.669	1.6379	8.42	10.81
63	0.225	0.649	1.5836	5.18	7.1

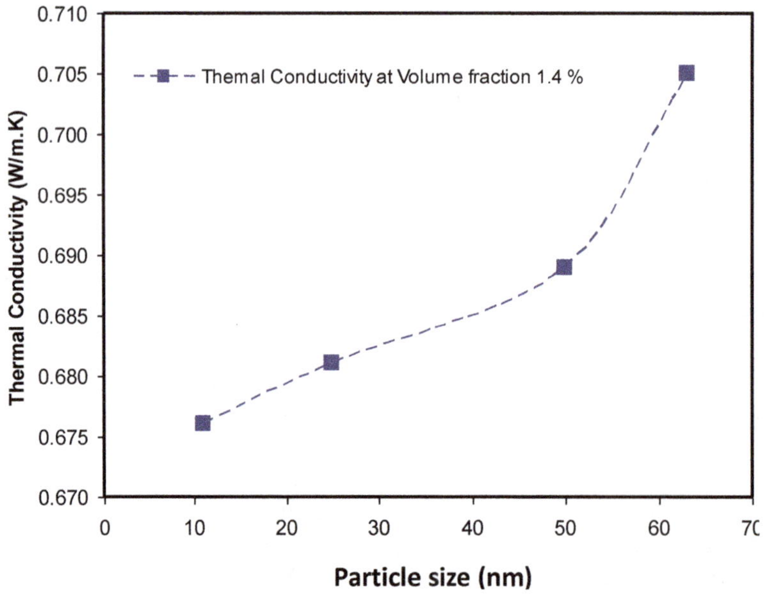

Figure 7. Variation of thermal conductivity of Al_2O_3 nanofluid with particle size obtained for 1.4 % volume fraction.

Furthermore, the thermal conductivity and thermal diffusivity of the smallest nanoparticles are lower than that of the largest nanoparticles. This would be attributed to phonon scattering at the solid–liquid interface. The

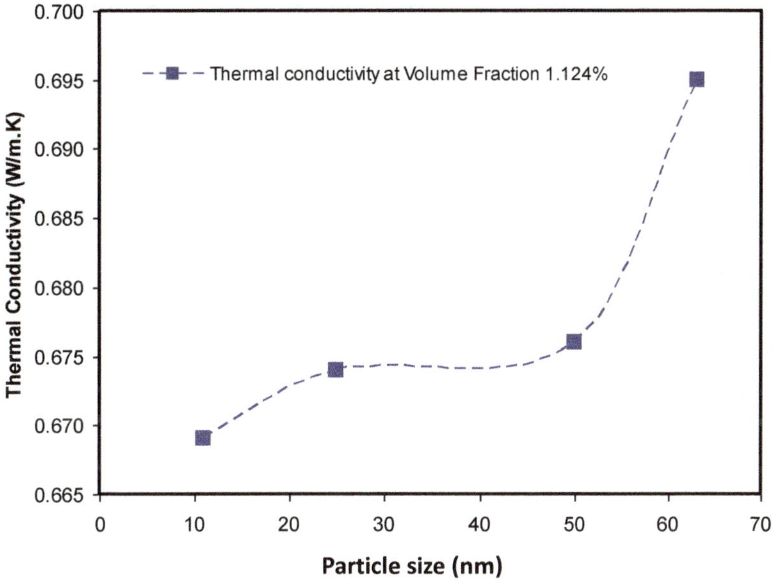

Figure 8. Variation of thermal conductivity of Al_2O_3 nanofluid with particle size obtained for 1.124 % volume fraction.

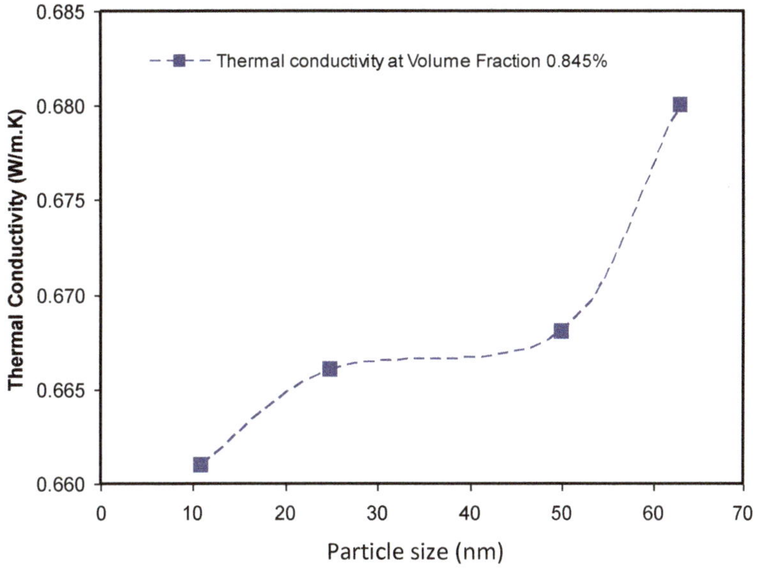

Figure 9. Variation of thermal conductivity of Al_2O_3 nanofluid with particle size obtained for 0.845% volume fraction.

phonon mean free path in small particles may be reduced by phonon–phonon scattering, scattering at the boundaries between nanoparticle and molecules, as well as, lattice imperfections. However, the relationship between the particle size and its thermal conductivity and thermal diffusivity are nonlinearly dependent for the particle size as shown in Figures 7 to 16.

The lattice imperfections in smaller nanoparticles cannot remarkably affect the thermal conductivity and thermal diffusivity. This indicates that, the phonon–phonon scattering at the boundary between nanoparticle and fluid interfacial can be more remarkable than that at the lattice imperfections. Though lattice imperfections are readily formed in small particles, the thermal conductivity and thermal diffusivity of nanoparticles is mainly subject to its size. Therefore, the results suggest that, the thermal conductivity of nanofluid is subject to a size-dependent effect. The results show that, the enhancement thermal

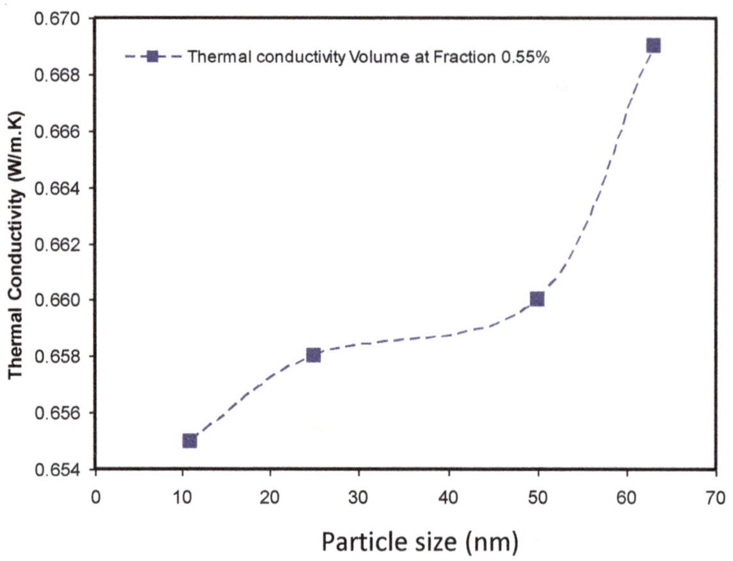

Figure 10. Variation of thermal conductivity of Al_2O_3 nanofluid with particle size obtained for 0.55 volume fraction.

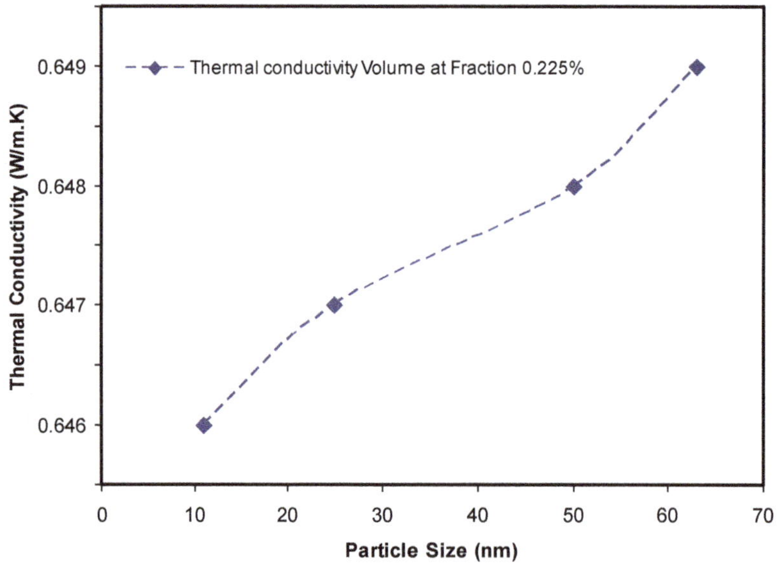

Figure 11. Variation of thermal conductivity of Al_2O_3 nanofluid with particle size obtained for 0.225% volume fraction.

conductivity were 9.56, 10.37, 11.6, 14.26% at particle size 11, 25, 50, and 63 nm at volume fraction 1.4%, respectively. While, the enhancement thermal diffusivity of Al_2O_3 nanofluid was 16.84, 17.34, 18.24, 21.31 at particle size 11, 25, 50, and 63 nm at volume fraction 1.4%, respectively. These results indicate that, the enhancement thermal conductivities and thermal diffusivities of Al_2O_3 nanoparticle suspension in distilled water have increased with increase in the particle size of nanoparticles. The results are listed in the Table 1.

Effect of volume fraction on thermal properties of nanofluids

Figures 17 to 24 shows the effect of volume fraction concentration on the thermal conductivity and thermal diffusivity of 11, 25, 50, 63 nm Al_2O_3 nanoparticles suspended in distilled water. The enhancement in thermal conductivity and thermal diffusivity increased with the increasing volume fraction concentration of nanoparticles. It is also observable that, the thermal

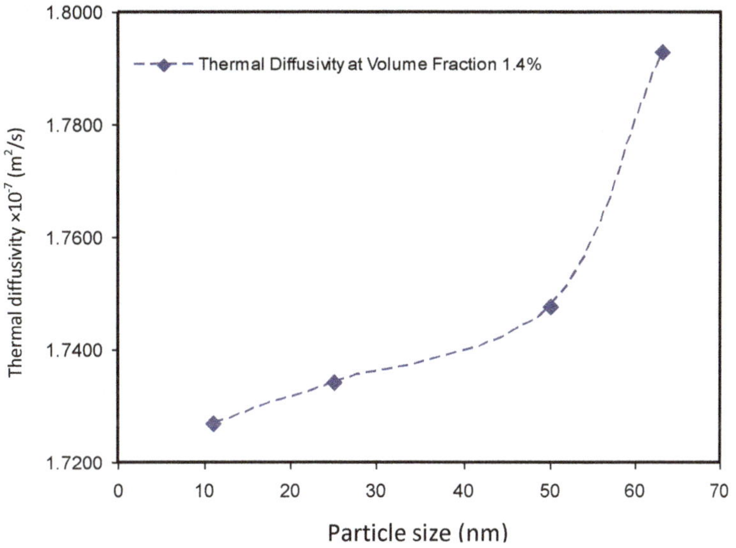

Figure 12. Variation of thermal diffusivity of Al_2O_3 nanofluid with particle size obtained for 1.4% volume fraction.

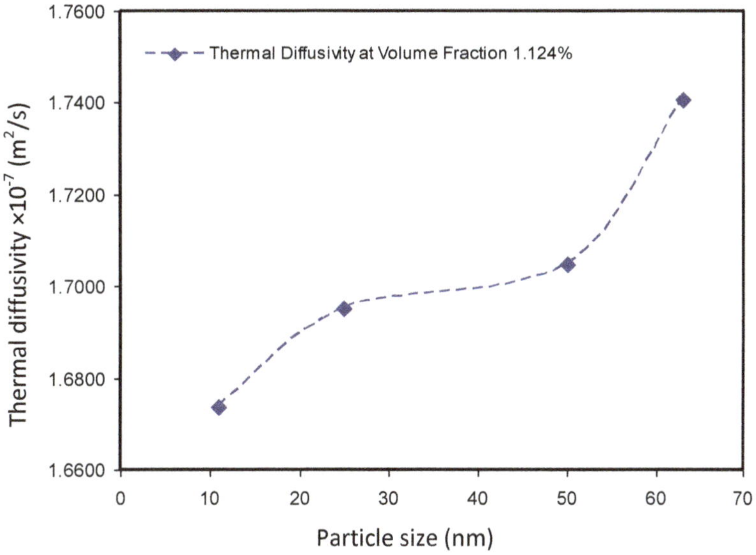

Figure 13. Variation of thermal diffusivity of Al_2O_3 nanofluid with particle size obtained for 1.124% volume fraction.

conductivity and thermal diffusivity increases linearly with the volume fraction concentration of nanoparticles. This observation can provide an insight into the mechanism of the thermal exchanger transport in nanofluids. We particularly mention the volume fraction concentration of nanoparticles dependence for thermal conductivity and thermal diffusivity in accordance with Figures 17 to 24, because the thermal conductivity and diffusivity would shows more enhancements if the nanoparticles formed suspensions in base fluids and this could be more profound in nanofluids containing a higher volume fraction of nanoparticles. Al_2O_3 (25 nm) nanofluid exhibit 10.37% enhancement with 1.4 % volume fraction of nanoparticles in distilled water while the nanofluid presents 9.23, 7.94, 6.64, 4.86% enhancement with 1.124, 0.845, 0.55, and 0.225 volume fraction nanoparticles in water, respectively. Al_2O_3 nanofluids exhibit 9.56, 8.42, 7.13, 6.32, and 4.70% enhancements with 1.4, 1.124, 0.845, 0.55, and 0.225% volume fraction concentration suspensions in distilled water, respectively. All results of Al_2O_3 of (11, 25, 50, and 63 nm) are listed in Table 1.

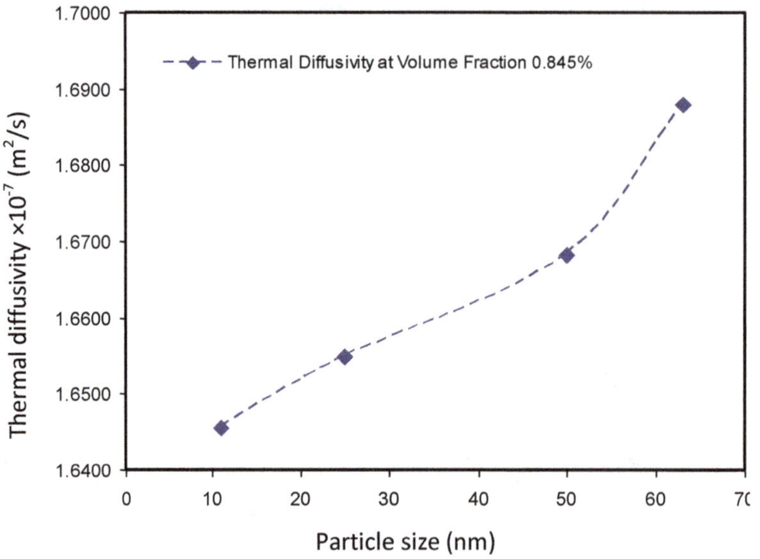

Figure 14. Variation of thermal diffusivity of Al_2O_3 nanofluid with particle size obtained for 0.845% volume fraction.

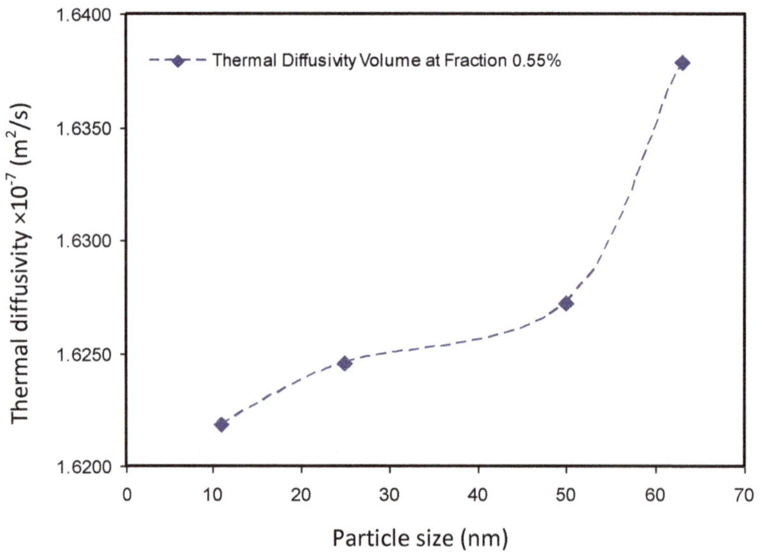

Figure 15. Variation of thermal diffusivity of Al_2O_3 nanofluid with particle size obtained for 0.55% volume fraction.

The nanoparticles suspension demonstrates some unique and novel thermal properties when compared to the traditional heat transfer of fluids. There are several mechanisms that will enhancement the thermal properties of nanofluids; Brownian motion of nanoparticles, interfacial layer between solid\fluids, nanoparticle structuring/aggregation, and effects of nanoparticle clustering (Liqiu, 2009). First, Brownian motion of nanoparticles could contribute to the thermal conduction enhancement through two ways, direct contribution due to motion of nanoparticles that transports heat, and indirect contribution due to micro-convection of fluid surrounding individual nanoparticles. The direct contribution of Brownian motion has been taken by comparing the time scale of particle motion with that of heat diffusion in the fluid. Equivalently, we can compare the time required for particle to move by the distance equal to its size with time required for heat to move in the liquid by the same distance (Liqiu, 2009; Keblinski et al., 2002). The indirect contribution has also been shown to

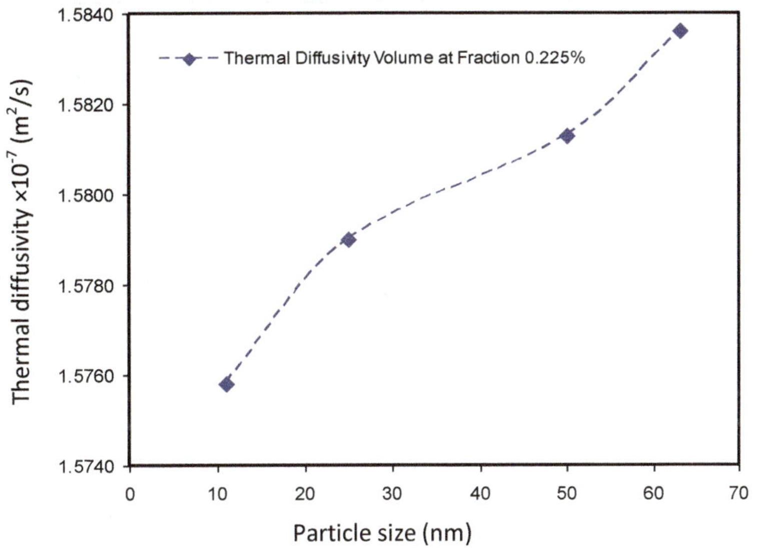

Figure 16. Variation of thermal diffusivity of Al_2O_3 nanofluid with particle size obtained for 0.225% volume fraction.

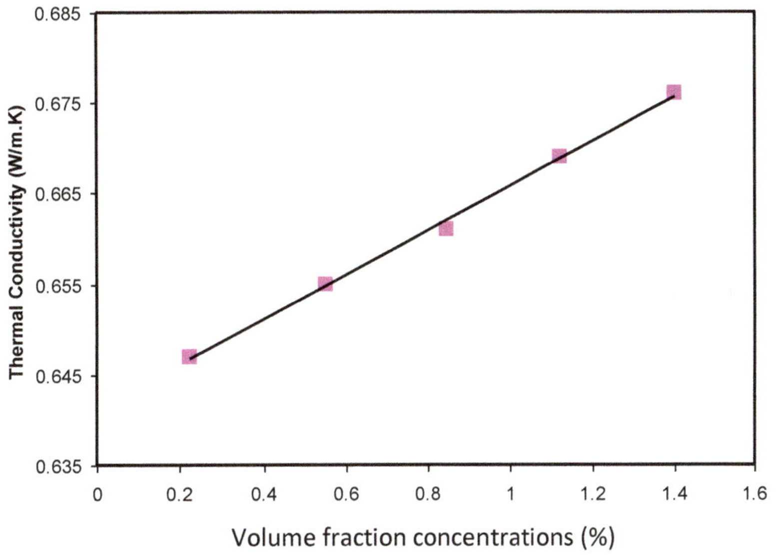

Figure 17. Thermal conductivity of Al_2O_3 (11nm) nanofluid as a function of Volume fraction concentration.

play a minute role through the same reasoning for the direct contribution and also through molecular modeling (Evans et al., 2006). Secondly, interfacial thermal resistance would lead to an increase in the estimate value of thermal conductivity of nanofluid and an increase thermal conductivity with increase particle size, which satisfy the experimental results.

An interface effect that could enhance thermal conductivity is the layering of the fluid at the solid interface, by which the atomic structure of the liquid layer is significantly more ordered than that of bulk liquid. Given that, crystalline solids (which are obviously ordered) display much better thermal transport than liquids, such liquid layering at the interface would be expected to lead to a higher thermal conductivity (Keblinski et al., 2002). To evaluate an upper limit for the effect of the interfacial layer, let us suppose that, the thermal conductivity of this interfacial fluid is exactly same as that of the solid nanoparticle. The resultant larger effective volume of the nanoparticle-layered-fluid

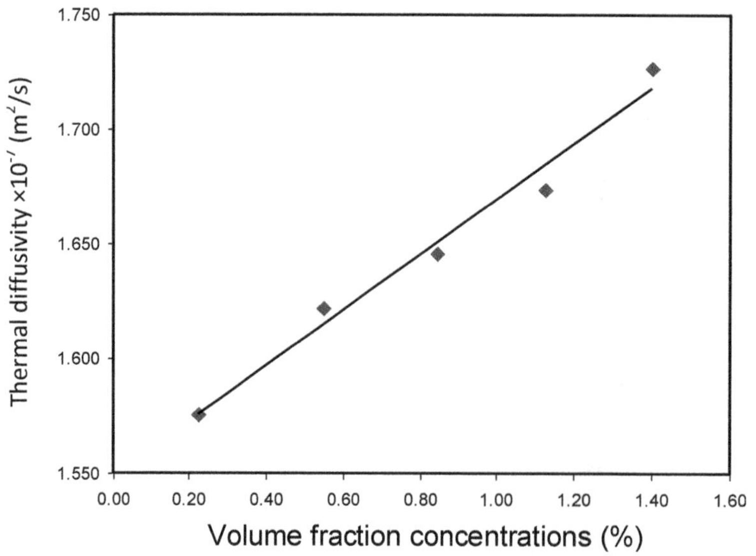

Figure 18. Thermal conductivity of Al_2O_3 (11nm) nanofluid as a function of volume fraction concentration.

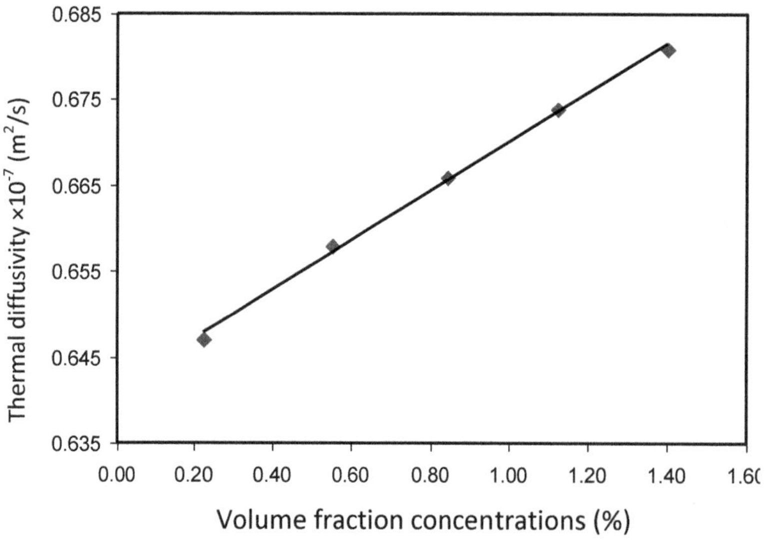

Figure 19. Thermal conductivity of Al_2O_3 (25 nm) nanofluid as a function of Volume fraction concentration.

structure could enhance the thermal conductivity of nanofluid. Third, nanoparticle structuring/aggregation have dominated mechanism for the thermal conductivity enhancement of nanofluids, due to interconnected nanoparticles in the fluid enhances the thermal conduction.

Lastly, by creating paths of lower thermal conductivity resistance, clustering of particles into percolating would have major effect on the effective thermal conductivity. The effective volume of cluster, that is, the volume from which other clusters are excluded, can be much larger than the physical volume of the particles. Since within such clusters, heat can move very rapidly, the volume fraction of highly conductive phase is larger than the volume of solid which is significantly increase thermal conductivity. The effect of clustering is depending on the ratio of the volume of the solid particles in the cluster to the total volume of the cluster. With decreasing packing fraction, the effective volume of the cluster increases, thus enhancing thermal conductivity. A further dramatic increase of enhancement thermal conductivity can take place if the nanoparticles do not need to be in physical

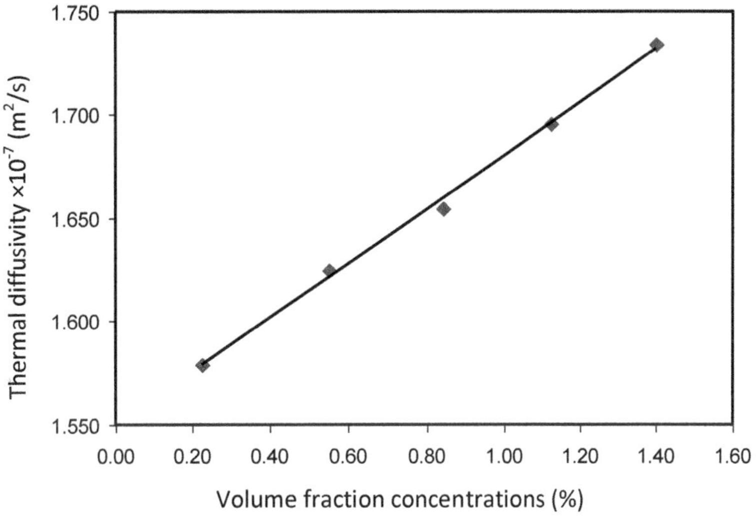

Figure 20. Thermal diffusivity of Al_2O_3 (25 nm) nanofluid as a function of Volume fraction concentration.

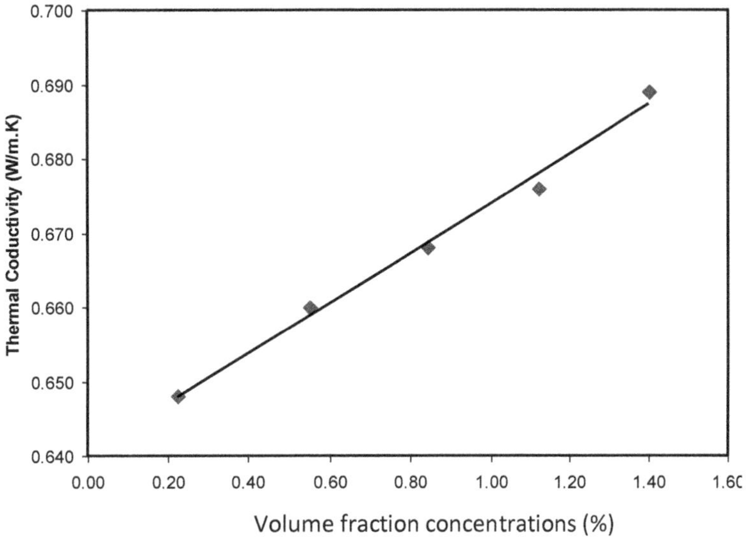

Figure 21. Thermal conductivity of Al_2O_3 (50 nm) nanofluid as a function of volume fraction concentration.

contact, but just within a specific distance, allowing rapid heat flow between them. Such fluid-mediated clusters exhibit a very low packing fraction and thus, a very large effective volume. The principal is capable of explaining the unusually large experimental observed in enhancement of the thermal conductivity (Keblinski et al., 2002).

The results shows significant enhancement in the thermal conductivity and thermal diffusivity of Al_2O_3 nanofluid. For example, the nanofluid with the particle size of 11 nm, the enhancement of thermal conductivity and thermal diffusivity were recorded and it varies from 4.70 at 0.225% volume fraction to 9.56 at 1.4%, and 6.61 at 0.225% volume fraction to 16.84 at 1.4% volume fraction respectively. However, for nanofluid with the particle size 63 nm, the thermal conductivity varies from 5.18 at 0.225% volume fraction to 14.26 at 1.4% volume fraction. While, thermal diffusivity varies from 7.1 at 0.225% to 21.31 at 1.4% volume fraction.

Conclusion

The coupled transient heat conduction equations of the

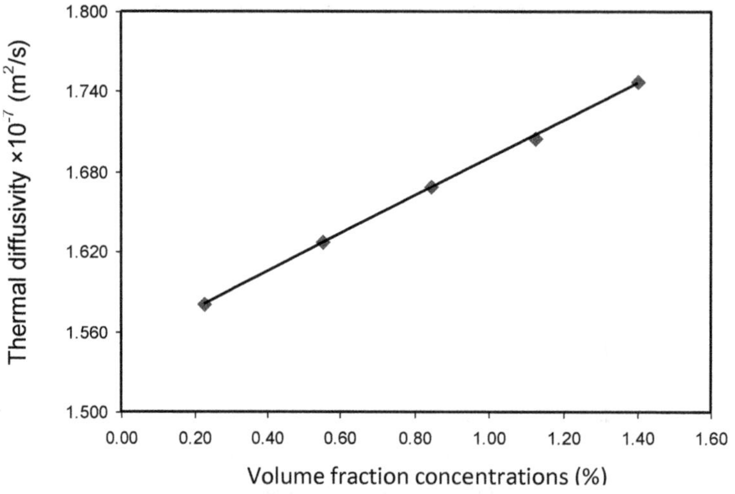

Figure 22. Thermal diffusivity of Al_2O_3 (50 nm) nanofluid as a function of volume fraction concentration.

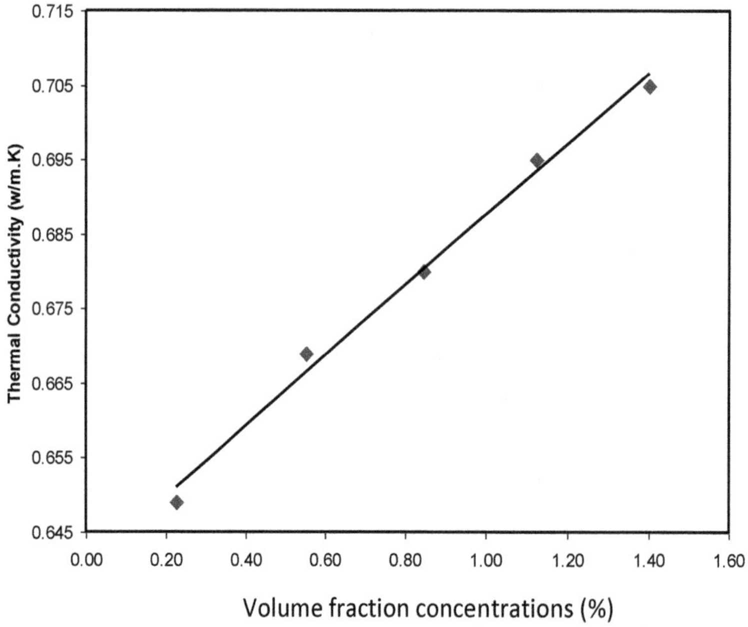

Figure 23. Thermal conductivity of Al_2O_3 (63 nm) nanofluid as a function of volume fraction concentration.

heating wire and the nanofluid were solved simultaneously using the finite difference Method based on Crank- Nicolson algorithm to get thermal conductivity and thermal diffusivity of Al_2O_3 (11, 25, 50, and 63 nm) nanofluids. Crank-Nicolson algorithm was used to discretize the coupled transient heat conduction equations. New experimental data of the thermal conductivity and thermal diffusivity of Al_2O_3 (11, 25, 50, and 63 nm) nanoparticles suspension in distilled water were obtained in this study.

Results show that, the thermal conductivity and thermal diffusivity are increase with increasing the particle size and volume fraction concentration. Since our measurements were carried at room temperature the Brownian motion effect can be ignored, thus, our results show that, interfacial layer between solid/fluids, nanoparticle structuring/aggregation particle clustering, particle size, and the volume fraction concentration of particles in the water base fluid give a significant effect on thermal conductivity and thermal diffusivity of nanofluids.

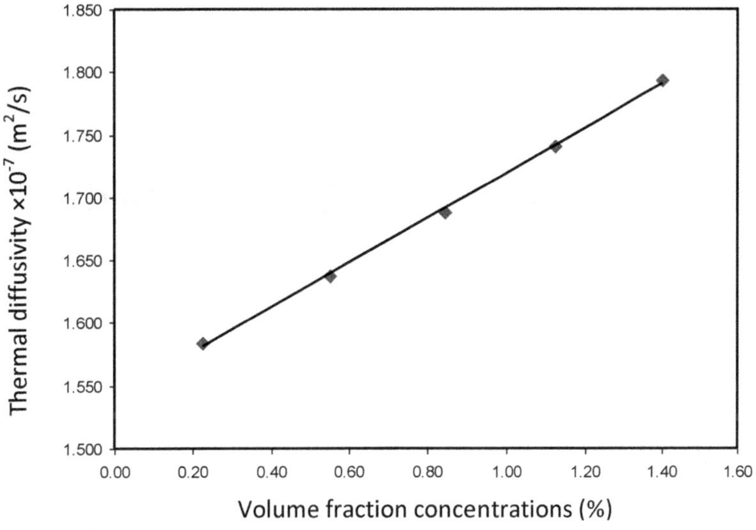

Figure 24. Thermal diffusivity of Al_2O_3 (63 nm) nanofluid as a function of volume fraction concentration.

ACKNOWLEDGEMENTS

We gratefully acknowledge the Department of Physics, UPM for providing the research facilities to enable us to carry out this research. We also like to acknowledge Ministry of Education for the financial support through Fundamental research grant (01-11-08-664FR/5523664) and University research grant (05-02-12-2179RU).

REFERENCES

Calvin HL, Peterson GP (2007). The Effect of Particle Size on the Effective Thermal Conductivity of Al2O3-Water Nanofluids. J. Appl. Phys. 101(4):044312-044316.

Calvin HL, Peterson GP (2010). Experimental Studies of Natural Convection Heat Transfer of Al_2O_3/DIWater Nanoparticle Suspensions (Nanofluids). Adv. Mech. Eng. pp. 1-10.

Choi SUS (1998). Nanofluid technology: Current Status and Future research, Korea-US Technical Conf. on Strategic Technologies, Vienna, V.A.

Choi SUS, Zhang Z, Yu W, Lockwood FE, Grulke EA (2001). Anomalous Thermal Conductivity Enhancement in Nanotube Suspensions. Appl. Phys. Lett. 79:2252-2254.

Eastman J, Choi SUS, Li S, Thompson LJ, Lee S (1997). Enhanced Thermal Conductivity Through The Development of Nanofluids. Mater. Res. Soc. Symptoms. 457:3-11.

Evans W, Fish J, Keblinski P (2006). Role of Brownian motion Hydrodynamics on Nanofluids Thermal Conductivity. Appl. Phys. Lett. 88(9):093116-093118.

Jonathan DS, Richard ER, Robert JS (1990). Collinear Photothermal Deflection Spectroscopy With Light Scattering Samples. Appl. Optics. 29(28):4225-4234.

Keblinski P, Phillpot SR, Choi SUS, Eastman JA (2002). Mechanisms Of Heat Flow in Suspensions of Nano-Sized. Int. J. Heat Mass Transfer. 45:855-863.

Kim SH, Choi SR, Kim D (2007). Thermal Conductivity of Metal-Oxide Nanofluids: Particle Size Dependence and Effect of Laser Irradiation. J. Heat Transfer. 129:298-307.

Lee S, Choi SUS, Li S, Eastman J (1999). Measuring thermal conductivity of fluids containing oxide nanoparticles. J. Heat Transfer ASME. 121(2):280-289.

Liqiu W (2009). Advances in Transport Phenomena, Springer-Verlag Berlin Heidelberg.

Maiga S, Nguyen C, Galanis N, Gilles R, Thierry M, Mickael C (2006). Heat Transfer Enhancement in Turbulent Tube Flow Using Al_2O_3 Nanoparticle Suspension. Int. J. Numer. Methods Heat Fluid Flow. 16:275-292.

Michael PB, Yanhui Y, Pramod W, Amyn ST (2009a). The Thermal Conductivity of Alumina Nanofluids in Water, Ethylene Glycol, and Ethylene Glycol + Water Mixtures. J. Nanopart. Res. 12(4):1469-1477.

Michael PB, Yanhui Y, Pramod W, Amyn ST (2009b). The Effect of Particle Size on the Thermal Conductivity of Alumina Nanofluids. J. Nanopart. Res. 11:1129-1136.

Omid A, Ahmad F (2009). Numerical Investigation of Natural Convection of Al_2O_3 Nanofluid in Vertical Annuli. Heat Mass Transfer. 46(1):15-23.

Pang C, Kang YT (2012). Stability and Thermal Conductivity Characteristics of Nanofluids (H_2O/CH_3OH + NaCl + Al_2O_3 Nanoparticles) for CO_2 Absorption Application. International Refrigeration and Air Conditioning Conference at Purdue.

Walvekar R, Faris IA, Khalid M (2012). Thermal Conductivity of Carbon Nanotube Nanofluid – Experimental and Theoretical Study. Heat Transfer – Asian Res. 41:145-163.

Wang X, Xu X, Choi SUS (1999). Thermal Conductivity of Nanoparticle-Fluid Mixture. J. Thermophy. Heat Transfer. 13:474-480.

Wen D, Ding Y (2006). Natural Convective Heat Transfer of Suspensions of Titanium Dioxide Nanoparticles (Nanofluids). IEEE Trans. Nanotechnol. 5(3):220-227.

Wen QL, Qing MF (2008). Study for the Particle's Scale Effect on Some Thermophysical Properties of Nanofluids by a Simplified Molecular Dynamics Method. Eng. Anal. Bound. Elem. 32:282-289.

Wongwises S, Weerapun D (2007). A Critical Review of Convective Heat Transfer of Nanofluids, Renew. Sustain. Energy Rev. 11:797-817.

Yimin X, Wilfried R (2000). Conceptions for Heat Transfer Correlation of Nanofluids, Int. J. Heat Mass Transfer. 43:3701-3707.

5

Metamaterials in microwave applications: A selective survey

Vipul Sharma[1], S. S. Pattnaik[2] and Tanuj Garg[1]

[1]Department of Electronics and Communication Engineering, Gurukul Kangri University, Haridwar, India.
[2]Department of ETV, NITTTR, Chandigarh, India.

Recently, metamaterial has grabbed prime focus of research fraternity working in the field of antennas, microwave filters and components design. Many different structures of split ring resonators (SRR) for their respective applications have been reported till date. In all of these findings in some way or other, using metamaterial, drastic reduction in size of the component has been achieved. Thus, the metamaterial can be viewed as a powerful miniaturization tool in the field of radio frequency (RF) and microwave. Since its inception in 1999, many variants of SRR have come up leading to more efficient designs with better RF properties. This review paper presents an application based review of such variants and will be useful for new researchers exploring its use in their research problems.

Key words: Metamaterial, antenna miniaturization, negative refractive index metamaterial (NRIM),split ring resonators (SRR).

INTRODUCTION

In a pioneering paper, written in 1968, Prof. V.G. Vaselego coined a term Metamaterial, a hypothetical material having negative value of electrical and magnetic permeabilities and theoretically he derived the properties (Vaselego, 1968). For almost thirty years after its inception, there was hardly any substantial work reported this field. In 1999, Pendry et al. first realized negative permeability by introducing periodic array of nonmagnetic conducting units. They demonstrated that periodic array of nonmagnetic conductor split rings show negative effective permeability at high frequency side of resonance (Pendry et al., 1999). It was first time in year 2000, when Smith et al. came up with physical realization of simultaneously negative values of effective magnetic permeability and electrical permittivity and thus a negative value of refractive index (NRIM) in microwave regime (Smith et al., 2000). The structure they called left handed medium comprised of periodic array of interspaced conducting nonmagnetic split ring resonators and continuous wires. The array conducting wire provided electric field resonance and thus negative value of electrical permittivity. Since then many variants of split ring structure have been reported by researchers for different engineering applications. But complete potential of metamaterial is yet to be explored. The main objective of this review paper is to discuss different variants of metamaterial structure and their applicability in microwave engineering problems.

METAMATERIALS IN MICROWAVE APPLICATIONS

Metamaterial in the form of various structures has been extensively used by the microwave researchers to design miniaturized microwave components, filters and antennas. Sabah et al. (2010) presented a method of mechanical tuning of metamaterial resonant frequency by varying substrate thickness. In the paper, a metamaterial structures whose unit cell has triangular split ring resonator (TSRR) and wire strip (WS) have been analyzed for S- and C- microwave bands. Retrieval method has been used for calculation of effective material parameters. Simulation results show that the new metamaterials exhibits double negative properties in the frequency region of interest and their resonant frequency can be tuned by varying substrate thickness. In Sulaiman et al. (2010), the authors proposed two types of metamaterial composed of patch antennas, one antenna using metamaterial as substrate and the other one using metamaterial as cover. Perfectly electric conductor in omega shape on RT 5880 substrate has been used as metamaterial unit. When metamaterial unit is used as substrate of patch antenna, there exist drastic reduction in size of the patch, and becomes comparable to the metamaterial unit. The reduction in size is also accompanied by improvement in return loss. The price paid is the bandwidth. This antenna is found suitable for narrow band applications. When planar array slab consisting of metamaterial units is used as a cover of conventional patch antenna, significant improvement in directivity was also observed.

In Ziolkowski et al. (2009), the authors present design, fabrication and simulation of electrically small, coaxially fed metamaterial inspired Z antennas having ability to change resonant frequency by changing value of lumped element inductor. The structure consists of a coaxially fed monopole printed on one side of RT5880 substrate, while Z element printed on the other side. The Z element is composed by joining two “J” elements connected by a lumped element inductor. The authors have tested the design for two Z-antennas one at 570 MHz and another at 300 MHz. The results show that by changing value of lumped element inductor joining the two “J” elements, the resonant frequency of the antenna can be successfully changed.

Further, Palandoken et al. (2009) also presented a novel application of metamaterial for antenna miniaturization. Metamaterial is composed of unit cells having spiral rings connected to metal strips on both sides of FR4 substrate, arranged in 2×3 planar array form. The antenna comprises of a printed dipole directly connected to three of six unit cells. Rectangular slot in truncated ground plane allows better impedance matching. It has been shown that loaded dipole gives broad bandwidth about 63% with drastic reduction in size.

Ziolkowski et al. (2009) also explored the metamaterial property to design antenna at 300 MHz. This paper demonstrates electrically small, metamaterial inspired 3D magnetic EZ antenna, consisting of extruded capacitively loaded loop (CLL) driven by coaxially fed semi loop antenna. Resonant frequency of the antenna is found tunable by amount of Quartz filled capacitive gap of CLL) element. The antenna is found to be nearly matched without any external circuit and shows more than 95% efficiency with a fractional bandwidth of 1.66%. Sharma et al. (2011) introduced elliptical split ring resonator (ESRR structure to get metamaterial property. The structure is dual fed in offset and shows negative value refractive index at multiple frequencies. When tested for radiation properties the ESRR gives highly directional radiation pattern. In the design of metamaterial antennas some researchers also addressed the issue of optimization in design. Vidyalakshmi et al. (2009) proposed a hybrid of genetic algorithm (GA) and artificial neural network (ANN) called GA ANN for optimization of metamaterial split ring resonator to achieve given resonant frequency. Size of outer square ring, conductor width, dielectric spacing between outer and inner rings are inputs to the optimization program and error between desired resonant frequency and calculated resonant frequency has been taken as cost function to be minimized. ANN has been used to obtain fitness of chromosomes. Performance GA ANN when compared with stand alone GA and stand alone ANN, GA ANN converges more fast and gives higher accuracy. Sumanta et al. (2012) introduced a new structure, N-sided Regular Polygon Split Ring Resonator (NRPSRR). They have also nicely presented the mathematical analysis and derivation of the structure. Further my increasing the limit of N to infinity, they have obtained resonant frequency for cicular SRR. Genetic Algorithm has been successfully used to obtain simulation model of the structure.

Conclusion

The paper presents a selective review of findings reported by various researchers in the field of metamaterial research and its application in microwave structures. Various microwave structures using metamaterial for miniaturization, have been discussed. Mathematical modelling of metamaterial is a new dimension; the research fraternity is focussing now. Less research has been reported in this particular field till date. The aspect has been addressed in this survey. This paper will go a long way for the new scientists interested to use metamaterial in microwave application and miniaturization problems.

Conflict of Interests

The author(s) have not declared any conflict of interests.

REFERENCES

Palandoken M, Grede A, Heino H (2009). Broadband Microstrip Antenna With Left-Handed Metamaterials", IEEE Trans. Ant. Prop. 57(2):331-338.

Pendry JB, Holden AJ, Robbins DJ, Stewart WJ (1999). Magnetism from conductors and enhanced nonlinear phenomena. IEEE Trans. Microwave Theory Technique, 47:2075–2084.

Sabah C (2010). Tunable Metamaterial Design Composed of Triangular Split Ring Resonator and Wire Strip for S- and C- Microwave bands", PIER B 22:341-357.

Smith DR Padilla Willie J, Vier DC, Nemat-Nasser SC, Schultz S (2000). Composite Medium with Simultaneously Negative Permeability and Permittivity, Phys. Rev. Lett. 84(18):4184-4187.

Sulaiman AA, Othman A, Jusoh MH, Baba NH, Awang RA, Ain MF (2010). Small Patch Antenna on Omega Structure Metamaterial. Europ. J. Sci. Res. 43(4):527-537.

Sumanta Bose M. Ramaraj Dr. S (2012). Raghavan, Swadhyaya Kumar, Mathematical Modeling, Equivalent Circuit Analysis and Genetic Algorithm Optimization of an N-sided Regular Polygon Split Ring Resonator (NRPSRR), 2nd International Conference on Communication, Computing. Security [ICCCS], Procedia Technology 6:763-770.

Vidyalakshmi MR, Raghavan S (2010). Comparison of Optimization Techniques for Square Split Ring Resonator, Int. J. Microwave. Optical Technol. 5(5):280-286.

Ziolkowski RW, Jin P, Nielsen JA, Tanielian MH, Holloway CL (2009). Experimental Verification of Z Antennas at UHF Frequencies", IEEE Antenna and Wireless Propagation Lett. 8:1329-1333.

Ziolkowski RW, Lin CC, Nielson ja, Tanielian MH, Holloway CL (2009). Design and Experimentation Verification of a 3D Magnetic EZ Antenna at 300 MHz", IEEE Ant Wireless Prop Letters. P. 8.

6

Effects of heat absorption and chemical reaction on a three dimensional MHD convective flow past a porous plate

N. Ahmed and K. Kr. Das

Department of Mathematics, Gauhati University, Guwahati-781014, India.

An attempt has been made to investigate the effects of heat sink and chemical reaction on a three dimensional Magneto hydrodynamics (MHD) convective flow with mass transfer of an incompressible viscous electrically conducting fluid past a porous vertical plate with transverse sinusoidal suction velocity. A magnetic field of uniform strength is assumed to be applied transversely to the direction of the main flow. The magnetic Reynolds number is considered to be small that induced magnetic field can be neglected. The governing equations are solved by regular perturbation technique. The expression for velocity field, temperature field, species concentration, current density, the skin friction, Nusselt number and Sherwood number at the plate are obtained in non dimensional forms. The effect of Hartman number, chemical reaction parameter, heat sink parameter on the velocity field, zeroth order skin friction and the amplitude of the first order skin friction, first order Nusselt number and the first order Sherwood number at the plate are discussed graphically. It is seen that chemical reaction and heat sink have significant effects on the flow and on the heat and mass transfer characteristics.

Key words: Three-dimensional convective flow, heat transfer, incompressible viscous fluid, wall shear stress, heat sink.

INTRODUCTION

The investigation of magneto hydrodynamics (MHD) convection with mass transfer problems in presence of transverse magnetic field have attracted the attention of a number of scholars because of its wide application in many branches of science and technology such as geophysics, astrophysics, plasma physics, missile technology, etc. Engineers employ MHD principles in the design of heat exchangers, pumps and flow meters, thermal protection, etc. From technological point of view, MHD convection flow problems are also very significant in the fields of stellar and planetary magnetospheres, aeronautics, chemical engineering and electronics. MHD is also stabilizing a flow against the transition from laminar to turbulent flow and in reduction of turbulent drag and suppression of flow separation. The application of MHD principles in medicine and biology are of paramount interest owing to their significance in biomedical engineering in general and in the treatment of various pathological states in particular. Applications in biomedical engineering include cardiac magnetic resonance imaging (MRI), electro cardio gram (ECG) etc. The principle of dynamo and motor is a classical example of MHD convection.

The problems of above phenomena of MHD convection have been studied by many authors. Ferraro and Plumpton (1966), Cramer and Pai (1973) and Sanyal and Bhattacharya (1992) are some of them. The problem of the convection flows arising in fluids as a result of interaction of the force of gravity and density difference caused by simultaneous diffusion of thermal energy and chemical species have been investigated by Bejan and Khair (1985), Raptis and Kafousias (1982) and Ahmed et al. (2005).

The effect of three dimensional flow caused by the periodic suction perpendicular to the main flow when the difference between the wall temperature and free steam temperature gives rise to buoyancy force in the direction of the free steam on heat transfer characteristic was investigated by Ahmed and Sarma (1997), Singh et al. (1998) and Choudhury and Chand (2002). Recently Jain and Gupta (2006) have investigated the effect of transverse sinusiodal injection velocity distribution on three dimensional free convective Couette flow of a various incompressible fluid in slip flow regime under the influence of heat sink. An analytical solution to the problem of the three dimensional free convective flow of an incompressible viscous fluid past a porous vertical plate with transverse sinusoidal suction velocity taking into account the presence of species concentration was obtained by Ahmed et al. (2006).

In many times it has been observed that foreign mass reacts with the fluid and in such a situation chemical reaction plays an important role in chemical industry. Theoretical descriptions of non-linear chemical dynamics have been presented by Epstein and Pojman (1998) and Gray and Scott (1990). The effects of chemical reaction and mass transfer on MHD flow past a semi-infinite plate was analysed by Devi and Kandasamy (2000). The effects of mass transfer, Soret effect and chemical reaction on an oscillatory MHD free convective flow through a porous medium have been investigated by Ahmed and Kalita (2010).

In view of the importance of the combined effect of chemical reaction and heat absorption, it is proposed to study a problem of three dimensional MHD convective flows past a porous vertical infinite plate with chemical reaction and heat absorption. The infinite plate assumption is one such classical idealization of great practical importance. Although the flow over a flat plate is the simplest case of boundary layer development in external flow, yet its significance cannot be undervalued because of its relevance to numerous engineering applications. Several configurations such as flow over airfoils, turbine blades, ship hulls, etc. can initially be estimated as flow past flat plates (Scheme 1). The justification of considering the three dimensional flow is that most of the fluid flows that occur in nature are three dimensional. Of course we have chosen a simple model of a three dimensional flow caused by transverse sinusoidal suction velocity.

The objective of the present work is to investigate the effect of chemical reaction as well as heat sink on a three dimensional convective flow past a porous plate. Our work is a generalization to the work done by Ahmed and Sarma (2010).

BASIC EQUATIONS

The equations governing the steady motion of an incompressible viscous electrically conducting fluid in presence of a magnetic field are:

The equation of continuity: $\operatorname{div}\vec{q} = 0$ (1)

The Gauss's law of magnetism: $\operatorname{div}\vec{B} = 0$ (2)

The momentum equation:

$$\left(\vec{q}.\vec{\nabla}\right)\vec{q} = -\frac{1}{\rho}\vec{\nabla}p + \frac{\vec{J}\times\vec{B}}{\rho} + \nu\nabla^2\vec{q} + \vec{g} \tag{3}$$

The energy equation:

$$\rho C_p\left[\left(\vec{q}.\vec{\nabla}\right)\bar{T}\right] = k\nabla^2\bar{T} + \varphi + \frac{\vec{J}^2}{\sigma} + Q_0\left(\bar{T}_\infty - \bar{T}\right) \tag{4}$$

The species continuity equation:

$$\left[\left(\vec{q}.\vec{\nabla}\right)\bar{C}\right] = D_M\nabla^2\bar{C} + D_T\nabla^2\bar{T} + \bar{K}\left(\bar{C}_\infty - \bar{C}\right) \tag{5}$$

The Ohm's law: $\vec{J} = \sigma\left[\vec{E} + \vec{q}\times\vec{B}\right]$ (6)

We now consider the steady convective flow of an incompressible viscous electrically conducting fluid in presence of heat sink taking into account the species concentration and chemical reaction past a vertical porous plate with transverse sinusoidal suction velocity as mentioned earlier by making the following assumptions:

(i) All the fluid properties except the density in the buoyancy term are constant.

(ii) A magnetic field of uniform strength B_0 is applied transversely to the direction of the main steam.

(iii) The magnetic Reynolds number is so small that the induced magnetic field can be neglected.

(iv) The viscous dissipation and magnetic dissipation energy are negligible.

(v) $\bar{T}_w > \bar{T}_\infty$ and $\bar{C}_w > \bar{C}_\infty$.

We introduce a co-ordinate system $\left(\bar{x},\bar{y},\bar{z}\right)$ with X-axis vertically upwards along the plate, Y-axis perpendicular to it and directed into the fluid region and Z-axis along the width of the plate. Let $\vec{q} = \bar{u}\hat{i} + \bar{v}\hat{j} + \bar{w}\hat{k}$ be the fluid velocity at the point $\left(\bar{x},\bar{y},\bar{z}\right)$ and $B_0\hat{j}$ be the applied magnetic field, $\hat{i},\hat{j},\hat{k}$ being the unit vectors along +ve X-axis, Y-axis and Z-axis respectively. The suction velocity is taken as follows:

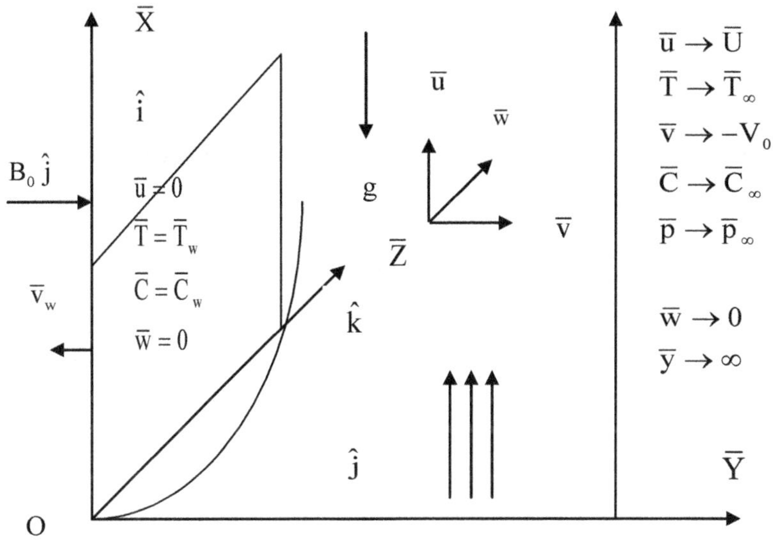

Scheme 1. Flow configuration.

$$\bar{v}_w(\bar{z}) = -V_0\left[1+\varepsilon\cos\frac{\pi\bar{z}}{L}\right] \tag{7}$$

which consists of a basic steady distribution $-V_0$ with a superimposed weak distribution $-\varepsilon V_0\cos\left(\frac{\pi\bar{z}}{L}\right)$. Since the plate is infinite in length in X-direction, therefore all the quantities except possibly the pressure are assumed to be independent of $\bar{x}$. With the foregoing assumptions and under usual boundary layer and Boussinesq approximation, Equations 1, 3, 4 and 5 are reduced to Equation of continuity:

$$\frac{\partial\bar{v}}{\partial\bar{y}}+\frac{\partial\bar{w}}{\partial\bar{z}}=0 \tag{8}$$

Momentum equations:

$$\bar{v}\frac{\partial\bar{u}}{\partial\bar{y}}+\bar{w}\frac{\partial\bar{u}}{\partial\bar{z}}=g\beta\left(\bar{T}-\bar{T}_\infty\right)+g\bar{\beta}\left(\bar{C}-\bar{C}_\infty\right)+\nu\left(\frac{\partial^2\bar{u}}{\partial\bar{y}^2}+\frac{\partial^2\bar{u}}{\partial\bar{z}^2}\right)+\frac{\sigma B_0^2}{\rho}\left(\bar{U}-\bar{u}\right) \tag{9}$$

$$\bar{v}\frac{\partial\bar{v}}{\partial\bar{y}}+\bar{w}\frac{\partial\bar{v}}{\partial\bar{z}}=-\frac{1}{\rho}\frac{\partial\bar{p}}{\partial\bar{y}}+\nu\left(\frac{\partial^2\bar{u}}{\partial\bar{y}^2}+\frac{\partial^2\bar{u}}{\partial\bar{z}^2}\right) \tag{10}$$

$$\bar{v}\frac{\partial\bar{w}}{\partial\bar{y}}+\bar{w}\frac{\partial\bar{w}}{\partial\bar{z}}=-\frac{1}{\rho}\frac{\partial\bar{p}}{\partial\bar{z}}+\nu\left(\frac{\partial^2\bar{w}}{\partial\bar{y}^2}+\frac{\partial^2\bar{w}}{\partial\bar{z}^2}\right)-\frac{\sigma B_0^2\,\bar{w}}{\rho} \tag{11}$$

Energy equation:

$$\bar{v}\frac{\partial\bar{T}}{\partial\bar{y}}+\bar{w}\frac{\partial\bar{T}}{\partial\bar{z}}=\alpha\left(\frac{\partial^2\bar{T}}{\partial\bar{y}^2}+\frac{\partial^2\bar{T}}{\partial\bar{z}^2}\right)+\frac{Q_0\left(\bar{T}_\infty-\bar{T}\right)}{\rho C_p} \tag{12}$$

Species continuity equation:

$$\bar{v}\frac{\partial\bar{C}}{\partial\bar{y}}+\bar{w}\frac{\partial\bar{C}}{\partial\bar{z}}=D_M\left(\frac{\partial^2\bar{C}}{\partial\bar{y}^2}+\frac{\partial^2\bar{C}}{\partial\bar{z}^2}\right)+D_T\left(\frac{\partial^2\bar{T}}{\partial\bar{y}^2}+\frac{\partial^2\bar{T}}{\partial\bar{z}^2}\right)+\bar{K}\left(\bar{C}_\infty-\bar{C}\right) \tag{13}$$

Equation 2 is satisfied by $\vec{B}=B_0\,\hat{j}$. The symbols are defined in the nomenclature. The relevant boundary conditions are:

$$\text{at } \bar{y}=0:\quad \bar{u}=0,\ \bar{v}=\bar{v}_w,\ \bar{w}=0,\ \bar{T}=\bar{T}_w,\ \bar{C}=\bar{C}_w \tag{14}$$

$$\text{at } \bar{y}\to\infty:\ \bar{u}=\bar{U},\ \bar{v}=-V_0,\ \bar{w}=0,\ \bar{T}=\bar{T}_\infty,\ \bar{C}=\bar{C}_\infty,\ \bar{p}=\bar{p}_\infty \tag{15}$$

We introduce the following non-dimensional quantities:

$$\left.\begin{array}{l} y=\frac{\bar{y}}{L},\ z=\frac{\bar{z}}{L},\ u=\frac{\bar{u}}{V_0},\ v=\frac{\bar{v}}{V_0},\ U=\frac{\bar{U}}{V_0},\ w=\frac{\bar{w}}{V_0},\ \theta=\frac{\bar{T}-\bar{T}_\infty}{\bar{T}_w-\bar{T}_\infty},\ Q=\frac{Q_0L}{\rho C_pV_0}, \\ \phi=\frac{\bar{C}-\bar{C}_\infty}{\bar{C}_w-\bar{C}_\infty},P_r=\frac{\nu}{\alpha},S_c=\frac{\nu}{D_M},S_r=\frac{D_T\left(\bar{T}_w-\bar{T}_\infty\right)}{\nu\left(\bar{C}_w-\bar{C}_\infty\right)},G_r=\frac{Lg\beta\left(\bar{T}_w-\bar{T}_\infty\right)}{V_0^2},K=\frac{\bar{K}L}{V_0} \\ G_m=\frac{Lg\bar{\beta}\left(\bar{C}_w-\bar{C}_\infty\right)}{V_0^2},M=\frac{\sigma B_0^2\nu}{\rho V_0^2},\ R_e=\frac{V_0L}{\nu},\ p=\frac{\bar{p}}{\rho\left(\frac{\nu}{L}\right)^2},\ p_\infty=\frac{\bar{p}_\infty}{\rho\left(\frac{\nu}{L}\right)^2} \end{array}\right\} \tag{16}$$

The non-dimensional forms of Equations 8, 9, 10, 11, 12 and 13

$$\frac{\partial v}{\partial y}+\frac{\partial w}{\partial z}=0 \tag{17}$$

$$v\frac{\partial u}{\partial y}+w\frac{\partial u}{\partial z}=G_r\,\theta+G_m\,\phi+\frac{1}{R_e}\left(\frac{\partial^2u}{\partial y^2}+\frac{\partial^2w}{\partial z^2}\right)+MR_e\left(U-u\right) \tag{18}$$

$$v\frac{\partial v}{\partial y}+w\frac{\partial v}{\partial z}=-\frac{1}{R_e^2}\frac{\partial p}{\partial y}+\frac{1}{R_e}\left(\frac{\partial^2v}{\partial y^2}+\frac{\partial^2v}{\partial z^2}\right) \tag{19}$$

$$v\frac{\partial w}{\partial y}+w\frac{\partial w}{\partial z}=-\frac{1}{R_e^2}\frac{\partial p}{\partial z}+\frac{1}{R_e}\left(\frac{\partial^2 w}{\partial y^2}+\frac{\partial^2 w}{\partial z^2}\right)-MR_e w \quad (20)$$

$$v\frac{\partial \theta}{\partial y}+w\frac{\partial \theta}{\partial z}=\frac{1}{P_r R_e}\left(\frac{\partial^2 \theta}{\partial y^2}+\frac{\partial^2 \theta}{\partial z^2}\right)-Q\theta \quad (21)$$

$$v\frac{\partial \phi}{\partial y}+w\frac{\partial \phi}{\partial z}=\frac{1}{S_c R_e}\left(\frac{\partial^2 \phi}{\partial y^2}+\frac{\partial^2 \phi}{\partial z^2}\right)+\frac{S_r}{R_e}\left(\frac{\partial^2 \theta}{\partial y^2}+\frac{\partial^2 \theta}{\partial z^2}\right)-K\phi \quad (22)$$

with relevant boundary conditions:

$$y=0\ :\ u=0\,,\ \bar{v}=-(1+\varepsilon \operatorname{Cos}\pi z)\,,\ w=0\,,\ \theta=1\ ,\ \phi=1 \quad (23)$$

$$y\to\infty\ :\ u=U\,,\ \bar{v}=-1\ ,\ \bar{w}=0\ ,\ \theta=0\ ,\ \phi=0\ ,\ p=p_\infty \quad (24)$$

METHOD OF SOLUTION

We assume the solution of Equations 17 to 22 to be of the form:

$$u=u_0(y)+\varepsilon u_1(y,z)+0(\varepsilon^2) \quad (25)$$

$$v=v_0(y)+\varepsilon v_1(y,z)+0(\varepsilon^2) \quad (26)$$

$$w=w_0(y)+\varepsilon w_1(y,z)+0(\varepsilon^2) \quad (27)$$

$$p=p_0(y)+\varepsilon p_1(y,z)+0(\varepsilon^2) \quad (28)$$

$$\theta=\theta_0(y)+\varepsilon \theta_1(y,z)+0(\varepsilon^2) \quad (29)$$

$$\phi=\phi_0(y)+\varepsilon \phi_1(y,z)+0(\varepsilon^2) \quad (30)$$

with $p_0=p_\infty$, $w_0=0$ (31)

Substituting these in Equations 17 to 22 and equating the harmonic terms and neglecting ε^2 we get the following set of the differential equations:

Zeroth-order equations:

$$\frac{dv_0}{dy}=0 \quad (32)$$

$$v_0\frac{du_0}{dy}=G_r\theta_0+G_m\phi_0+\frac{1}{R_e}\frac{d^2u_0}{dy^2}+MR_e(U-u) \quad (33)$$

$$v_0\frac{d\theta_0}{dy}=\frac{1}{P_r R_e}\frac{d^2\theta_0}{dy^2}-Q\theta_0 \quad (34)$$

$$v_0\frac{d\phi_0}{dy}=\frac{1}{S_c R_e}\frac{d^2\phi_0}{dy^2}+\frac{S_r}{R_e}\frac{d^2\theta_0}{dy^2}-K\phi_0 \quad (35)$$

First-order equations:

$$\frac{\partial v_1}{\partial y}+\frac{\partial w_1}{\partial z}=0 \quad (36)$$

$$-\frac{\partial u_1}{\partial y}+v_1\frac{du_0}{dy}=G_r\theta_1+G_m\phi_1+\frac{1}{R_e}\left(\frac{\partial^2 u_1}{\partial y^2}+\frac{\partial^2 u_1}{\partial z^2}\right)-MR_e u_1 \quad (37)$$

$$-\frac{\partial v_1}{\partial y}=-\frac{1}{R_e^2}\frac{\partial p_1}{\partial y}+\frac{1}{R_e}\left(\frac{\partial^2 v_1}{\partial y^2}+\frac{\partial^2 v_1}{\partial z^2}\right) \quad (38)$$

$$-\frac{\partial w_1}{\partial y}=-\frac{1}{R_e^2}\frac{\partial p_1}{\partial y}+\frac{1}{R_e}\left(\frac{\partial^2 w_1}{\partial y^2}+\frac{\partial^2 w_1}{\partial z^2}\right)-MR_e w_1 \quad (39)$$

$$-\frac{\partial \theta_1}{\partial y}+v_1\frac{d\theta_0}{dy}=\frac{1}{P_r R_e}\left(\frac{\partial^2 \theta_1}{\partial y^2}+\frac{\partial^2 \theta_1}{\partial z^2}\right)-Q\theta_1 \quad (40)$$

$$-\frac{\partial \phi_1}{\partial y}+v_1\frac{d\phi_0}{dy}=\frac{1}{S_c R_e}\left(\frac{\partial^2 \phi_1}{\partial y^2}+\frac{\partial^2 \phi_1}{\partial z^2}\right)+\frac{S_r}{R_e}\left(\frac{\partial^2 \theta_1}{\partial y^2}+\frac{\partial^2 \theta_1}{\partial z^2}\right)-K\phi_1 \quad (41)$$

Subject to boundary conditions:

$$y=0\ :\ u_0=0\,,\ v_0=-1\,,\ \theta_0=1\,,\ \phi_0=1\,,\ u_1=0\,,\ v_1=-\operatorname{Cos}\pi z$$
$$w_1=0\ ,\ \theta_1=0\ ,\ \phi_1=0 \quad (42)$$

$$y\to\infty\ :\ u_0=U\ ,\ v_0=-1\,,\ \theta_0=0\ ,\ \phi_0=0\ ,\ u_1=0\ ,\ v_1=0$$
$$,\ w_1=0\ ,\ p_1=0\ ,\ \theta_1=0\ ,\ \phi_1=0 \quad (43)$$

The solution of Equations 32 to 35 under the boundary conditions 42 and 43 are

$$v_0=-1 \quad (44)$$

$$\theta_0=e^{-ay} \quad (45)$$

$$\phi_0=(1-a_1)e^{-by}+a_1e^{-ay} \quad (46)$$

$$u_0=U+A_1e^{-ay}+A_2e^{-b_e y}+(-A_1-A_2-U)e^{-\lambda R_e y} \quad (47)$$

where

$$a=\frac{P_rR_e+\sqrt{P_r^2R_e^2+4P_rR_eQ}}{2}\ ,\ a_1=\frac{-a^2S_rS_c}{a^2-S_cRa-KS_cR_e}\ ,\ \lambda=\frac{1+\sqrt{1+4M}}{2}\ ,$$

$$b=\frac{S_cR_e+\sqrt{S_c^2R_e^2+4KS_cR_e}}{2}\ ,$$

$$A_1=\frac{-G_m a_1R_e}{a^2-R_ea-MR_e^2}-\frac{G_rR_e}{a^2-R_ea-MR_e^2}\ ,\ A_2=\frac{-G_m(1-a_1)R_e}{b^2-R_eb-MR_e^2}$$

We shall first consider the Equations 36, 38 and 39 for $v_1(y,z)$, $w_1(y,z)$ and $p_1(y,z)$ which are independent of main flow component u_1, temperature field θ_1 and concentration field ϕ_1. The suction velocity $v_w = -(1+\varepsilon \operatorname{Cos}\pi z)$ consists of a uniform distribution -1 with superimposed weak sinusoidal distribution $\varepsilon \operatorname{Cos}\pi z$. Hence the velocity components v, w and p are also separated into mean and small sinusoidal components v_1, w_1 and p_1. We assume v_1, w_1 and p_1 to be of the following forms:

$$v_1 = -\pi v_{11}(y)\operatorname{Cos}\pi z \tag{48}$$

$$w_1 = v'_{11}(y)\operatorname{Sin}\pi z \tag{49}$$

$$p_1 = R_e^2 p_{11}(y)\operatorname{Cos}\pi z \tag{50}$$

On substitution of Equations 48, 49 and 50, Equation 36 is satisfied and Equations 38 and 39 reduce to the following ordinary differential equations

$$v''_{11} + R_e v'_{11} - \pi^2 v_{11} = -\frac{R_e p'_{11}}{\pi} \tag{51}$$

$$v'''_{11} + R_e v''_{11} - (\pi^2 + MR^2)v'_{11} = -R_e \pi p_{11} \tag{52}$$

with relevant boundary conditions

$$y = 0 \quad : \quad v_{11} = \frac{1}{\pi} \quad , \quad v'_{11} = 0 \tag{53}$$

$$y \to \infty : v_{11} = 0 \; , \; v'_{11} = 0 \; , \; p_{11} = 0 \tag{54}$$

The solutions of these equations are:

$$v_{11} = \frac{1}{\pi(\bar{\xi} - \xi)}\left[\bar{\xi} e^{-\xi y} - \xi e^{-\bar{\xi} y}\right] \tag{55}$$

$$p_{11} = \frac{1}{R_e \pi^2(\bar{\xi} - \xi)}\left[(\pi^2 + MR_e^2 + R_e\bar{\xi} - \bar{\xi}^2)e^{-\bar{\xi}y} - (\pi^2 + MR_e^2 + R_e\xi - \xi^2)e^{-\xi y}\right]$$

$$= \frac{1}{R_e \pi^2(\bar{\xi} - \xi)}\left[\bar{\xi}_1 e^{-\bar{\xi} y} - \xi_1 e^{-\xi y}\right] \tag{56}$$

Where

$$\xi = \frac{R_e \lambda + \sqrt{R_e^2 \lambda^2 + 4\pi^2}}{2},$$

$$\bar{\xi} = \frac{R_e \bar{\lambda} + \sqrt{R_e^2 \bar{\lambda}^2 + 4\pi^2}}{2}, \quad \lambda = \frac{1 + \sqrt{1 + 4M}}{2},$$

$$\bar{\lambda} = \frac{1 - \sqrt{1 + 4M}}{2}, \quad \bar{\xi}_1 = (\pi^2 + MR_e^2 + R_e\bar{\xi} - \bar{\xi}^2),$$

$$\xi_1 = (\pi^2 + MR_e^2 + R_e\xi - \xi^2)$$

Hence the solutions for the velocity components v_1, w_1 and pressure p_1 are as follows:

$$v_1 = \frac{1}{\xi - \bar{\xi}}\left[\bar{\xi} e^{-\xi y} - \xi e^{-\bar{\xi} y}\right]\operatorname{Cos}\pi z \tag{57}$$

$$w_1 = \frac{\xi\bar{\xi}}{\pi(\xi - \bar{\xi})}\left[e^{-\bar{\xi} y} - e^{-\xi y}\right]\operatorname{Sin}\pi z \tag{58}$$

$$p_1 = \frac{R_e \xi\bar{\xi}}{\pi^2(\bar{\xi} - \xi)}\left[\bar{\xi}_1 e^{-\bar{\xi} y} - \xi_1 e^{-\xi y}\right]\operatorname{Cos}\pi z \tag{59}$$

SOLUTION FOR FIRST ORDER FLOW, CONCENTRATION AND TEMPERATURE FIELD

We now consider Equations 30, 33 and 34. The solutions for velocity component u, temperature field θ and concentration field ϕ are also separated into mean and sinusoidal components u_1, θ_1 and ϕ_1. To reduce the partial differential Equations 30, 33, 34 into ordinary differential equations, we consider the following forms for u_1, θ_1 and ϕ_1.

$$u_1 = u_{11}(y)\operatorname{Cos}\pi z \tag{60}$$

$$\theta_1 = \theta_{11}(y)\operatorname{Cos}\pi z \tag{61}$$

$$\phi_1 = \phi_{11}(y)\operatorname{Cos}\pi z \tag{62}$$

Using the expressions for v_1, u_1, θ_1, ϕ_1 in Equations 37, 40 and 41 we get the following differential equations:

$$u''_{11} + R_e u'_{11} - (\pi^2 + MR_e^2)u_{11} = -\pi R_e v_{11} u'_0 - R_e G_r \theta_{11} - R_e G_m \phi_{11} \tag{63}$$

$$\theta''_{11} + P_r R_e \theta'_{11} - (\pi^2 + P_r R_e Q)\theta_{11} = -\pi P_r R_e v_{11} \theta'_0 \tag{64}$$

$$\phi''_{11} + S_c R_e \phi'_{11} - (\pi^2 + K)\phi_{11} = -\pi S_c R_e v_{11} \phi'_0 - S_c S_r(\theta''_{11} - \pi^2\theta_{11}) \tag{65}$$

with the boundary conditions

$$\left.\begin{aligned} y = 0 : u_{11} = 0, \; \theta_{11} = 0 \; , \; \phi_{11} = 0 \\ y \to \infty : u_{11} = 0 \; , \; \theta_{11} = 0 \; , \; \phi_{11} = 0 \end{aligned}\right\} \tag{66}$$

The solutions of Equations 64, 65 and 63 subject to boundary conditions (66) are

$$\theta_{11} = G_0 e^{-hy} + G_1 e^{-(\xi+a)y} + G_2 e^{-(\bar{\xi}+a)y} \tag{67}$$

$$\phi_{11} = H_0 e^{-my} + H_1 e^{-hy} + H_2 e^{-(\xi+a)y} + H_3 e^{-(\bar{\xi}+a)y} + H_4 e^{-(\xi+b_e)y} + H_5 e^{-(\bar{\xi}+b)y} \tag{68}$$

$$u_{11} = M_0 e^{-ny} + M_1 e^{-hy} + M_2 e^{-my} + M_3 e^{-(\xi+a)y} + M_4 e^{-(\bar{\xi}+a)y} + M_5 e^{-(\xi+b)y} + + M_6 e^{-(\bar{\xi}+b)y} + M_7 e^{-(\xi+\lambda R_e)y} + M_8 e^{-(\bar{\xi}+\lambda R_e)y} \tag{69}$$

where

$$G_1 = \frac{aP_rR_e\bar{\xi}}{(\bar{\xi}-\xi)\left\{(\xi+a)^2 - P_rR_e(\xi+a) - (\pi^2 + P_rR_eQ)\right\}},$$

$$G_2 = \frac{-aP_rR_e\xi}{(\bar{\xi}-\xi)\left\{(\bar{\xi}+a)^2 - P_rR_e(\bar{\xi}+a) - (\pi^2 + P_rR_eQ)\right\}}, \quad G_0 = -(G_1+G_2),$$

$$h = \frac{P_rR_e + \sqrt{P_r^2R_e^2 + 4(\pi^2 + P_rR_eQ)}}{2},$$

$$m = \frac{S_cR_e + \sqrt{S_c^2R_e^2 + 4(\pi^2 + K)}}{2},$$

$$E_1 = -S_cS_rG_0(h^2-\pi^2),\ E_2 = -S_cS_rG_1\left\{(\xi+a)^2 - \pi^2\right\},\ E_3 = -S_cS_rG_2\left\{(\bar{\xi}+a)^2 - \pi^2\right\},$$

$$B_1 = \frac{(1-a_1)bS_cR_e\xi}{(\bar{\xi}-\xi)},\ B_2 = \frac{aa_1S_cR_e\bar{\xi}}{(\bar{\xi}-\xi)},\ B_3 = \frac{-S_cR_eb\xi(1-a_1)}{(\bar{\xi}-\xi)},\ B_4 = \frac{-aa_1S_cR_e\xi}{(\bar{\xi}-\xi)},$$

$$H_1 = \frac{E_1}{h^2 - S_cR_eh - (\pi^2+K)},\ H_2 = \frac{B_2+E_2}{(\xi+a)^2 - S_cR_e(\xi+a) - (\pi^2+K)},$$

$$H_3 = \frac{B_4+E_3}{(\bar{\xi}+a)^2 - S_cR_e(\bar{\xi}+a) - (\pi^2+K)},\ H_4 = \frac{B_1}{(\xi+b)^2 - S_cR_e(\xi+b) - (\pi^2+K)},$$

$$H_5 = \frac{B_3}{(\bar{\xi}+b)^2 - S_cR_e(\bar{\xi}+b) - (\pi^2+K)},\ H_0 = -\sum_{i=1}^{5} H_i$$

$$,\ n = \frac{R_e + \sqrt{R_e^2 + 4(\pi^2 + MR_e^2)}}{2},$$

$$K_1 = \frac{aA_1R_e\bar{\xi}}{(\xi-\bar{\xi})},$$

$$K_2 = \frac{A_2bR_e\bar{\xi}}{(\bar{\xi}-\xi)},\ K_3 = -\frac{\bar{\xi}\lambda R_e^2(A_1+A_2+U)}{(\bar{\xi}-\xi)},\ K_4 = \frac{-aA_1R_e\xi}{(\bar{\xi}-\xi)},$$

$$K_5 = -\frac{A_2bR_e\xi}{(\bar{\xi}-\xi)},\ K_6 = \lambda R_e^2(A_1+A_2+U)\xi,$$

$$L_1 = R_e(G_rG_0 + G_mH_1),\ L_2 = -G_mH_0R_e$$

$$L_3 = K_1 - G_rG_1R_e - G_mH_2R_e,\ L_4 = K_4 - G_mH_3R_e - R_eG_rG_2,\ L_5 = K_2 - G_mH_4R_e$$

$$L_6 = K_5 - G_mH_5R_e,\ M_1 = \frac{L_1}{h^2 - R_eh - (\pi^2 + MR_e^2)},\ M_2 = \frac{L_2}{m^2 - R_em - (\pi^2 + MR_e^2)},$$

$$M_3 = \frac{L_3}{(\xi+a)^2 - R_e(\xi+a) - (\pi^2 + MR_e^2)},\ M_4 = \frac{L_4}{(\bar{\xi}+a)^2 - R_e(\bar{\xi}+a) - (\pi^2 + MR_e^2)},$$

$$M_5 = \frac{L_5}{(\xi+b)^2 - R_e(\xi+b) - (\pi^2 + MR_e^2)},\ M_6 = \frac{L_6}{(\bar{\xi}+b)^2 - R_e(\bar{\xi}+b) - (\pi^2 + MR_e^2)},$$

$$M_7 = \frac{K_3}{(\xi+\lambda R_e)^2 - R_e(\xi+\lambda R_e) - (\pi^2 + MR_e^2)},$$

$$M_8 = \frac{K_6}{(\bar{\xi}+\lambda R_e)^2 - R_e(\bar{\xi}+\lambda R_e) - (\pi^2 + MR_e^2)},$$

$$M_0 = -\sum_{i=1}^{8} M_i$$

Skin friction at the plate

The non-dimensional skin-friction at the plate in direction of the free steam is given by

$$\tau = \frac{\left.\mu\frac{\partial\bar{u}}{\partial\bar{y}}\right)_{\bar{y}=0}}{\rho V_0^2} = -\frac{1}{R_e}\left[u_0'(0) + \varepsilon u_{11}'(0)\operatorname{Cos}\pi z\right] = \tau_0 + \varepsilon Q_1 \operatorname{Cos}\pi z \tag{70}$$

where

$$\tau_0 = -\frac{1}{R_e}u_0'(0) = \frac{aA_1}{R_e} + \frac{bA_2}{R_e} - \lambda(A_1 + A_2 + U) \tag{71}$$

and

$$Q_1 = -\frac{1}{R_e}u_{11}'(0)$$

$$= \frac{1}{R_e}\left[\begin{array}{l} nM_0 + hM_1 + mM_2 + (\xi+a)M_3 + (\bar{\xi}+a)M_4 + (\xi+b)M_5 \\ +(\bar{\xi}+b)M_6 + (\lambda R_e + \xi)M_7 + (\lambda R_e + \bar{\xi})M_8 \end{array}\right] \tag{72}$$

The co-efficient of rate of heat transfer

The heat flux from the plate to the in terms of Nusselt number Nu is given by

$$Nu = -\frac{k}{\rho V_0 C_p(\bar{T}_w - \bar{T}_\infty)}\left(\frac{\partial\bar{T}}{\partial\bar{y}}\right)_{y=0} = -\left.\frac{1}{P_rR_e}\frac{\partial\theta}{\partial y}\right]_{y=0} = Nu_0 + \varepsilon Q_2 \operatorname{Cos}\pi z \tag{73}$$

Where

$$Nu_0 = -\frac{\theta_0'(0)}{P_r R_e} = \frac{a}{P_r R_e} \quad (74)$$

And

$$Q_2 = -\frac{\theta_{11}'(0)}{P_r R_e} = \frac{1}{P_r R_e}\left[hG_0 + (\xi + a)G_1 + (\bar{\xi} + a)G_2\right] \quad (75)$$

The coefficient of mass transfer

The mass flux at the wall y = 0 in terms of Sherwood number Sh is given by

$$Sh = \frac{-D_M}{V_0(\bar{C}_w - \bar{C}_\infty)}\left(\frac{\partial \bar{C}}{\partial \bar{y}}\right)_{y=0} = -\frac{1}{S_c R_e}\frac{\partial \phi}{\partial y}\Big]_{y=0} = -\frac{1}{S_c R_e}\left[\phi_0'(0) + \varepsilon\phi_{11}' \cos\pi z\right]$$

$$= Sh_0 + \varepsilon Q_3 \cos\pi z \quad (76)$$

Where

$$Sh_0 = -\frac{1}{S_c R_e}\left[b(1 - a_1) + a a_1\right] \quad (77)$$

and

$$Q_3 = -\frac{1}{S_c R_e}\phi_{11}'(0) = \frac{1}{S_c R_e}\begin{bmatrix} mH_0 + hH_1 + (\xi + a)H_2 + (\bar{\xi} + a)H_3 \\ (\xi + b)H_4 + (\bar{\xi} + b)H_5 \end{bmatrix} \quad (78)$$

Current density

The current density $\vec{J}$ is given by

$$\vec{J} = \sigma\vec{q} \times \vec{B} = \sigma B_0\left(-\hat{i}\bar{w} + \hat{k}\bar{u}\right) \quad (79)$$

The magnitude of $\vec{J}$ is given

by $|\vec{J}| = \sigma B_0\sqrt{\bar{w}^2 + \bar{u}^2} = \sigma B_0 V_0\sqrt{u^2 + w^2} \quad (80)$

The current density (in magnitude) in non dimensional form is given by:

$$J_c = \frac{|\vec{J}|}{\sigma B_0 V_0} = \sqrt{u^2 + w^2} = u\sqrt{1 + \left(\frac{w}{u}\right)^2} = u$$

(since $\frac{w}{u} << 1$) (81)

That is, the magnitude of the non dimensional current density is proportional to the boundary layer velocity.

RESULTS AND DISCUSSION

In order to study the effects of heat sink parameter (Q), Reynolds number (R_e), and chemical reaction parameter (K), we have carried out the data tabulations for u, θ_0, ϕ_0, Q_1, Q_2, Q_3 and τ_0 which are respectively the dimensional velocity, zeroth order temperature, zeroth order species concentration, amplitudes of the first order skin friction, Nusselt number and Sherwood number; the zeroth order skin friction at the plate y = 0 and their values are demonstrated in the graphs. Throughout our discussion P_r (Prandtl number) is considered to be equal to 71 which corresponds to air. Since the water vapor is used as a diffusing chemical species of common interest in air therefore the values of S_c is taken to be 0.60 (water vapor). The values of the Grashof number G_r for heat transfer has been chosen as 10 (externally cooled plate) whereas the values of Grashof number G_m for mass transfer is considered to be 15, the free steam velocity is selected to be 1 and the small reference parameter ε is chosen as 0.001 and the remaining parameters namely chemical reaction parameter (K), heat sink parameter (Q), Reynolds number (R_e), Soret number S_r are chosen arbitrarily.

Figures 1, 2 and 3 exhibit the variation of velocity profile u against y for different values of chemical reaction parameter (K), heat sink parameter (Q) and Hartmann number (M). It is seen from these figures that the velocity quickly increases up to some thin layer of the liquid adjacent to the plate and after this, fluid velocity decreases asymptotically towards 1 as $y \to \infty$; that is, in the free steam. This shows that the buoyancy effects (due to concentration and temperature differences) are significant near the hot plate.

It is observed from Figure 1 that the fluid motion is retarded (that is, the fluid velocity decreases) on account of chemical reaction. This shows that the consumption of chemical species leads to fall in the concentration field which in turn diminishes the buoyancy effects due to concentration gradients. Consequently, the flow field is decelerated. It is also inferred from Figures 2 and 3 that the heat sink parameter (Q) as well as Hartmann number (M) impedes the fluid motion. In other words, fluid motion is retarded due to application of transverse magnetic field. This phenomenon clearly agrees with the fact that Lorentz force that appears due to interaction of the magnetic field and fluid velocity resists the fluid motion.
Figure 4 demonstrates the variation of zeroth order fluid temperature θ_0 against y under the effect of heat sink parameter (Q). It is clear from this figure that zeroth order fluid temperature θ_0 asymptotically falls from 1 to zero as $y \to \infty$. The same figure further indicates that the heat sink parameter results in a steady decrease in the zeroth order fluid temperature.

The variation of zeroth order species concentration ϕ_0 versus y under the influences of heat sink parameter (Q) and chemical reaction parameter (K) have been presented in Figures 5 and 6. These figures show that

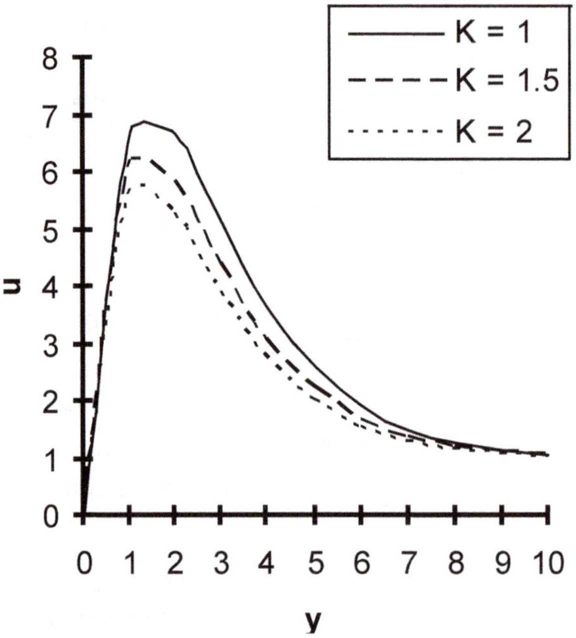

Figure 1. Velocity distribution versus y for K when Q =1, M = 1, R_e = 0.5, S_r = 0.5.

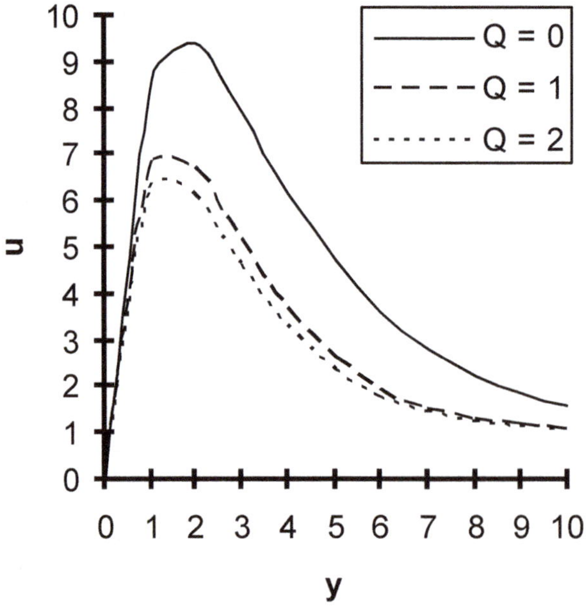

Figure 2. Velocity distribution versus y for Q when K =1, M = 1, R_e = 0.5, S_r = 0.5.

zeroth order concentration of the fluid fall under the effect of heat sink parameter (Q) and chemical reaction parameter (K). Moreover, it is noticed from these figure that ϕ_0 asymptotically decreases from maximum value ϕ_0 = 1 to its minimum value ϕ_0 = 0 as one moves far away the plate ($y \rightarrow \infty$).

Figures 7, 8 and 9 depict the variation of amplitude of the perturbed part of skin-friction Q_1 versus Reynolds number R_e. From these figures we observe that magnetic field effect as well as heat sink effect causes Q_1 to decrease whereas Q_1 increases for the increasing values of chemical reaction parameter. There is an

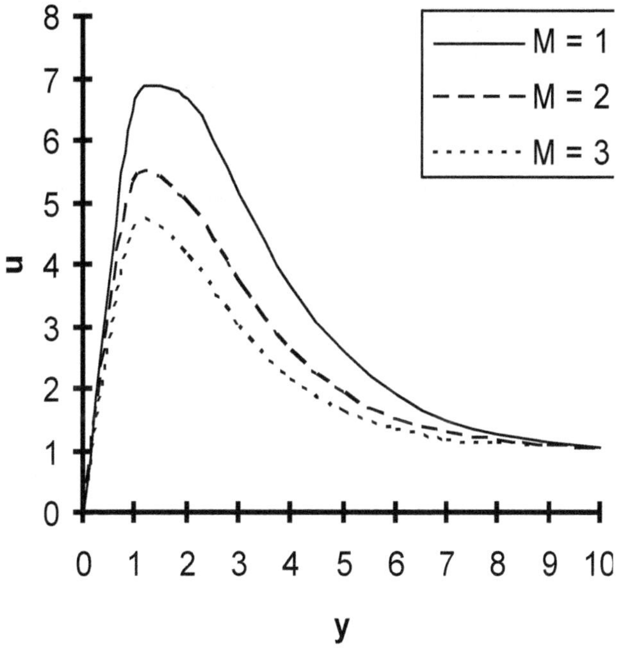

Figure 3. Velocity distribution versus y for M when Q =1, K = 1, R_e =0 .5, S_r = 0.5.

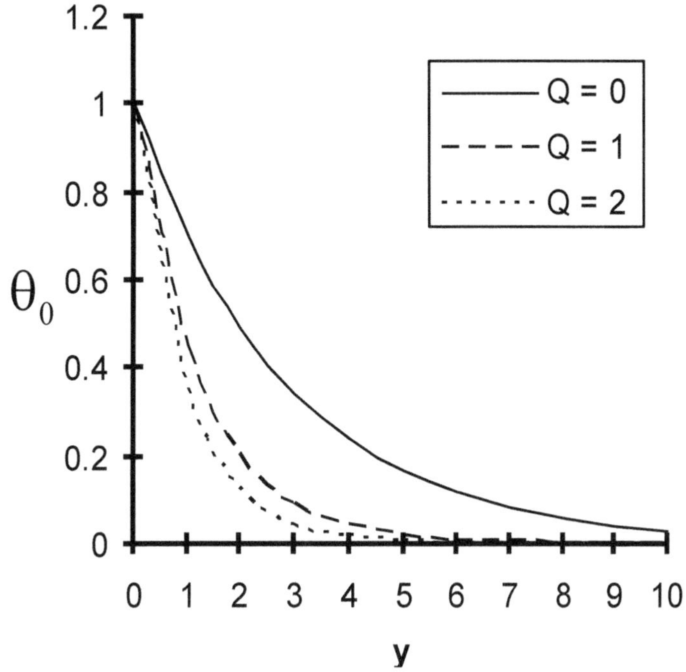

Figure 4. Zeroth order temperature distribution versus y for Q when R_e = 0.5.

indication from these figures that Q_1 falls as R_e increases. That is for low viscosity Q_1 is not significantly affected by heat sink parameter (Q), chemical reaction parameter (K) or by Hartmann number (M).

The influence of heat sink parameter (Q) on the amplitude of Q_2 of the perturbed part of the Nusselt number is displayed in Figure 10. It is noticed from the figure that an increase in the value of Reynolds number

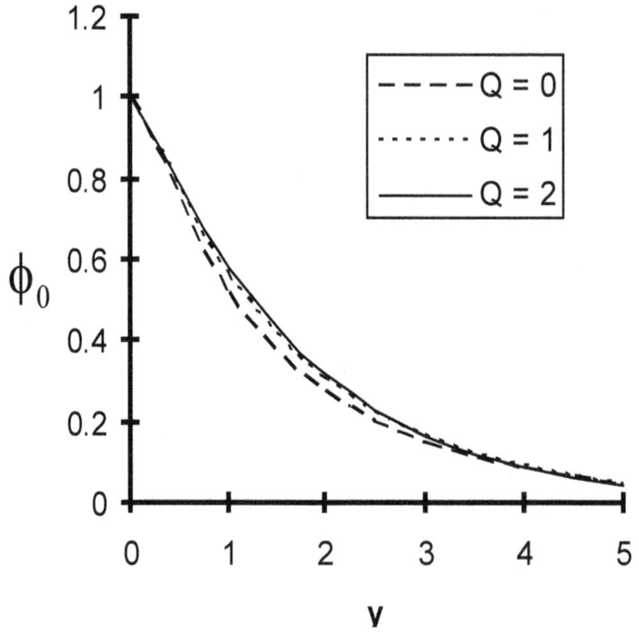

Figure 5. Zeroth order concentration profile versus y for Q when K = 1, R_e = 0.5, S_r = 0.5.

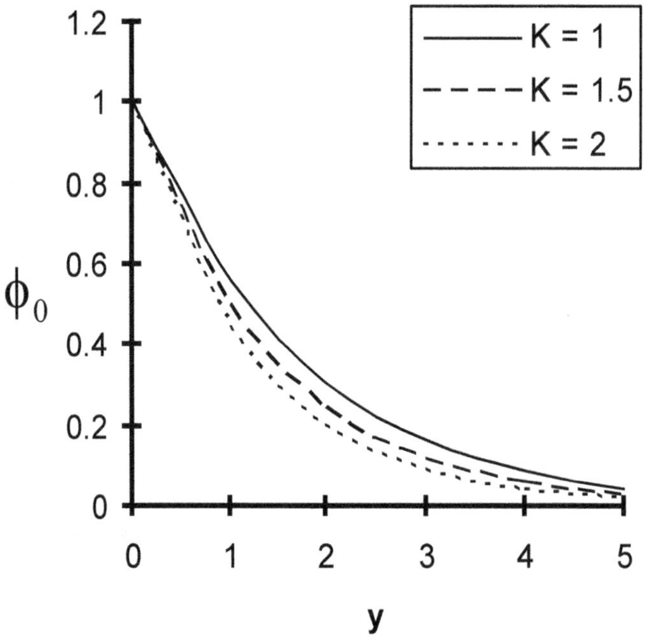

Figure 6. Zeroth order concentration profile versus y for K when Q = 1, R_e = 0.5, S_r = 0.5.

(R_e) or heat sink parameter (Q) causes Q_2 to increase; that is, Q_2 drops due to high viscosity or low strength of heat sink.

Figures 11 and 12 exhibits the change in behaviour of amplitude, of perturbed part, and of the Sherwood number Q_3 under the influence of the Reynolds number R_e, the chemical reaction parameter (K) and

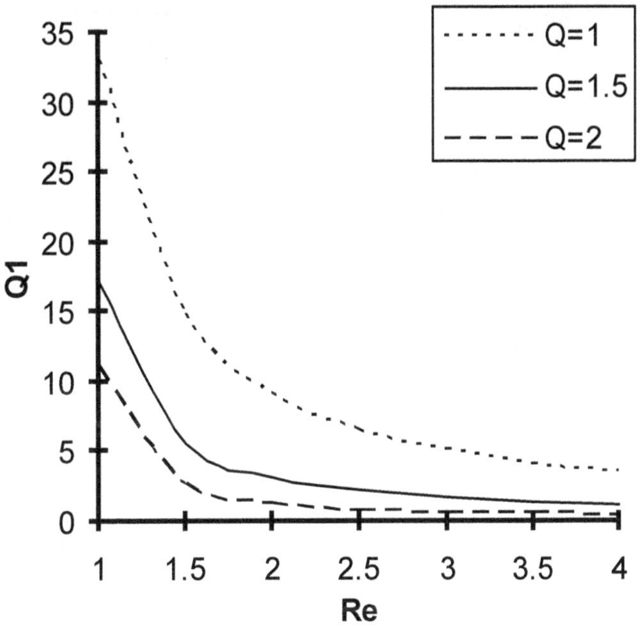

Figure 7. The amplitude Q_1 of the first order skin friction versus R_e for K=1, M=1, S_r=0.5.

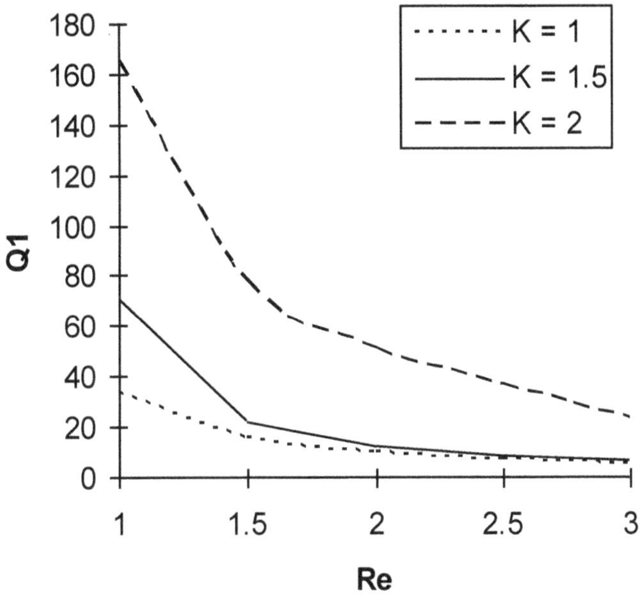

Figure 8. The amplitude Q_1 of the first order skin friction versus R_e for Q=1, M=1, S_r= 0.5.

heat sink parameter (Q). These figures show that Q_3 is increased due to chemical reaction effect where as there is a steady decline in Q_3 when heat sink parameter (Q) is increased.

The variation of the zeroth order skin friction τ_0 at the plate y = 0 under the influence of chemical reaction parameter (K), heat sink parameter (Q) and Hartmann

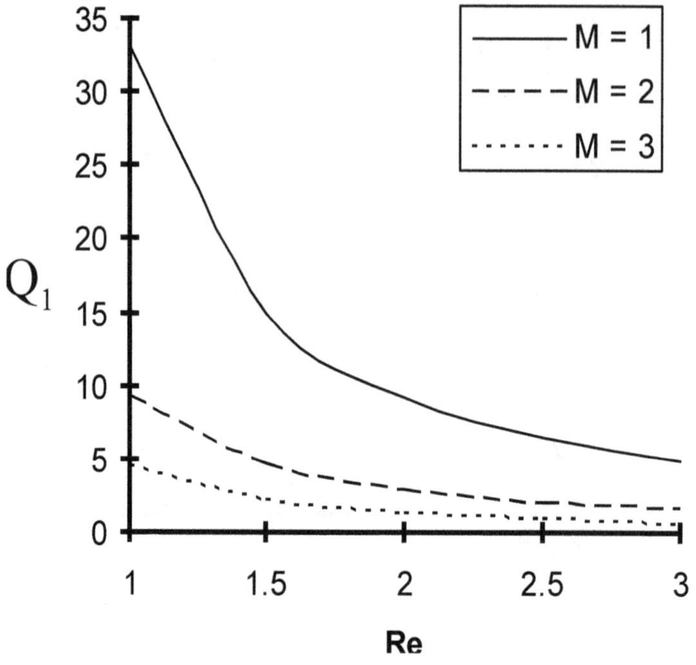

Figure 9. The amplitude Q_1 of the first order skin friction versus R_e for K = 1, Q = 1, S_r = 0.5.

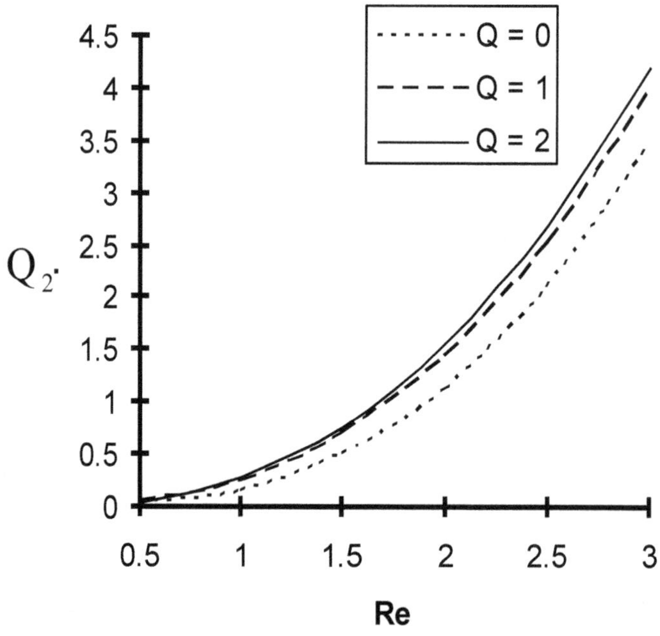

Figure 10. The amplitude Q_2 of the first order Nusselt number versus R_e for K = 1, S_r = 0.5.

number (M) are presented respectively in Figures 13, 14 and 15. It is noticed from these figures that the magnitude of viscous drag at the plate decreases due to the chemical reaction parameter (K), heat sink (Q) and

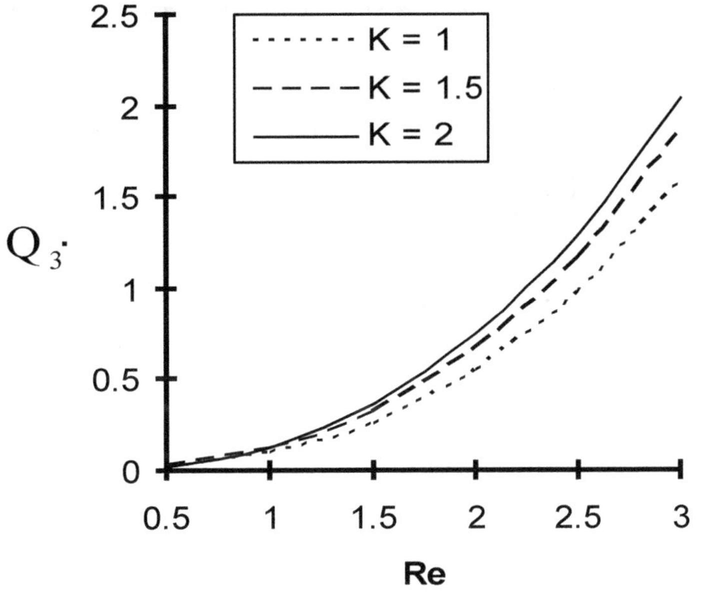

Figure 11. The amplitude Q_3 of the first order Sherwood number versus R_e for Q=1, S_r=0.5.

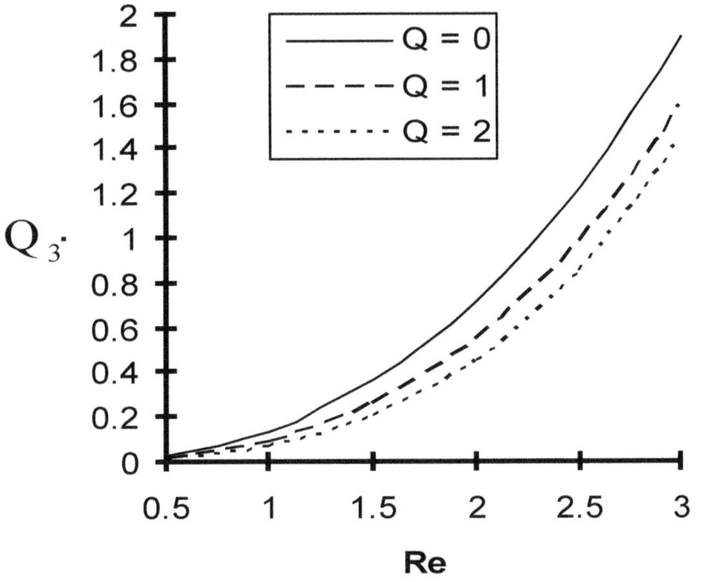

Figure 12. The amplitude Q_3 of the first order Sherwood number versus R_e for Q =1, S_r=0.5.

magnetic field (M).

Conclusion

1. The chemical reaction, heat sink and magnetic field lead the fluid motion to retard. Thus the chemically reacting fluid motion may be controlled with the application of heat sink and magnetic field.
2. The heat sink results in a steady decrease in the fluid temperature. Hence the fluid temperature may be controlled by using a suitable heat sink.

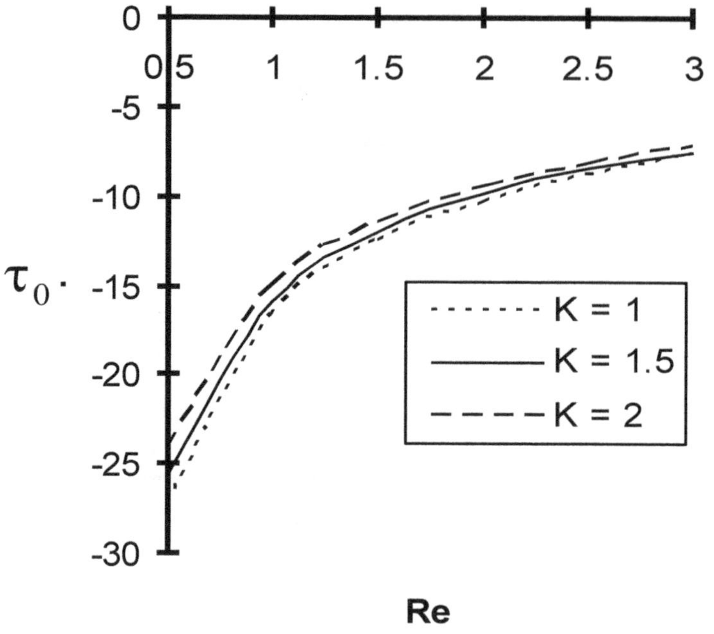

Figure 13. The zeroth order skin friction τ_0 at the plate versus R_e for Q = 1, M=1, S_r=0.5.

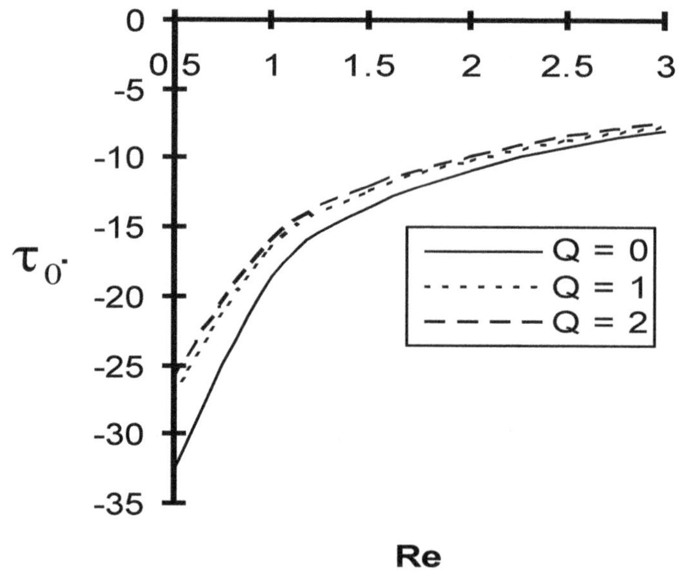

Figure 14. The zeroth order skin friction τ_0 at the plate versus R_e for Q = 1, M=1, S_r=0.5.

3. The concentration of the fluid rises under the effect of heat sink whereas it falls due to the effect of chemical reaction.
4. Magnitude of the first order skin friction increases due to chemical reaction effect and it decreases under the effects of absorption heat sink and the applied transverse magnetic field.
5. The first order Nusselt number drops due to high viscosity or low strength heat sink.
6. The heat absorbing sink leads the first order Sherwood

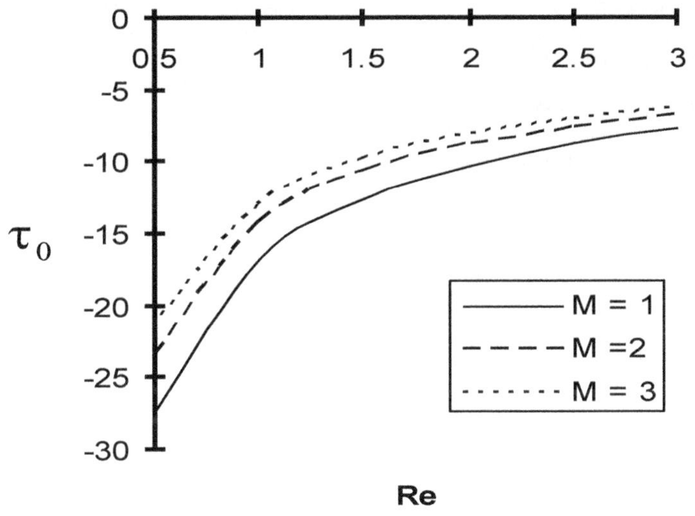

Figure 15. The zeroth order skin friction τ_0 at the plate versus R_e for Q = 1, K=1, S_r=0.5.

number to fall but it rises under the effect of chemical reaction parameter.

7. Magnitude of the zeroth order skin friction diminishes due to chemical reaction effect, magnetic field as well as the heat sink.

Nomenclature: $\vec{B}$, magnetic induction vector; B_0, strength of applied magnetic field; $\bar{C}_\infty$, species concentration in the free stream; $\bar{C}_w$, species concentration at the plate; $\bar{C}_p$, specific heat at constant pressure; D_M, chemical molecular diffusivity; D_T, chemical thermal diffusivity; $\vec{E}$, electric field; $\vec{g}$, gravitational acceleration; **g,** acceleration due to gravity; G_r, Grashof number for heat transfer; G_m, Grashof number for mass transfer; $\vec{J}$, electric current density; **k,** thermal conductivity; $\bar{K}$, first order chemical reaction; **K,** chemical reaction parameter; **L,** wave length of the periodic suction; **M,** Hartmann number; $\bar{p}$, pressure; $\bar{p}_\infty$, pressure in the free steam; **p,** non dimensional pressure; p_∞, non dimensional pressure in the free steam; $\vec{q}$, velocity vector; $\bar{Q}$, first order heat sink; **Q,** non dimensional first order heat sink; R_e, Reynolds number; S_r, Soret number; P_r, Prandtl number; S_c, Schmidt number; $\bar{T}$, temperature in the boundary layer; $\bar{T}_w$, temperature at the plate; $\bar{T}_\infty$, fluid temperature at the free steam; $\bar{U}$, free steam velocity; U, non dimensional free steam velocity; $(\bar{u},\bar{v},\bar{w})$, components of the fluid velocity; (u,v,w), non dimensional components of the fluid velocity; V_0, mean suction velocity; $(\bar{x},\bar{y},\bar{z})$, coordinate system; $\hat{i},\hat{j},\hat{k}$, unit vectors in the increasing direction of $\bar{x},\bar{y},\bar{z}$; $\vec{J}\times\vec{B}$, Lorentz force per unit volume; α, thermal diffusivity; β, coefficient of volume expansion for heat transfer; $\bar{\beta}$, coefficient of volume expansion for mass transfer; σ, electrical conductivity; ν, kinematic viscosity; ρ, density of the fluid; ε, small reference parameter; θ, non dimensional temperature; ϕ, non dimensional concentration; φ, viscous dissipation of energy per unit volume; μ, coefficient of viscosity.

REFERENCES

Ahmed N, Sarma D, Sarma D (2005). MHD free and force convective flow and mass transfer through a porous medium. Far East J. Appl. Math. 21(3):271.

Ahmed N, Sarma D (1997). Three dimensional free convective flow and heat transfer through a porous medium, Indian J. Pure Appl. Math. 26(10):1345.

Ahmed N, Sarma HK (2010). Effect on thermal diffusion on a three-dimensional MHD mixed convection with mass transfer flow past a porous vertical plate, J. energy, Heat Mass Transfer 32:199-221.

Ahmed N, Kalita H (2010). Oscillatory MHD free convective flow through a porous medium with mass transfer, Soret effect and chemical

reaction, Indian J. Sci. Technol. 3(8):919-924.
Ahmed N, Sarma D, Barua DP (2006). Three-dimensional free convective flow and mass transfer along a porous vertical plate, Bulletin of the Allahabad Mathematical Society. 21:125-141.
Bejan A, Khair KR (1985). Heat and mass transfer in a porous medium, Int. J. Heat Transfer. 28:902-918.
Choudhury RC, Chand T (2002). Three-Dimensional Flow and Heat Transfer through Porous Medium. Int. J. Appl. Mech. Eng. 7(4):1141-1156.
Cramer KR, Pai SI (1973). Magneto Fluid Dynamics for Engineers and Applied physicists, McGraw Hill Book Company, New York.
Devi SPA, Kandasamy R (2000). Effect of chemical reaction, heat transfer and mass transfer on MHD flow past a semi-infinite plate, Z. Angew. Math. Mech. 80:697-701.
Epstein IR, Pojman JA (1998). An introduction to non linear chemical dynamics, Oxford University Press, Oxford.
Ferraro VCA, Plumpton C (1966). An introduction to magneto fluid mechanics, Clarandon Press, Oxford.
Gray P, Scott SK (1990). Chemical oscillation and instabilities: Nonlinear chemical kinetics, Oxford University Press, Oxford.
Jain NC, Gupta P (2006). Three-Dimensional Free Convection Couetee Flow with Transpiration Cooling, J. Zhejiang University Sci. A. 7(3):340-346.
Raptis A, Kafousias N (1982). Magneto hydrodynamic free convective flow and mass transfer through a porous medium bounded by an infinite vertical porous plate with constant heat flux. Can. J. Phys. 60:1724-1729.
Sanyal DC, Bhattacharya S (1992). Similarity solutions of an unsteady incompressible thermal MHD boundary layer flow by group theoretic approach, Indian J. Eng. Sci. 30:561-569.

7

The effects of heat treatment on some mechanical properties of laminated black pine (*Pinus nigra*)

Mustafa ORDU[1], Mustafa ALTINOK[2] Abdi ATILGAN[3], Murat OZALP[1] and Huseyin PEKER[3]

[1]Dumlupinar University, Simav Technical Education Faculty, Simav-Kutahya, Turkey.
[2]Gazi University, Technical Education Faculty,Anakara, Turkey
[3]Artvin Coruh University, Faculty of Forestry, Artvin, Turkey.

Black pine (*Pinus nigra*) has high industrial usage potential and large plantation in Turkey. Larch wood was treated with heat to determine its mechanical properties, such as bending strength and pressure and order to determine its adhesion strength. Laminating adhesives are used for the construction of polyvinyl acetate and polyurethane. Heat treatment was applied to the samples at 100-150°C for 4 h in drying oven. In the result of the experiments, heat treated samples give better results due to temperature increase.

Key words: Heat treatment, bending strength, compression strength, bonding strength, black pine.

INTRODUCTION

Laminated wood products are becoming popular in various applications such as construction, furniture units and indoor decorations due to the utilization of raw materials with alternative approach. Compared with sawn lumber, laminated lumber can be used to create wood products that are free from defects and are much larger in size than pieces of wood sawn. There are also advantageous in that they are dimensionally more stable, have a variable cross-section and are more attractive than wood products manufactured from solid wood (Dilik, 1997; Kurtoglu, 1978).

Today, wood age is still in common use. It is used by human and insects and the external effects can last longer than the using of the wood. Drying process is done according to the properties of the earth.

Most economical way of drying within the shortest possible time can be done with a minimum defects and loss. Researchers desire to develop alternative methods. One of these is pure superheated steam drying method. The industrial world mostly applied shorter drying time compared to conventional drying method. The technology of wood protection using heat leads to: natural durability of wood, stabilization of wood-dimension as well as improvement and increase without any preservatives (Kantay and Kartal, 2008a). In addition to better durability, other advantages of heat treated wood are: reduced hygroscopic and improved dimensional stability (Jamsa and Viitaniemi, 2001). Heat treated pine and spruce are mainly used for outdoor constructions, for example garden furniture, windows, doors and wall or fence board. When better weather and decay resistance is desired, the temperatures used for heat treatment must be over 200 (Syrjanen and Oy, 2001). Wood was subjected to heat treatment in the first scientific study in 1930 by German scientists Stamm and Hansen and an American scientist in 1940. Bavendam (1950) and Runkel and Buro (1950) continued to work on this issue. Kollman and Schneider (1960) and BurmesterRusch (1970) worked on this issue again.

In the 1990s, the Netherlands, Finland and French

scientists have done a lot of work on it (Viitaniemi, 2000). Heat-treated wood has led to decrease in the amount of moisture balance. High temperature (220°C) in heat treatment is applied as a result of the normal amount of moisture balance, in almost half of the wood treated with heat. Heat treatment radially and tangentially significantly reduces the amount of shrinkage (Anon, 2003). Environmental pollution caused by reducing the cost of modification of wood and alternative wood protection method considered as heat treatment are applied for several purposes: first, to reduce the moisture of wood, that is, give stabilization to the wood size and enhance biological resistance to wood-destroying organisms.

In addition, heat treatment reduces the amount of moisture balance in the wood, increases its permeability, such as CCA and CCB impregnated materials (varnish and paint-surface) to improve its performance (Yildiz, 2002). Many fungi and insects are checked by the use of biological degradation, open-air effects, burning and more importantly, reduction of dimensional instability (Rowell, 2005). Heat treated wood is the most important feature of the balance due to the decrease in the amount of moisture in the wood (Kantay, 1993). In this study, the effects of wooden materials treated with heat on hardness, brilliance and resistance to stickiness of varnishes were investigated. For this purpose, firstly Scotch pine (*Pinussylvestris* L.) and chestnut (*Castanea sativa* M.) wooden samples were kept in temperatures of 100, 150 and 200°C for 2, 4 and 6 h and after that they were varnished by the water-based varnish.

According to the test results, it was determined that while the hardness, brightness and resistance to stickiness were improved in both wooden types which were kept for 2 h in 100, 150 and 200°C, but they deteriorated when kept for 4 and 6 h in same temperatures (Ozalp and Gezer, 2009). In a study of wood of Scotch pine (*Pinus sylvestris* L.), it underwent thermal processing for 4, 6 and 8 h at 150, 170 and 190ºC. When scotch pine was treated with heat in the experiments, its bending elasticity module (EM), bending resistance (ED), pressure resistance (BD), weight loss (AK), total color change (ΔE *) and volumetric swelling (HS) values were determined. In the test results according to the EM and ED values of the heat treatment, reduction in Scotch pine led to increase in BD values. ED has been mostly affected by mechanical resistance.

Heat treatment resulted in the dark color of Scotch and made the swelling volume to decrease by approximately 50%. When the temperature of heat treatment and application time increases, the amount of change in all of these features increases (Ozçiftçi et al., 2009). Modified thermal (heat) properties of the materials used in treated wood are directly connected to low and high temperatures, resulting in the use of different materials. Thermal processing with low temperature can be applied in wood building elements such as furniture, garden furniture, sauna elements and in door-window frames, shutters etc. Thermal processing with high temperature can be applied to equipment outside the door and fence, exterior cladding, sauna and bathroom elements, flooring materials, garden furniture and as sound barriers (Kantay and Kartal, 2008b).

Application of inadequate glue to the surface of wooden material will lead to weak joint; and when high pressure is applied on the specimen mechanic adhesion decreases leading to a weak joint (McNamara and Waters, 1970). Woods of diffuse porous show different sticky characteristics than ring-porous wood. The resistance of glue line of ring-porous wood increases relative to the density. Marra (1992) conducted a study on beech wood samples kept in 100 and 150°C for 4 h, and after that they were laminated with polyurethane and polyvinyl acetate adhesives. The test results of heat-treated laminated beech wood and control samples showed that mechanical properties including compression strength, bending strength and bonding strength were affected positively by heat treatment; and increase in temperature and duration further increased strength values of the laminated wood specimens (Altinok et al., 2010).

The aim of the present study is to investigate the effects of heat treatment on same mechanical properties of laminated black pine. This study also aspired to determine the ideal heat and adhesive.

MATERIALS AND METHODS

Black pine wood was used in all sample experimental studies. The Black pine wood was taken from the area of Simav Mountain in Simav-Kutahya Province of Turkey. Samples are prepared from whole sap wood trunk. Density of dry wood is 0.53 g/cm^3.

All tests in this study were carried out in Department of Furniture and Decoration Education Laboratory on universal testing device. Bending strength tests were carried out according to TS 2474 standards (TS 2474, 1976). The samples having 2×2×30 cm dimensions with 3 layers were used for bending strength. The thickness of the layers is 6.5 mm. 10 samples have been used for each experiment. The samples prepared are acclimatized up to 12% moisture at 20°C and 65% relative humidity conditions. Test specimens are determined through measurement from the middle parts by ± 1% mm sensitivity micrometer. Span to thickness ratio for bending is 24 cm. Bending strength test mechanism is given in Figure 1.

The following equations were used in the calculation of bending strength (σ_e):

$$ó_e = \frac{3.P.L_s}{2.b.h^2} \quad (1)$$

Where σ_e is bending strength (N/mm^2); P is maximum force at the moment of breaking (N); Ls is distance between points of support

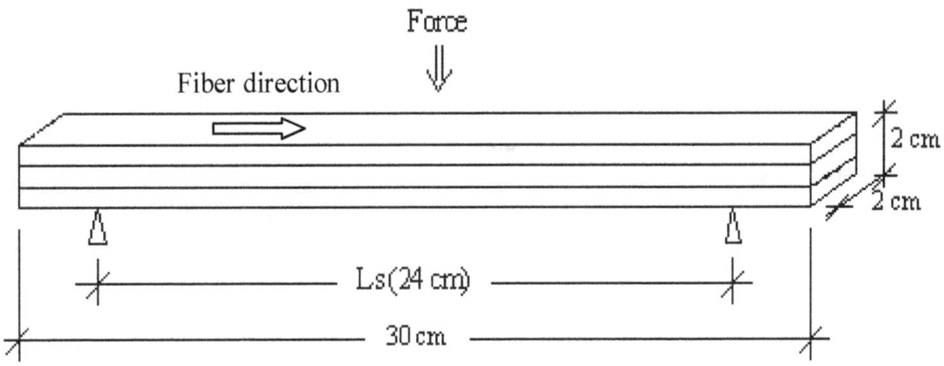

Figure 1. The dimensions of bending strength test specimens.

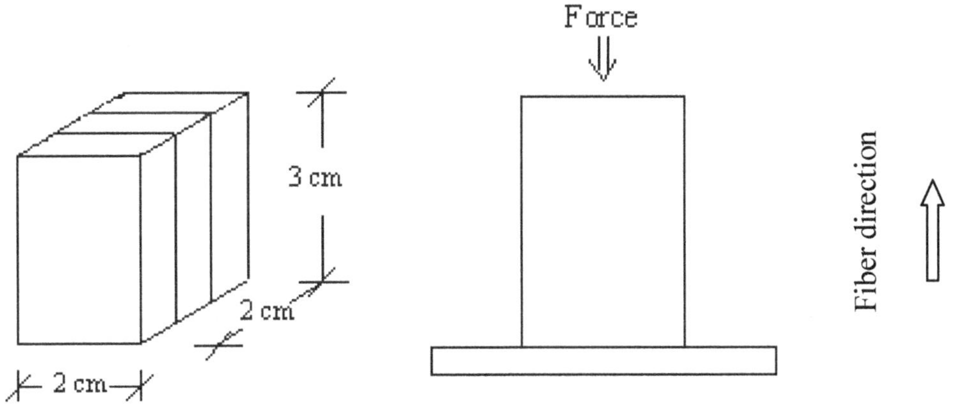

Figure 2. The dimensions of compression strength test specimens.

(mm); b is width of sample piece (mm); h is thickness of sample piece (mm).

Compression strength tests were carried out according to TS 2595 standards (TS 2595, 1976). The thickness of the layers is 6.5 mm. 10 samples have been used for each experiment. The samples prepared are acclimatized up to 12% moisture at 20°C and 65% relative humidity conditions. Test specimens are determined through measurement from the middle parts by ± 1% mm sensitivity micrometer. Span to thickness ratio for bending is 24 cm. Compression strength test mechanism is given in Figure 2. The following equations are used for the calculation of compression strength (σ_b);

$$ó_b = \frac{F_{max}}{a.b} \tag{2}$$

Where σ_b is compression strength (N mm^2); F_{max} is maximum force at the moment of breaking (N); *a* is width of sample breadth cross-section (mm); *b* is lengthof sample breadth cross-section (mm).

The specimens were prepared according to TS 53 and TS 2470 TS 2474-53 (1976). Bonding strength test was done according to DIN 53255 standards. The samples having 15x20x150 mm dimensions with 2 layers were used for bonding strength. The thickness of layers is 7.5 mm. 10 samples have been used for each experiment. The samples prepared are acclimatized up to 12% moisture at 20°C and 65% relative humidity conditions. Test specimens are determined through measurement from the middle parts by ± 1% mm sensitivity micrometer. Bonding strength test mechanism is given in Figure 3. The following equations are used for the calculation of bonding strength (σ),

$$ó = \frac{F_{max}}{b.l} \tag{3}$$

Where σ is bonding strength (N mm^{-2}); F_{max} is maximum force at the moment of breaking (N); *b* is width of bonding surface (mm); l is length of bonding surface (mm).

Adhesives used

In this study, polyurethane and polyvinyl acetate (PVAc) adhesives are used for preparing all samples. The amount of the adhesives applied to the surface of the samples is 235.4 (g/m^2) on average. General properties of polyurethane and polyvinyl acetate adhesives used are given in Tables 1 and 2.

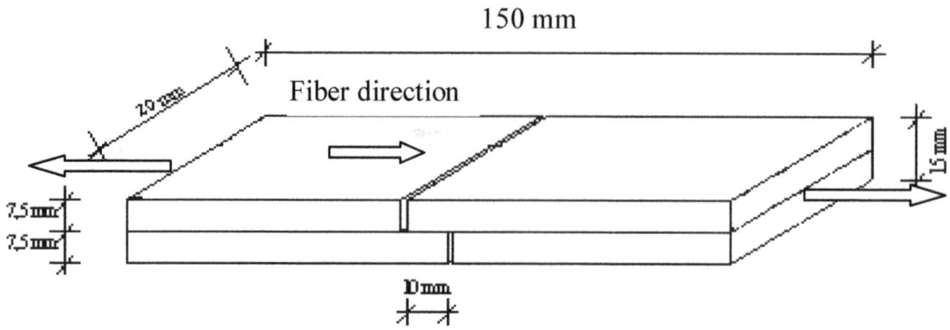

Figure 3. The dimensions of bonding strength test specimens.

Table 1. General properties of polyurethane adhesive.

General properties	Polyurethane adhesive (PU)
Commercial type	Liquid-solid
Storing period (month)	6–9
Colour	Bold brown
Effect on health	Harmfull above 60°C
Wood humudity (%)	Max. 15
Dry material (%)	20–90
Amount of using (gr/m^2)	200–250
Fitting time (hour)	0.5–1
Pressure (N/mm^2)	0.3-0.8
Temperature (°C)	10–60
Resist to water	Resist to boiled water
Resist to microorganism	Extra

Table 2. General properties of polyvinyl acetate.

General properties (°C)	Polyvinyl acetate (PVAc)
Amount of using	150–200(g/m^2)
Density (20)	1.1 (g/m^3)
Viscosity (20)	160-200 cps
pH (20)	5
Pressing time (20)	20 min

Heating process practices

Heat treatment was applied to the samples at two controlled (±1°C) temperatures (100-150°C) for 4 h under atmospheric pressure at drying oven. After heat treatment, treated and untreated samples were conditioned to 12% moisture contents (MC) in a conditioning room at 20 ± 2°C and with 65% (±5) relative humidity (RH) according to TS 642 ISO 554. Then they were laminated with polyurethane and polyvinyl acetate adhesives.

Press pressure and pressure period

All laminated samples were pressed and prepared at 20°C temperature, with 65% relative humidity and under 0.4 N/mm^2 pressure with 3 h time period in hydraulic press.

RESULTS AND DISCUSSION

Bending strength results

The values of average bending strength for laminated Black pine (3 layers) are given in Table 3. The increase of temperature in the heat treatment affected the bending strength properties of laminated black pine positively. The interaction between these obtained values has been examined with the variance analysis and the results obtained are given in Table 4. According to the results of the variance analysis conducted, the effects of heat treatment and type of adhesive on the bending strength of laminated Black pine are found significant with 5% error.

Compression strength results

The values of average compression strength for laminated Black pine (3 layers) are given in Table 5. The increase of temperature in the heat treatment affected the compression strength properties of laminated Black pine positively. The interaction between these obtained values has been examined with the variance analysis and the results obtained are given in Table 6. According to the results of the variance analysis conducted, the effects of heat treatment and type of adhesive on the compression strength of laminated black pine have been found significant with 5% error.

Bonding strength results

The values of average bonding strength for Black pine are given in Table 7. The increase of temperature in the

Table 3. The values of average bending strength results (N/mm^2).

Used Adhesives	The values of average bending strength results (N/mm^2)			
		Non-treatment	100°C	150°C
Polyvinyl acetate (PVAc)	Min	57.013	59.555	61.282
	Max	62.412	66.604	66.855
	Average	59.132	63.214	63.232
	δ_x	1.74	2.19	2.12
	n	10	10	10
Polyurethane (PU)	Min	54.523	56.829	59.384
	Max	60.231	62.89	65.361
	Average	57.807	59.084	62.382
	δ_x	1.65	1.71	1.92
	N	10	10	10

Table 4. Bending strength variance analysis results.

Variance source	Df	Sum of squares	Mean squares	F test	P
Type of adhesive	1	66.341	66.341	16.33	0.000
Heat treatment	2	191.471	95.735	23.57	0.000
Error	56	227.436	4.061		
Total	59	485.248			

Table 5. The values of average compression strength results (N/mm^2).

Used Adhesives	The values of average compression strength results (N/mm^2)			
		Non-treatment	100°C	150°C
Polyvinyl acetate (PVAc)	Min	22.16	23.83	29.76
	Max	27.48	27.85	33.53
	Average	24.59	25.80	30.89
	δ_x	1.47	1.22	1.19
	n	10	10	10
Polyurethane (PU)	Min	26.95	28.52	31.63
	Max	32.21	32.68	36.36
	Average	29.74	30.37	33.96
	δ_x	1.54	1.27	1.59
	n	10	10	10

Table 6. Compression strength variance analysis results.

Variance source	Df	Sum of squares	Mean squares	F test	P
Type of adhesive	1	272.65	272.65	131.43	0.000
Heat treatment	2	315.23	157.61	75,98	0.000
Error	56	116,17	2.07		
Total	59	704.04			

Table 7. The values of average bonding strength results (N/mm^2).

Used adhesives	The values of average bonding strength results (N/mm^2)			
		Non-treatment	100°C	150°C
Polyvinyl acetate (PVAc)	Min	11.21	12.21	13.94
	Max	12.04	14.66	16.07
	Average	11.52	13.63	15.15
	δ_x	0.26	0.81	0.72
	n	10	10	10
Polyurethane (PU)	Min	11.39	13.20	14.12
	Max	12.13	16.01	16.80
	Average	11.81	14.70	15.12
	δ_x	0.20	0.80	0.81
	n	10	10	10

Table 8. Bonding strength variance analysis results.

Variance source	Df	Sum of squares	Mean squares	F test	P
Type of adhesive	1	3.435	3.435	7.05	0.010
Heat treatment	2	127.228	63.614	130.58	0.000
Error	56	27.282	0,487		
Total	59	157.945			

heat treatment affected the bonding strength properties of laminated Black pine positively. The interaction between these obtained values has been examined with the variance analysis and the results obtained are given in Table 8. According to the results of the variance analysis conducted, the effects of heat treatment and type of adhesive on the bonding strength of laminated Black pine have been found significant with 5% error. As a result of the researches conducted, it is observed that as the heat treatment increased from 100 to 150°C, bending strength increased by 6. 93% in samples laminated with polyvinyl acetate adhesive and by 7.65% in samples laminated with polyurethane adhesive. Compression strength increased by 25.62% in samples laminated with polyvinyl acetate adhesive and by 14.18% in samples laminated with polyurethane adhesive. Bonding strength increased by 31.51% in samples laminated with polyvinyl acetate adhesive and by 28.02% in samples laminated with polyurethane adhesive.

The results of this study (bending strength, compression strength, bonding strength) are compatible with studies done by Altinok et al. (2010).

Conclusion

Polyvinyl acetate and polyurethane wood glue is used to laminate Black pine at 100 and 150°C for 4 h in heat treatment. Through thermal processing, the popular Black pine wood gave better results, as reflected in its mechanical properties. Degree of increase of heat treatment has a positive impact on its mechanical properties. Through heat treatment wood species with no commercial value can be used in areas where they had no use previously. As a result, application of heat treatment is suggested to improve its strength properties before laminating Black pine.

REFERENCES

Altinok M, Ozalp M, Korkut S (2010). The Effects Of Heat Treatment On Some Mechanical Properties Of Laminated Beech Wood (*FagusOrientalisL.*). Wood Res. 55(3):131-141.

Anon (2003). Thermoodwood Handbook. Finnish Thermowood Association, Helsinki, Finland.

Dilik T (1997). Production Of Window Profiles From Laminated Lumber and Determination Of Manufacturing Factors. Istanbul University; Ph.D. Thesis, Istanbul, Turkey.

Jamsa S, Viitaniemi P (2001). Heat Treatmet of Wood -Better Durability Without Chemicals. Proceedings of Special Seminar held in Antibes, France on 9 February, 2001: 21. http://www.holzfragen.de/bilder2/dgfh_hitze_eng.pdf.

Kantay R (1993). "Lumber Drying and Steaming", Forestry Educational and Cultural Foundation Publishing number 6, Istanbul.

Kantay R, Kartal N (2008a). Istanbul University, Faculty of Forestry Department of Forest Industrial Engineering. http://www.ahsaponline.net.arsiv/dergi/35/termalmod.htm.

Kantay R, Kartal SN (2008b). Properties of Heat Treated Wood Materials. Istanbul University, Faculty of Forestry Department of ForestryForestIndustryEngineering, Department of BiologyandWoodProtectionTechnologywood magazine 33. http://www.ahsaponline.net.arsiv/dergi/33/termawoo.htm

Kurtoglu A (1978). Moisture Distribution in Laminated Layered Loud Bearing Thick Wood Materials. Rev. Fac. For. Series-A. 28(1).

Marra AA (1992). Technology of wood bonding, New York.

McNamara VS, Waters D (1970). Comparison of the rate glue-line strength development for oak and maple. F.P.J. 20(3).

Ozalp M, Gezer I (2009). The investigation of heat treatment with water-based varnish single component in varnish applications of wood material. Afr. J. Biotechnol. 8(8):1689-1694.

Ozçiftçi A, Altun S, Yapıcı F (2009). International Symposium on Advanced Technologies (IATS'09), 13-15 May 2009, Karabük, Türkiye.

Rowell RM (2005). Chemical modification of wood. In: Handbook of Wood Chemistry and Wood Composites. Ed: Roger M. Rowell. CRC Press, Boca Raton, London, Newyork, Washintgon DC. ISBN 0-8493-1588-3.

Syrjanen T, Oy K (2001). Productıon And Classıficatıon Of heat Treated Wood In Fınland Proceedings of Special Seminar held in Antibes, France on 9 February 2001.

TS 2470 (1976). Wood-sampling methods and general requirements for physical and mechanical tests, Ankara, (in Turkish).

TS 2474-53 (1976). Wood-determination of ultimate strength in static bending, Ankara, (in Turkish).

TS 2595 (1976). Wood-determination of ultimate stress in compression parallel to grain, Ankara, (in Turkish).

TSE 53 (1976). Wood-sampling and test methods-determination of physical properties, Ankara, (in Turkish).

Viitaniemi P (2000). New Properties for Thermally-Treated Wood.Industrial Horizons, P. 9.

Yildiz S (2002). Physical, mechanical, technological and chemical properties of beech and spruce wood treated by heating. Ph.D. dissertation, Karadeniz Technical University, Trabzon.

8

Combustion properties of Rowan wood impregnated with various chemical materials

Hakan Keskin[1], Neslihan Süzer Ertürk[2], Mustafa Hilmi Çolakoğlu[3] and Süleyman Korkut[4]

[1]Department of Woodworking Industry Engineering, Technology Faculty, Gazi University, 06500 Beşevler, Ankara, Turkey.
[2]Department of Industrial Technology, Industrial Arts Education Faculty, Gazi University, 06830 Gölbaşı, Ankara, Turkey.
[3]Technology Development Foundation of Turkey (TTGV) 06800 Bilkent, Ankara, Turkey.
[4]Department of Forest Industrial Engineering, Forestry Faculty, Düzce University, 81620 Düzce, Turkey.

The common use of Rowan wood (*Sorbus aucuparia* Lipsky) in the forest products industry of Turkey, was evaluated for some of the combustion properties after impregnation with various chemical materials. For this purpose, the wood test samples prepared from Rowan wood materials according to TS 345 were treated with Tanalith-E, Vacsol-Azure, Imersol-Aqua and Boron compounds (Borax and Boric asid) by the vacuum impregnation process in accordance with ASTM D-1413 standards and directives of the manufacturer. After impregnation, each sample was tested for observation of retention amount and combustion properties (ASTM E 160-50). It has been proven that combustion properties of the impregnated Rowan wood materials were affected by impregnated materials. As a result, combustion temperature was measured as highest in Imersol Aqua (458.686°C) and the lowest in Borax (439.023°C). The impact of impregnation material decreases after the ignition and the main impact was in flame sourced combustion. In addition, boric acid reduces the material loss in combustion, ensuring that the material is more resistant to sudden collapse and destruction. As a result, it appears that impregnation of wood material with appropriate boron treatment will further increase combustion temperatures and provide additional fire resistance and a degree of security.

Key words: Rowan wood, combustion properties, Tanalith-E, Vacsol-Azure, Imersol-Aqua, Boron compounds.

INTRODUCTION

If the wood materials are used without processing by preservative chemicals (with regard to the area of usage), fungal stains, insect infectation, humidity, fire etc. damage the wood. As a result of these damages, the woods require to be repaired, maintained or replaced before its economic life ends (Richardson, 1987). For this reason, in most places the wood materials should be impregnated with some proper chemicals. For this purpose, ammonium sulfate, ammonium chloride, borax, boric acid, phosphoric acid, etc., are used mostly. In the case of wood not impregnated but only painted and varnished, the prevention on the surfaces is limited to a maximum of two years (Evans et al., 1992).

In a research which was conducted by Örs et al. (1999a), wood samples prepared from Scotch pine (*Pinus sylvestris* L.) and chestnut (*Castanea sativa* Mill.) were impregnated with tanalith–CBC, water repellent (WR) solutions + synthetic varnish and WR + polyurethane varnish. Afterwards, synthetic and polyurethane varnishes were applied to the surfaces. Impregnation with T-CBC was not retarding combustion in both types of wood, but the weight loss occurs as 20% in chestnut and

13% in yellow pine. Varnishing after the impregnation did not affect combustion properties. Furthermore, the alder wood was impregnated with boron compounds to keep from the biotic and abiotic and observed that boron compounds decreases the combustion in great extent (Uysal, 1998). On the ther hand, Douglas wood was impregnated with boron compounds and PEF-400 to observe the changes in the combustion properties (Yalınkılıç et al., 1997) and polyethylene glycol groups was affected negatively but boron compounds have positive impact on combustion.

Wood and wood-based materials consist of organic compounds and are composed mainly of carbon and hydrogen; for this reason, they are combustible. For ignition, oxygen, flame source and flammable material are necessary. The combustibility of the wood is favorable when it is used as a fuel, but unfavorable as a building material. It is impossible to make wood incombustible but it is possible to make it fire resistant (Baysal, 2011). Wood has excellent natural fire resistance as a result of its remarkably low thermal conductivity and the fact that wood char is formed when wood is burned. In order to reduce flammability and provide safety, wood is treated with fire-retardant chemicals. In other words, the combustibility of wood may be reduced with flameretardant or fire-retardant materials (Özçifçi, 2001a; Örs et al., 1999b). The most commonly used fire retardant chemicals in the wood industry are inorganic salts including ammonium and diammonium phosphate, ammonium chloride, ammoniumsulfate, borax, boric acid, phosphoric acid, and zinc chloride. Fire-retardant chemicals drastically reduce the rate at which flames travel across the wood surface, thereby reducing the capacity of the wood to contribute to a fire (Atar and Keskin, 2007; Atar et al., 2011).

Combustion properties of wood are important because of safety issues since it is one of the more commonly available flammable materials. Its combustion process is well studied particularly for modified woods which have been treated to resist burning (Baysal et al., 2003; Peker et al., 2004; Terzi, 2008; Seferoğlu, 2008).

In addition, wood-based materials such as plywood, OSB and MDF panels (Aslan and Özkaya, 2004), and laminated panels (Özen et al., 2001a; Uysal and Özçifçi, 2000; Özçifçi, 2001b; Okcu, 2006; Özkaya, 2002; Uysal and Kurt, 2006) have been investigated.

This study was performed to determine the impacts of vacuum impregnation process with various chemical materials (Tanalith-E, Vacsol-Azure, Imersol-Aqua, Borax and Boric asid) on the combustion properties of Rowan wood.

MATERIALS AND METHODS

Wood materials

Test samples were obtained from Kastamonu Forestry Regional Directory, Küre Directory, Kösreli Department number 200. Test samples were cut from the trees in accordance with TS 4176 (1984) standard. Recently use of Rowan wood is getting more popular in Turkey and surrounding countries due to its high demand. Test samples are prepared in accordance with TS 2470 (1976) and TS 53 (1981). Accordingly, non-deficient, knotless, normally growth (without zone line, without reaction wood and without decay and insect mushroom damage) wood materials was selected. Test samples cut to 70×70×800 mm were air-dried at 20±2°C temperature and 65±3% relative humidity conditions reaching up to 12% humidity level.

Impregnation material

Tanalith-E

Tanalith-E is an impregnation material used against the attacks of agent, yeast, insect and termite. It is a new generation of impregnation material consisting of copper and organic biocide (triazole) and not harmful to plant, animal and human health. Tanalith-E, light green in color, odor, pH 7, 104 g/cmP^3 density, smooth and completely water-soluble, water-based, non-corrosive to metal parts are available in the form of ready solution. Tanalith-E was applied to woods used in fences, railings, garden furniture, barns, silos, farm buildings, the wood used in children's play areas by vacuum - pressure method (Hickson's Timber Impregnation Co. (GB), 2000).

Vacsol Azure

Vacsol Azure a product of a new technology developed by using active ingredients, used in the process, ground wood materials on the level of fungi, insects (Propiconazole and tebuconazole), and termites (permethrin) to prevent decay by protecting against transparent impregnating agent. This solvent-based material is water-insoluble, pale yellow in color, flammable, density 0.806 g/cm^3 at 20±2°C, contains 64% of Volatile Organic Compounds (VOC) (Hickson's Timber Impregnation Co. (GB), 2000).

Imersol-Aqua

Imersol-AQUA, used as an impregnation material in this study was supplied from Hemel (Hemel-Hickson Timber Products Ldt.), Istanbul. Imersol-AQUA is non-flammable, odorless, fluent, water-based, completely, soluble in water, non-corrosive material with a pH value of 7 and a density of 1.03 g/cm^3. It is available as a ready-made solution. It contains 0.5 % w/w tebuconazole, 0.5 % w/w propiconazole, 1% w/w 3-Iodo-2-propynl-butyl carbonate and 0.5% w/w cypermethrin. Before the application of Imersol-AQUA on the wood material, all kinds of drilling, cutting, turning and milling operations should be completed and the relative humidity should be in equilibrium with the test environment. In the impregnation process, dipping duration should be at least 6 min and the impregnation pool must contain at least 15 L of impregnation material for 1 m^3 of wood. The impregnated wood should be left to dry for at least 24 h (Hickson's Timber Impregnation Co. (GB), 2000).

Preparation of the test samples

Using one type of wood, five different types of impregnation materials (Tanalith-E, Imersol Aqua, Vacsol Azure, Boric acid and Borax), eight types of combustion (flame sourced combustion temperature, without flame source combustion temperature, glowing stage combustion temperature, flame sourced combustion

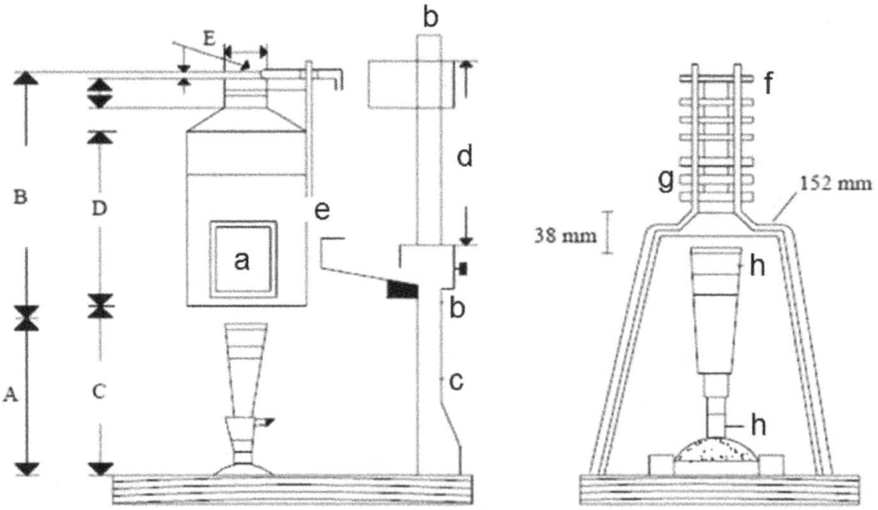

Figure 1. Combustion test apparatus. a: Mica glass b. End of slide, c: Flame burner guide, d: Slide, e: Potansiyometer or Milivoltmeter inlet, f: Test specimens, g: Wire cage, A: 270 mm., B: 430 mm., C: 295 mm., D: 305 mm., E: 38 mm.

light intensity, without flame sourced combustion light intensity, glowing stage combustion light intensity, combustion losses, combustion durations) and one control sample, a total of 450 (13×13×76 mm) samples were prepared with ten samples for each parameter.

Combustion tests

Combustion tests were done in combustion test devices according to ASTM E 160-50 (1975) standards. Accordingly, before the combustion test, impregnated samples were conditioned at 27°C and 30% relative humidity in a conditioning room until reaching 7% relative humidity. Every sample group was weighed before the test and stored on a wire stand. Samples on every stand were placed vertically on the stand with respect to the samples below and above. Distance between samples and the fire-flame outlet was fixed at 25±1.3 cm when the device was empty and gas pressure was fixed at 0.5 kg/cm^2 in the manometer. When ignited, the test temperature was set at 315±8°C in the funnel, using a calibrated thermocouple. The flame source was centered below the sample pile and the flame source combustion was continued for three minutes. After extinguishing the flame source, subsequent evaluation of flame spread (charring) without open-flame source combustion was carried out. Temperatures during combustion (°C) were determined with a thermometer (Figure 1). Without flame source cumbustion duration is the duration of combustion after the flame source turned-off up to flame source combustion. The glowing stage combustion duration is the duration from flame sourced combustion to dispersion of test samples.

Data analysis

SPSS 15.0 for Windows program is used in the statistical analysis of the combustion properties of the wood material tested. Multiple variance analysis was used to determine the difference between the combustion properties of the impregnated samples. In the case of significant difference between the groups (α = 0.05) confidence level was compared with Duncan's test. Factor that led to the success of the experiment was each other's rankings; least significant difference (LSD) was determined by leave of critical groups according to the value of homogeneity.

RESULTS

Combustion properties

Combustion temperatures

Mean values of combustion temperatures, light density, weight loss and combustion duration for different combustion types are given in Table 1.

The highest combustion temperature for burning types was found in without flame burning (570.160°C) and the lowest in glowing stage burning (259.511°C). The highest light density was measured in flame combustion (342.179 lux) and the lowest glowing stage (285.683 lux). Weight loss for different combustion types were found as equal (84.890%). Combustion duration was the highest in glowing stage combustion (9.055 min) and the lowest in without flame combustion (2.558 min).

Mean values of combustion temperatures, light density, weight loss and combustion duration for different impregnation materials are given in Table 2.

The highest combustion temperature for different impregnation materials was measured in Imersol Aqua (458.686°C) and the lowest in Borax (439.023°C). The highest light density was measured in Borax (343.337 lux) and the lowest in Vacsol Azure (294.102 lux) impregnation. The weigth loss was the highest in Imersol Aqua (91.667%) and the lowest in boric acid (78.333) impregnation. Combustion duration was the highest in Borax (6.888 min) and the lowest in Vacsol Azure (4.112 min).

Table 1. Mean values of combustion temperatures, light density, weight loss and combustion duration for different combustion types.

Combustion type	Temperature (°C)*	Ligtht density (lux)**	Weight loss (%)***	Combustion duration (min)****
AKY	520.216 A	342.179 A	84.890 A	3.500 B
KKY	570.160 A	312.318 A	84.890 A	2.558 C
KHY	259.511 B	285.683 A	84.890 A	9.055 A

*Different letters in the columns refer to significant changes in the combustion temperatures at 0.05 confidence level ($LSD_{0.5}$= 60.920); **different letters in the columns refer to significant changes in the light density at 0.05 confidence level ($LSD_{0.5}$= 54.120); ***different letters in the columns refer to significant changes in the weight loss at 0.05 confidence level ($LSD_{0.5}$= 1.401), ****different letters in the columns refer to significant changes in the combustion duration at 0.05 confidence level ($LSD_{0.5}$= 55.650).

Table 2. Mean values of combustion temperatures, light density, weight loss and combustion duration for different impregnation materials.

Impregnation material	Temperature (°C)*	Ligtht density (lux)**	Weight loss (%)***	Combustion duration (min)****
Control (Co)	436.840 A	293.282 A	84.667 B	3.110 D
Vacsol Azure (Va)	454.204 A	294.102 A	84.000 B	4.112 CD
Tanalith E (Te)	458.442 A	329.055 A	86.000 B	5.0 BC
Imersol Aqua (Ia)	458.686 A	320.690 A	91.667 A	5.777 AB
Borik Asit (Ba)	452.578 A	299.894 A	78.333 C	5.337 BC
Borax (Bx)	439.023 A	343.337 A	84.667 B	6.888 A

*Different letters in the columns refer to significant changes in the combustion temperatures at 0.05 confidence level ($LSD_{0.5}$= 86.160); **different letters in the columns refer to significant changes in the light density at 0.05 confidence level ($LSD_{0.5}$= 76.530); ***different letters in the columns refer to significant changes in the weight loss at 0.05 confidence level ($LSD_{0.5}$= 1.981), ****different letters in the columns refer to significant changes in the combustion duration at 0.05 confidence level ($LSD_{0.5}$= 78.700).

Multiple variance analysis (MANOVA) results of combustion temperature, light density, weight loss and combustion duration for different impregnation materials are given in Table 3.

According to this result, the impact of impregnation material type on combustion temperature, weight loss and combustion duration were found statistically meaningful ($\alpha < 0.05$). Accordingly, the impact of impregnation material type on light density was found as statistically unmeaningful ($\alpha < 0.05$).

Duncan's test results of interaction between impregnation material and combustion type is given in Table 4.

According to Duncan's test, the interaction between the impregnation material and combustion type was the highest in Ba+KKY (618.600°C) and the lowest in Ba+KHY (212.387°C). Duncan's test results of impregnation material and combustion type interaction was the highest for Ia+AKY, Ia+KKY, Ia+KHY (91.670%) and the lowest for Ba+AKY, Ba+KKY, Ba+KHY (78.330%). Duncan's test results for the interaction of impregnation material and combustion type was found as the highest in Bx+KHY (11.833 min) and the lowest in Va+KKY (1.505 min).

DISCUSSION

The amount of retention according to the type of impregnation material was found statistically meaningful. This may be due to difference in concentrations of impregnation solution. The highest amount of retention was obtained in Vacsol Azure, followed by Imersol Aqua, Tanalith-E, borax and boric acid. According to the type of impregnation material, the highest retention amounts were measured in Vacsol Azure samples (151.044 kg/m^3) followed by Imersol-Aqua (127.045 kg/m^3), Tanalith-E (104.083 kg/m^3), Borax (86.393 kg/m^3), boric acid (64.887 kg/m^3). The highest amount of retention in Vacsol Azure impregnation may be due to the difference in concentration and the impregnation capacity.

The impact of impregnation material on combustion temperature was found statistically meaningful ($\alpha<0.05$). Combustion temperature was measured as 454.204°C with Vacsol Azure, 458.442°C with Tanalith-E, 458.686°C with Imersol Aqua, 452.578°C with Boric acid, 439.023 °C with Borax and 436.840°C with control samples. Boron compounds decreased the combustion temperature as mentioned in the literature. The combustion temperatures was measured as 520.216°C with flame sourced combustion, 570.160°C without flame sourced combustion and 259.511°C with glowing stage combustion. The effect of impregnation material decreases after ignition has occurred, thus the main effect of impregnation material is during the flame sourced combustion.

The light intensity according to the type of impregnation materials was measured as the highest with

Table 3. Multiple variance analysis results of combustion temperature, light density, weight loss and combustion duration for different impregnation materials.

Factor	Degree of fredom	Sum of squares				Mean of squares				F value				P significance value (α < 0.05)			
		Combustion temperature	Light density	Weight loss	Combustion duration	Combustion temperature	Light density	Weight loss	Combustion duration	Combustion temperature	Light density	Weight loss	Combustion duration	Combustion temperature	Light density	Weight loss	Combustion duration
Impregnation material (A)	5	4182.272	19385.521	819.333	279693.722	836.454	3877.104	163.867	55938.744	0.100	0.591	37.336	8.073	0.000	0.126	0.000	0.000
Temperatures (B)	2	1001786.907	28757.947	0.000	1597640.333	500893.453	14378.974	0.000	798820.167	60.327	2.194	0.000	115.296				
Interaction (AB)	10	21178.695	44421.733	0.000	81567.444	2117.870	4442.173	0.000	8156.744	0.255	0.678	0.000	1.177				
Error	36	298905.916	235855.975	158.000	249423.333	8302.942	6551.555	4.389	6928.426								
Total	53	1326053.790	328421.177	977.333	2208324.833												

Table 4. Duncan's test results of interaction between impregnation material and combustion type.

Impregnation material + Combustion type	Temperature (°C)*	Weight loss (%)**	Combustion duration (min)***
Co+AKY	514.275 ^{A}	86.670 ^{B}	2.666 ^{FGH}
Co+KKY	548.244 ^{B}	86.670 ^{B}	0.500 ^{H}
Co+KHY	248.000 ^{A}	86.670 ^{B}	6.166 ^{DE}
Va+AKY	531.025 ^{A}	84.000 ^{B}	3.166 ^{FG}
Va+KKY	572.853 ^{B}	84.000 ^{B}	1.505 ^{GH}
Va+KHY	258.733 ^{A}	84.000 ^{B}	7.666 ^{CD}
Te+AKY	509.388 ^{A}	86.000 ^{B}	3.666 ^{EFG}
Te+KKY	563.472 ^{A}	86.000 ^{B}	2.333 ^{FGH}
Te+KHY	302.467 ^{B}	86.000 ^{B}	9 ^{BC}
Ia+AKY	533.111 ^{A}	91.670 ^{A}	2.833 ^{FGH}
Ia+KKY	570.955 ^{A}	91.670 A	4.333 ^{EF}
Ia+KHY	271.991 ^{B}	91.670 ^{A}	10.166 ^{AB}
Ba+AKY	526.747 ^{A}	78.330 ^{B}	4 ^{EFG}
Ba+KKY	618.600 ^{A}	78.330 ^{B}	2.511 ^{FGH}
Ba+KHY	212.387 ^{B}	78.330 ^{B}	9.500 ^{ABC}
Bx+AKY	506.750 ^{A}	84.670 ^{C}	4.666 ^{EF}
Bx+KKY	546.833 ^{A}	84.670 ^{C}	4.166 ^{EF}
Bx+KHY	263.487 ^{B}	84.670 ^{C}	11.833 ^{A}

*Different letters in the columns refer to significant changes in the combustion temperatures at 0.05 confidence level ($LSD_{0.5}$= 149.200); **different letters in the columns refer to significant changes in the weight loss at 0.05 confidence level ($LSD_{0.5}$= 3.431), ***different letters in the columns refer to significant changes in the combustion duration at 0.05 confidence level ($LSD_{0.5}$= 136.300).

Borax (343.337 lux) and the lowest with Vacsol Azure at (294.102 lux) respectively. The effect of the type of impregnation material on light intensity was found to be statistically insignificant (α <0.05). The light intensity according to the type of impregnation materials was found as 294.102 lux (Vacsol Azure), 329.055 (Tanalith-E), 320.690 lux (Imersol Aqua), 299.894 Lux (Boric acid), 343.337 (Borax) and 293.282 (control samples), respectively. Accordingly, the highest light intensity was measured in Borax as 343.337 lux and the lowest in Vacsol Azure as 294.102 lux. Boron compounds show a positive effect on the density of the smoke as a source of poisoning but not such a risk with other impregnation materials.

Light density according to type of combustion was measured as 342.179 lux with flame sourced combustion, 312.318 lux without flame sourced combustion and 285.683 lux in glowing stage combustion. Smoke does not occur in glowing stage combustion but at the same level with and without flame sourced combustions. Duncan's test results for the interaction of impregnation material and combustion type was measured as the highest with Bx+AKY as 350.639 lux and the lowest with Ba+KHY as 226.740 lux.

The impact of impregnation material type on weight loss was found statistically meaningful (α < 0.05). Weght loss was measured as 84.000% with Vacsol Azure, 86.000% with Tanalith-E, 91.667% with Imersol Aqua, 78.333% with Boric acid, 84.667% with Borax 84.667% with control samples. Thus it was highest with Imersol Aqua (91.667%) and lowest with Boric acid (78.333%). Weight loss for different combustion types was measured as the same (84.890%). Boric acid is more resistant to material loss, decreasing the risk of a sudden collapse and destruction in fire.

Duncan's's test results for the interaction of impregnation material and combustion type was measured as highest with Ia+AKY, Ia+KKY, Ia+KHY (91.670%) and as lowest with Ba+AKY, Ba+KKY, Ba+KHY (78.330%).

The impact of impregnation material on combustion type was found statistically meaningful (α<0.05). The combustion duration for different impregnation materials was measured as 4.112 min with Vacsol Azure, 5 min with Tanalith-E, 5.777 min with Imersol Aqua, 5.337 min with Boric acid, 6.888 min with Borax and 3.110 min with control samples. So, it was highest with Borax (6.888 min) and lowest with Vacsol Azure (4.112 min). Combustion duration according to type of combustion was measured as 3.500 min with flame sourced combustion, 2.558 min without flame sourced combustion and 9.055 min in glowing stage combustion. Thus, combustion duration was highest with glowing stage combustion (9.055 min) and lowest without flame sourced combustion (2.558 min). Duncan's test results for the interaction of impregnation material and combustion type were measured and the highest was 11.833 min with Bx+KHY and the lowest 1.505 min with Va+KKY.

Consequently, the boron compounds showed a positive effect on combustion properties of Rowan wood.

In this study, the combustion properties of Rowan wood were determined by impregnating it with various impregnation materials (Tanalith-E, Vacsol-Azure, Imersol-Aqua and Boron compounds (Borax and Boric asid)). These properties can be compared with the results of other studies in literature which are related to the effects of various impregnation materials on combustion properties of different tree species.

Yalınkılıç et al. (1998) investigated the combustion properties of Douglas wood impregnated with different agents to keep it from biotic and abiotic effects. They observed that boron compounds have impacts to decrease the combustion.

Örs et al. (1999c) researched the combustion increase impact of poly ethilenglicol (PEG-400) and WR. So yellow pine wood was treated with paraffin, styrene, methil metacrilate and izosiyanat after being impregnated with boric acid, borax and sodium per borate solutions with water or PEG-400. It was observed that boron compounds increases the combustion resistance and decreases the combustion impact of water repellents.

Özen et al. (2001b) impregnated yellow pine with sodium per borate, sodium tetra borate, Imersol-Aqua and Tanalith–CBC by dipping method and studied combustion properties of D-VTKA glued 3-ply laminates in accordance with ASTM E69 standard (ASTM, 1976). They observed sodium tetra borate and sodium per borate as the agents that negatively affect the combustion process.

Örs et al. (2002) investigated the combustion properties of heaven wood impregnated with Tanalith-CBC, Borax, Boric acit, Boric acid + Borax, Vacsol-WR, Imersol-WR 2000, Polietilenglikol-400 and Stiren. As a result of the tests, the boron compounds with vacuum process increased the fire reterdant of the heavenwood species growing in Turkey. It was observed that one of the vinil monomers Styrene and Vacsol-WR has shown a fire retardant effect.

Baysal (2003) carried out the combustion properties of yellow pine impregnated with boric acid, borax and tanning materials. Natural extractives showed unfavourable effects on fire parametres. Also, they showed the same or more badly burning properties compare to control specimen. However, boric acid and borax applied as secondary treatment agents over natural extractives possivitely affected some fire properties of Scots pine in significant level (P≤ 0.01).

ACKNOWLEDGMENTS

This study is a part of M.Sc.Thesis prepared by Neslihan SÜZER ERTÜRK, Institute of Science and Technology, Gazi University, Ankara, Turkey.

REFERENCES

Aslan S, Özkaya K (2004). Investigation of Combustion Resistance of Wood-based Panels Treated with Different Chemicals. Süleyman Demirel Univ. Fac. For. J. Ser: A. Number. 2:122-140.

ASTM E160-50 (1975). Standart Test Method for Combustible Properties of Terated Wood by the Crib Test. ASTM Standards, U.S.A.

ASTM-D 1413-76 (1976). Standard Test Method of Testing Wood Preservatives by Laboratory Soilblock Cultures. Annual Book of ASTM Standards, pp. 452-460.

Atar M, Keskin H (2007). Impacts of Coating with Various Varnishes After Impregnation with Boron Compounds on the Combustion Properties of Uludag Fir. J. Appl. Polym. Sci. 106:4018–4023.

Atar M, Keskin H, Korkut S, Sevim Korkut D (2011). Impact of impregnation with boron compounds on combustion properties of oriental beech (*Fagus orientalis* Lipsky) and varnishes. Afr. J. Biotechnol. 10(15):2867-2874.

Baysal E (2003). Fire Properties of Scots Pine Impregnated With Borates and Natural Extractives. J. Inst. Sci. Technol. Erciyes Univ. 19(1-2):59-69.

Baysal E, Peker H, Çolak M, Tarımer İ (2003). Fire Properties of Varnish Coated Wood and Fire Retardant Effect of Pretreatment with Borates. Sci. Eng. J. Fırat Univ. 15(4):645-653.

Baysal E (2011). Combustion properties of wood impregnated with commercial fertilizers. Afr. J. Biotechnol.10(82):19255-19260.

Evans PD, Michell AJ, Schmalzl K (1992). Studies of the Degradetion and Protection of Wood Surfaces. Wood Sci.Technol. 26:151-163.

Hickson's Timber Impregnation Co. (GB) (2000). Into the 21st. Century, Imersol-Aqua Brochure. Hickson Timber Treatments Datasheet 6214:1-4,

Okcu O (2006). Stiking and combustion properties of laminated vaneer lamber (LVL) impregnated with saqme impregnation materials. Science Expertisig Thesis, Department of Furniture and Decoration, Graduate School of Natural and Applied Sciences, Zonguldak Karaelmas Universty, Karabük, P. 70.

Örs Y, Atar M, Peker H (1999a). The Effects of Different Preservation Chemicals and Finishing on Wood Combustion Properties in *Pinus sylvestris* L. and *Castanea sativa* Mill., Turk. J. Agric. For. 23(5): 541-549.

Örs Y, Sönmez A, Uysal B (1999b). Fire Reterdant Chemicals Affecting Combustion Resistance of Wood. Turk. J. Agric. For. 23(EK2):389-394.

Örs Y, Atar M, Peker H (1999c). The Effect of Some Boron Compounds and Water Repellents on the Fire Resistance Properties of Scotch Pine Wood. Turk. J. Agric. For. 23(5):501-509.

Örs Y, Atar M, Peker H (2002). Çeşitli Maddelerle Emprenye Edilmiş Sakallı Kızılağaç (C. A. Mey.) Yalt. Odunun Yanma Özellikleri. Gazi Üniversitesi Fen Bilimleri Enstitüsü Dergisi 15(3):687-697.

Özçifçi A (2001a). The Effects of Bleaching Chemicals on Combustion Properties of European Oak (*Quercus Sessiliflora* Salisb.), Z.K.Ü.K.T.E.F. Teknoloji J. 4(3-4):63-72.

Özçifci A (2001b). Technical properties of laminated wood materials ompregnated with some chemicals. Ph. D. Thesis, Gazi University Institute of Sciene and Technology, Ankara, P. 97.

Özen R, Özçifci A, Uysal B (2001a). The Combustion Properties of Glued Laminated Timber (Glulam) Prepared with PVAc Adhesive. Z.K.Ü.K.T.E.F. Teknoloji J. 4(1-2):139-148.

Özen R, Özçifci A, Uysal B (2001b). The Combustion Properties of Laminated Wood Materials Prepared from Scotch Pine (*Pinus sylvestris* L.). Pamukkale Univ. Eng. Coll. J. Eng. Sci. 7(1):131-138.

Özkaya K (2002). Research on determination of combustion characteristics of wood-based board materials treated by different chemical materials. M. Sc. Thesis, Hacettepe University Institute of Science, p. 124. Ankara.

Peker H, Tan H, Baysal E (2004). Fire Resistance of Spruce Wood Treated With Some Chemicals. Sci. Eng. J. Fırat Univ. 16(1):163-175.

Richardson BA (1987). Wood Preservation. The Construction Press: Lancaster, England.

Seferoğlu D (2008). Determination of the effects of surfacing processes on combustion resistance of wood materials. M. Sc. Thesis, Department of Furniture and Decoration, Graduate School of Natural and Applied Sciences, Karabük University, Karabük, p. 93.

Terzi E (2008). Fire properties of wood materials treated with amonia compounds. M. Sc. Thesis, İstanbul University Institute of Sciences, p. 126.

TS 4176 (1984). Wood - Sampling Sample Trees and Long for Determination of Physical and Mechanical Properties of Wood in Homogeneous Stands. T.S.E., Ankara.

TS 2470 (1976). Wood - Sampling Methods and General Requirements for Physical and Mechanical Tests. Türk Standartları Enstitüsü, Ankara.

TS 53 (1981). Wood - Sampling and test methods - Determination of physical properties, T.S.E., Ankara.

Uysal B (1998). The Effects of Various Water Repellent and Fire Retardant Chemicals on Fire Resistance of Alder Wood. Z.K.Ü.K.T.E.F. Teknoloji J. 1(2):43-52.

Uysal B, Özçifci A (2000). Ihlamur (*Morus alba* L.) Odunundan PVAc Tutkalı ile Üretilen Lamine Ağaç Malzemenin Yanma Özellikleri. Gazi Üniversitesi Fen Bilimleri Enstitüsü Dergisi 13(4):1023-1035.

Uysal B, Kurt Ş (2006). Combustion Properties of Laminated Veneer Lumbers Bonded with PVAC, PF Adhesives and Impregnated with Some Chemicals. Doğuş University J. 7(1):112-126.

Yalınkılıç MK, Örs Y, Ay N, Baysal E, Demirci Z (1997). Impregnation Treatments and Anatomical Properties of the Wood of Regional Grown Douglas Fir (*Pseudotsuga menziesii* (Mirb) Franco). Turk. J. Agric. For. 21(5):433-444.

Yalınkılıç MK, Demirci Z, Baysal E (1998). Fire Resistance of Douglas Fir [*Pseudotsuga Menziesii* (Mirb.) Franco] Wood Treated with some Chemicals. Pamukkale Univ. Eng. Coll. J. Eng. Sci. 4(1-2):613-624.

9

Analysis and supervision of the water extraction of a thermal power plant

M. N. Lakhoua

University of Carthage, UR: Mechatronics Systems and Signals, ESTI, Tunis, Tunisia.

The aim of this paper is to analysis the water treatment process in a thermal power plant (TPP). In fact, we present an application of a supervisory control and data acquisition (SCADA) system. Thus, an example of a SCADA system of the center of RADES in Tunisia was presented. Our contribution in this work consists in the analysis of the water loss in the TPP on the one hand and the supervision of the water extraction circuit using the SCADA system, on the other hand.

Key words: Water extraction, water loss; thermal power plant, supervisory control and data acquisition (SCADA).

INTRODUCTION

Every day, we use the electric energy without even to be conscious of it. The electric energy serves in all domains including those where we think that it is not used (central heating of gas, thermal motor-driven vehicles...). Means of production of this energy are very various; we classify them today depending on whether they are based on renewable energies or fossil energies. With regards to these last, reserves not being inexhaustible, we tries to replace them by the renewable energies that have for main advantage to be less polluting.

The Tunisian Society of Electricity and Gas (STEG) is a society whose main function is to produce electricity in order to satisfy needs of its customers. Among electricity production centers of the STEG, we mention the center of RADES (near to Tunis, Tunisia) (Annual Report, 2012). It is one of the most important centers of the point of view the power installed (700 MWS). It has been inaugurated in 30 of May, 1986. It is about a thermal power plant (TPP) producing electricity while using dry water steam to drag the alternator in rotation, this steam is generated in a furnace that transforms the chemical energy of the fuel (natural gas, heavy fuel-oil) in calorific energy. By reason of the complex requirements and in order to avoid the maximum loss of production, it is extremely important to master all aspects having linked to the security and the profitability of the highest level.

In fact, the electricity production in a TPP is based on a set of energies transformations using water as support of energy. This water must have a noble quality in order to guarantee the installation security and to improve production groups' performances. It is therefore necessary to apply a rigorous water treatment and a control of its quality (Vitaly, 2008).

The process of the electricity production in the TPP of RADES is essentially based on the water distributed by the SONEDE (National Water Distribution Utility of

Tunisia). This water generally contains dissolved mineral salts and organic matters. The presence of these elements can generate problems bound to the furring, the corrosion and the different facilities contamination notably the furnace, the steam-powered turbine and water or steam circuits (Kagiannas et al., 2003; Ecob et al., 1995).

In order to assure the required quality of water in the water-steam cycle of the TPP, the water treatment process is necessary. Indeed, water passes by the filtration chain then introduced in the inverse osmosis station and thereafter in the demineralization station (Changling and Boon-Teck, 2006). In fact, the water of the SONEDE used in the TPP of RADES is unfit to the feeding of furnaces. It contains matters suspended and in various solutions in nature and in quantity of salts and gases dissolved. The matter suspended is constituted of the sand, of colloidal clays, insoluble mineral salts and of organic matters (products of animal and plant deterioration). These bodies give a certain coloration to water (turbidity) that requires a clarification treatment. This undesirable foulness can drive to the serious damages. Among which we mention notably corrosion and furring (or encrustations). In order to avoid these problems, it is necessary to:

(i) Eliminate gases (CO_2, O_2, N_2) of the water by the physical degassing or the chemical degassing by the injection of the oxygen reduction as N_2H_4...
(ii) Use of the destitute water of mineral salts for example water done demineralization with a conductivity ($\sigma < 0.2$ µS/cm) and a content in silica $SiO_2 < 30$ ppb.
(iii) Work with a sufficiently basic pH ($8.5 < pH < 9.5$).

The objective of this paper is to identify the water loss in a TPP and to control the water extraction circuit using a SCADA system. An example of a SCADA system of the TPP of RADES is presented.

PRESENTATION OF AN EXAMPLE OF A SCADA SYSTEM

The supervisory control and data acquisition (SCADA) term refers to a system that collects data coming from different sensors of an industrial or other process, these sensors can beings installed in the same site or distant (several Km), the introverted data are treated by a unit called processor power station (CPU, PCU, PC...), results are sent in real time to the Men / Machine interfacing that can be a computer with its peripherals (Baily and Wright, 2003).

The SCADA system assures the surveillance and the control of electric, mechanical or electronic equipment equipping all or a part of the network (Munro, 2008). It also allows operators to command and to control all facilities of the power station, as well as to offer all necessary information to the good conduct of a stage data of the power station (Carke et al., 2003; Horng, 2002). The intended role to the SCADA system is to collect data instantaneously of their sites and to transform them in numeric signals by following to send them through the network of communication toward the main and secondary stations (Wiles, 2008). This centralized supervision allows operators, since the control room of the TPP, to control facilities in their domain of exploitation and the different types of incidents (Ozdemir and Karaoc, 2006; Warcuse et al., 1997; Gergely et al., 2010). The center of RADES is equipped of a SCADA network. Stations belong to a network superior Ethernet (10 Mb/s). Mainly this network permits to do exchanges of files between the various stations (Annual Report, 2012). It avoids so the overcharge of the node network bus. Figure 1 shows an overall view of the TPP of RADES using a SCADA system. The SCADA system of the TPP of RADES orders and classifies all data for (Lakhoua, 2009a) (Lakhoua, 2009b) (Lakhoua, 2010a):

(i) Instantaneous impression.
(ii) Visualization on screen using data tables and tabular diagrams.
(iii) Registration of instantaneous exchanges of numeric and analogical data.
(iv) Instantaneous calculation for example corrections of gas debits, direct middle specific consumption, middle values.
(v) Storage of the analogical information of the process.
(vi) Calculation of outputs and losses of the process.
(vii) Surveillance of the SOE signals (entrances rapid contact 1ms)
(viii) Interfacing interactive Men / Machine for the surveillance of the system and the conduct of processes (tabular, curves view of alarm) (Figure 2).

The SCADA system of RADES is equipped of three communication networks (Lakhoua, 2010b):

(i) Field bus, 5 Mbits, permitting to do exchanges of the numeric data of the entrance card / exits (FBM) toward the central system (CP) via modules of communication (FCM);
(ii) Node bus, 10 Mbits, permitting to do exchanges of the numeric data of the central system (CP) via modules of communication (DNBT) toward the Men/Machine interfacing (workstations);
(iii) Ethernet TCP/IP, 100 Mbits, permitting to do exchanges of files between workstations of the Men/Machine interfacing. It avoids so the overcharge of the Nodebus network.

PRESENTATION OF THE INVERSE OSMOSIS AND THE DEMINERALIZATION STATIONS

Considering that the water of the SONEDE contains an

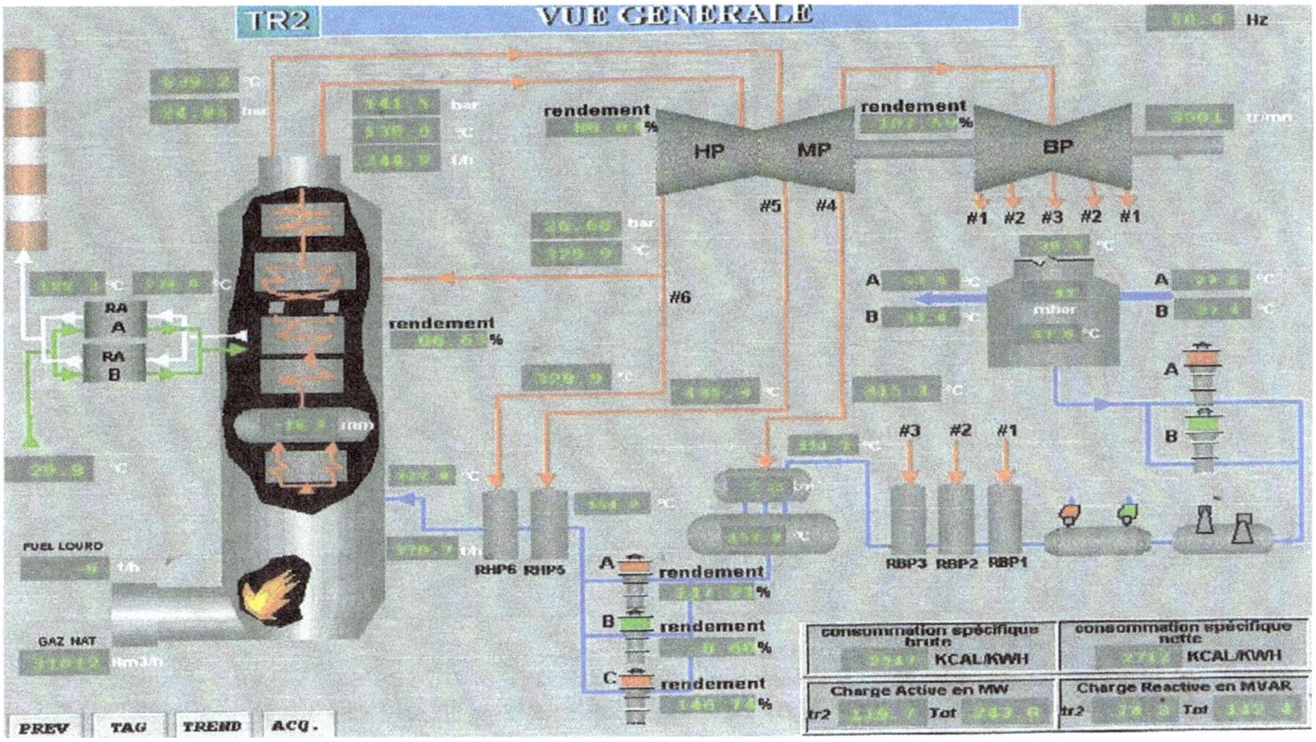

Figure 1. Overall view of the TPP of Radès with SCADA.

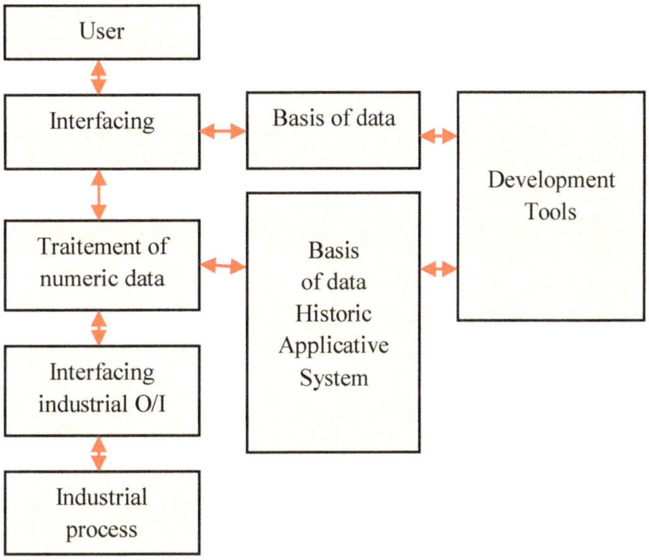

Figure 2. Principle of a SCADA system.

elevated rate in dissolved salts and in matter suspended, it is indispensable to adopt a stage of pretreatment to assure the good working of the inverse osmosis installation and to protect modules against risks of usuries, corrosion and especially membrane calmative (Tarja et al., 2006).

The pretreatment is constituted of two filtration chains each including a sand filter and an active coal filter. Thereafter, we present the two stations of the TPP: inverse osmosis and demineralization.

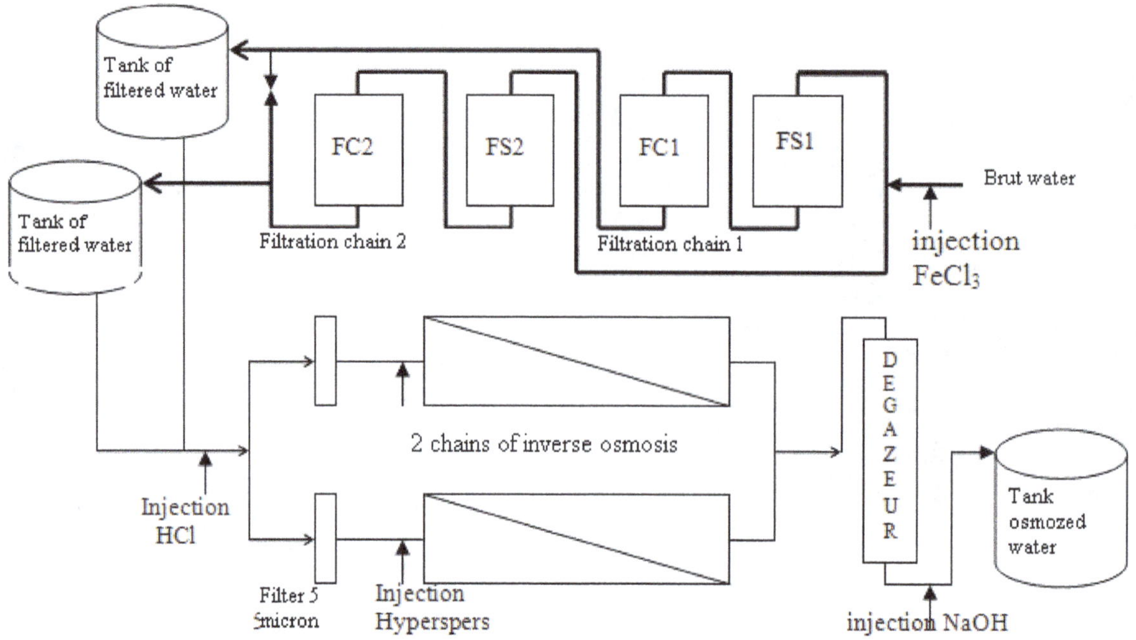

Figure 3. Functional diagram of the inverse osmosis station of the TPP of RADES. FS1: Sand filter of the filtration chain 1. FC1: Active coals filter of the filtration chain 1. FS2: Sand filter of the filtration chain 2. FC2: Active coals filter of the filtration chain 2.

The control of the water quality is an important task to maintain the efficiency and the sure and continuous working of the power station (Yubin et al., 2002). To guarantee the best water quality at the level of the water steam circuit, the TPP of RADES arranges an inverse osmosis station that permits to eliminate the majority of salts dissolved in the raw water before being treated in a demineralization station (Figure 3). This stage serves to minimize risks of failing by corrosion of the turbine or the loss of the efficiency and the power (Electricity and Gas Revue, 2011). The bold lines present the water circuit in the two filtration chains and the light lines present the water circuit in the two inverse osmosis chains. The basic principle of the ion exchange consists in withdrawing ions (remaining salts that are lower to 8%) in solution in water is to recover an ion of value, either to eliminate a harmful or bothersome ion for the ulterior utilization of water (Firoozshahi and Mengyang, 2010). The exchange of ions is a process which ions with a certain load contents in a solution are eliminated of this solution, and replaced in the same way by an equivalent quantity of other ions load gave out by the strong but the opposite load ions are not affected. In the demineralization chain, osmosis water passes by the following stages:

(i) A weak cationic exchanger (CF1);
(ii) A strong cationic exchanger (CF2);
(iii) A weak anionic exchanger (AF1);
(iv) A degasser;
(v) A strong anionic exchanger (AF2);
(vi) A strong cationic exchanger (CF3);
(vii) A strong anionic exchanger (AF3).

After the demineralization, the water must have a lower conductivity of 0.2 μS/cm, a pH between 6.5 and 7.5; silica < 30 ppb. Figure 4 shows the water treatment cycle in the demineralization station.

RESULTS OF THE IDENTIFICATION OF THE WATER LOSS AND THE SCADA APPLICATION

Demineralized water is distributed to the two production plants A and B, the laboratory and the unloading of the fuel station. Consumption of the latter two is negligible compared to the amount consumed by the process of generation of electricity (Figure 5). Demineralized water is distributed to the four stations of the TPP using two tanks. For each station, demineralized water is used primarily for the extra three following circuits:

(i) The water-steam circuit;
(ii) The water cooling circuit of the bodies of various rotating machinery (Noria circuit);
(iii) The secondary steam circuit used primarily for the fuel heating.

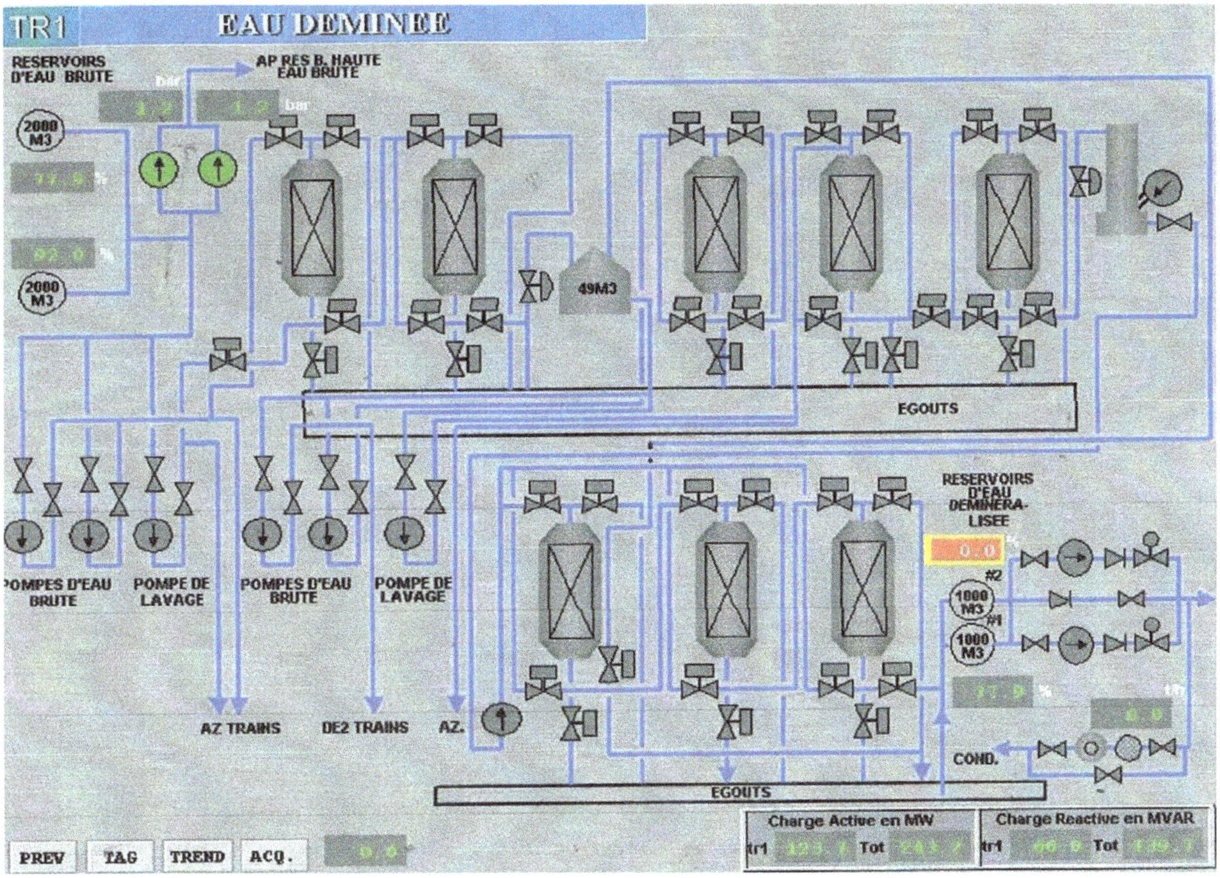

Figure 4. Demineralization station of the TPP of RADES.

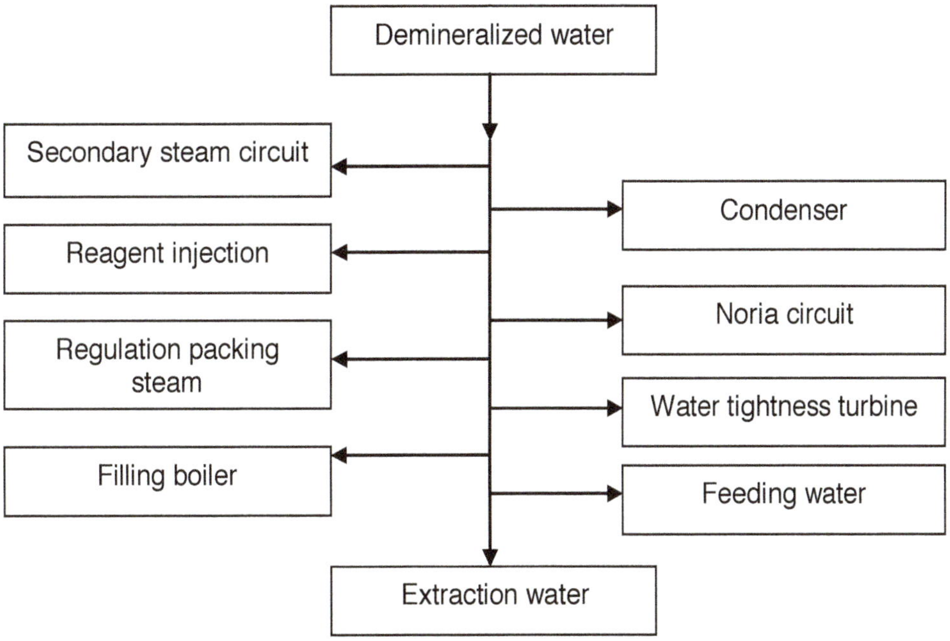

Figure 5. Demineralization station of the TPP.

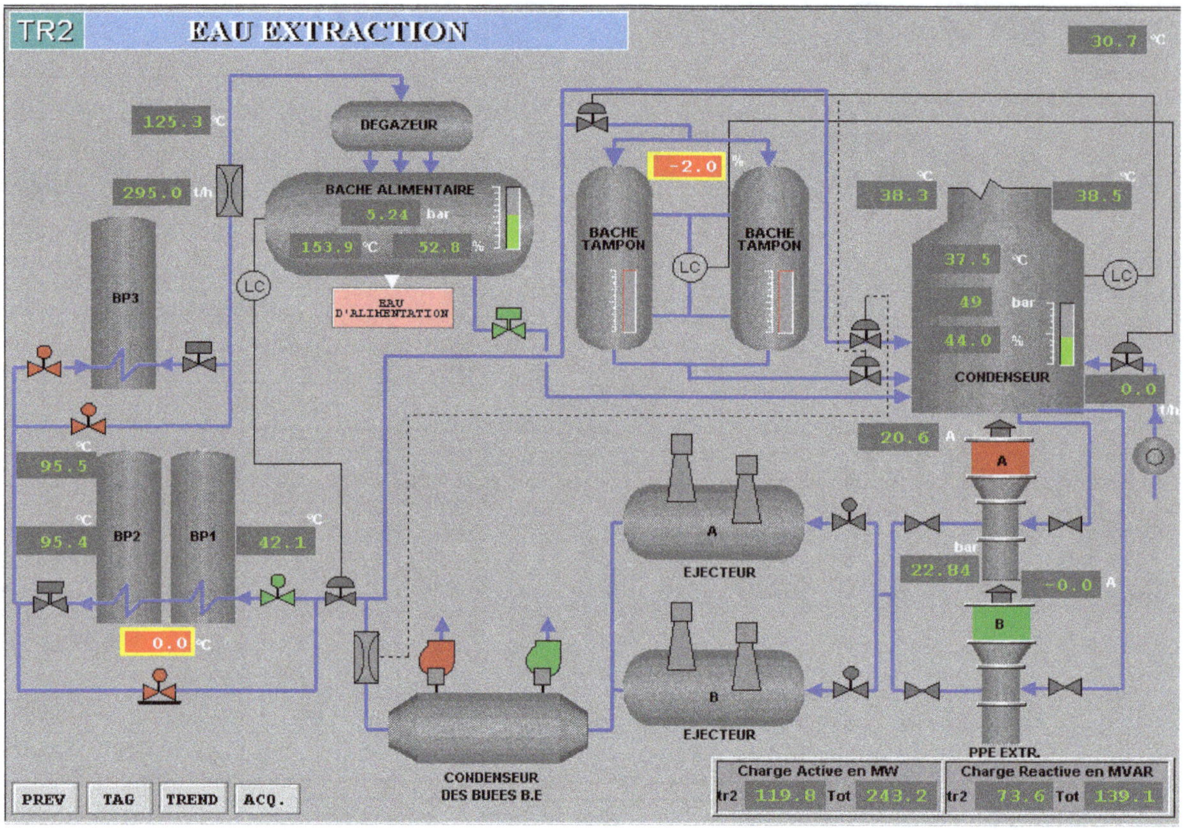

Figure 6. Display of the water extraction circuit of the TPP.

The ongoing purges are formed by the boiler purges which are necessary to maintain the quality of condenser water steam cycle. The ongoing purges volume varies between 50 and 70 m^3/day for one station. Sampling purges are carried out automatically by on line continuously analyses of the water of the boiler parameters (pH, conductivity...). The volume of these purges varies between 3 and 5 m^3/day for one station. The staple purges occur after judgments or in case of anomalies that require draining of the boiler or other bodies (ball boiler, covering food...). The sampling purges of the plant A allow extra Noria circuits of the two stations. The make-up of the water steam circuit varies between 70 and 80 m^3/day for one station.

In this application, we have installed an ultrasonic flow meter in the make-up water circuit and an electromagnetic flow meter in the osmosis position. Flow converted to the level of the ultrasonic flow meter will be issued as an electric signal 4 to 20 mA, towards the electronic room to treat by the input (UA 374 and UT 375) modules. Then, it will be forwarded to the ordinary and finally visualized in the control room. Figure 6 shows the display of the circuit of the water extraction of the TPP.

This application is declined in six stages:

Stage 1: Choosing the site of the signal (FBM module) that treats the deminiralized water.
Stage 2: Programming the AIN block for the supervision of the ultrasonic flow meter signals.
Stage 3: Testing the AIN block by injection of current.
Stage 4: Passing the cable between the electronic room and the SCADA room.
Stage 5: Programming three ACCUM blocks and two COUT blocks.
Stage 6: Improving the existing tabular of the water extraction circuit.

CONCLUSION

SCADA systems are used to control and monitor physical processes, examples of which are transmission of electricity, transportation of gas and oil in pipelines, water distribution, traffic lights, and other systems used as the basis of modern society. In this paper, we presented an identification of the water loss in a thermal power plant and an application of the SCADA system on the water extraction circuit. Moreover, we proved the importance on using a SCADA system for sustainable development in the supervision of the thermal power plants. Also the paper outlines the general concepts and required equipments for the supervision of such power plants. Some applications of SCADA system implementation in electrical companies over the world have been presented.

Conflict of Interest

The author(s) have not declared any conflict of interests.

REFERENCES

Annual Report (2012). Tunisian Society of Electricity and Gas. Tunisia.

Baily D, Wright E (2003). Practical SCADA for Industry. Elsevier. http://store.elsevier.com/Practical-SCADA-for-Industry/David-Bailey/isbn-9780750658058/

Carke G, Rynders D, Wright E (2003). Practical Modern SCADA Protocols. Elsevier.

Changling L, Boon-Teck O (2006). Frequency deviation of thermal power plants due to wind farms, IEEE Trans. Energy Conver. 21(3):708-716. http://dx.doi.org/10.1109/TEC.2006.874210

Ecob D, Williamson J, Hughes G, Davis J (1995). PLC's and SCADA - a water industry experience. IEE Colloquium on Application of Advanced PLC Systems with Specific Experiences from Water Treatment. pp. 601-610. http://dx.doi.org/10.1049/ic:19950742

Electricity and Gas Revue (2011). Tunisian Society of Electricity and Gas. N°16. Tunisia.

Firoozshahi A, Mengyang L (2010). Water treatment plant intelligent monitoring in large gas refinery. IEEE International Conference on Computational Technologies in Electrical and Electronics Engineering. pp. 785-789.

Gergely El, Coroiu L, Popentiu-Vladicescu F (2010). Analysis of the influence of the programming approach on the response time in PLC Control Programs, JCSCS. 3(1):61-64.

Horng JH (2002). SCADA system of DC motor with implementation of fuzzy logic controller on neural network. Adv. Eng. Software. pp. 361–364. http://dx.doi.org/10.1016/S0965-9978(02)00020-0

Kagiannas AG, Askounis D, Anagnostopoulos K, Psarras J (2003). Energy policy assessment of the Euro-Mediterranean cooperation, Energy Conver. Manage. pp. 2665-2686.

Lakhoua MN (2009a). Methodology for designing supervisory production systems: Case study of a counting system of natural gas, J. Elect. Eng. 9(N°3).

Lakhoua MN (2009b). Application of functional analysis for the design of supervisory systems: Case study of heavy fuel-oil tanks. Int. Trans. Syst. Sci. Applications. 5(N°1):21-33.

Lakhoua MN (2010a). Surveillance of pumps vibrations using a SCADA. Control Eng. Appl. Informatics. 12(N°1).

Lakhoua MN (2010b). SCADA applications in thermal power plants. Int. J. Phys. Sci. 5(N°7):1175-1182.

Munro K (2008). SCADA - A critical situation, Network Security. 1:4-6. http://dx.doi.org/10.1016/S1353-4858(08)70005-9

Ozdemir E, Karacor M (2006). Mobile phone based SCADA for industrial automation. ISA. Trans. 45(N°1):67-75. http://dx.doi.org/10.1016/S0019-0578(07)60066-4

Tarja AM, Ungureanu G, Capajana D, Covaciu FA (2006). SCADA System for Water Potential Management of a Hydropower Plants Cascade. IEEE Int. Conf. Automation. Quality Testing. Robotics. 1:410-414.

Vitaly A (2008). Alternative trends in development of thermal power plants. Appl. Thermal Eng. 28(2-3):190-194. http://dx.doi.org/10.1016/j.applthermaleng.2007.03.025

Warcuse J, Menz B, Payne JR (1997). Servers in SCADA applications. IEEE Trans. Ind. Appl. 9-2:1295-1334.

Wiles J (2008). Techno Security's Guide to Securing SCADA: A Comprehensive Handbook on Protecting the Critical Infrastructure. Elsevier. http://store.elsevier.com/Techno-Securitys-Guide-to-Securing-SCADA/Jack-Wiles/isbn-9780080569994/

Yubin X, Yinzhang G, Honggang W, Jianchao Z (2002). Distributed control system in water plant based on ControlNet. Proceedings of the 4th World Congress on Intelligent Control and Automation. 4:3113-3117.

Optimization of the louver angle and louver pitch for a louver finned and tube heat exchanger

Jiin-Yuh Jang and Chun-Chung Chen

Department of Mechanical Engineering, National Cheng Kung University, Tainan 70101, Taiwan.

The optimization of the louver angle (θ) and the louver pitch (Lp) for a louver finned and tube heat exchanger was investigated numerically along with a simplified conjugate-gradient method (SCGM). The area reduction ratio relative to a plain surface is the objective function to be maximized. A search for the optimum louver angle (θ) and louver pitch (Lp), ranging from 15°< θ < 40° and 2 mm < Lp < 3.2 mm, respectively, was performed. The results showed that the maximum area reduction ratios may reach 39~46% combined with the optimal design of (θ, Lp) at Re_D = 589~3533 (U_{in}= 0.5-3.0 m/s).

Key words: Optimization, louver pitch, louver angle, finned and tube heat exchanger.

INTRODUCTION

Fin-and-tube heat exchangers with louvered fins are widely employed in automobiles, air-conditioners and power generation, etc. The louvers act to interrupt the airflow and create a series of thin boundary layers that have lower thermal resistance. The first reliable published data on louvered fin surfaces was presented by Kays and London (1950). Davenport (1983) utilized smoke trace to study a standard variant of the corrugated louvered fin geometry and obtained heat transfer and friction correlations for corrugated louvered fin geometry. Achaichia and Cowell (1988) made an overall study of performance characteristics of flat-sided tube and louvered plate fin heat exchangers. They obtained the correlations for the louvered plate fin geometry. Sahnoun and Webb (1992) developed an analytical model to predict the heat transfer and friction characteristics of the corrugated louvered fin core. Sunden and Svantesson (1992) presented the investigations of heat transfer and pressure drop of standard louver fin and inclined louver fin. Their investigations illustrate that all the louvered surfaces are better efficient than the corresponding smooth surface. Wang et al. (1998) tested 17 samples of commercially available louver fin and tube heat exchangers for different geometrical parameters, including the number of tube row, fin pitch, and tube size.

In the 1990's, some investigators developed CFD code based on non-orthogonal, boundary-fitted meshes to calculate the flow over louvered fins. Suga et al. (1990) and Suga and Aoki (1991) used a rectangular flow domain filled with overlapping Cartesian meshes to calculate the flow and heat transfer over a finite-thickness fin. Hiramatsu et al. (1990) and Ikuta et al. (1990) utilized a block structured mesh with respective blocks for each louver.

Jang et al. (2001) numerically researched a three dimensional convex louver finned tube heat exchangers. The effects of different geometrical factor, containing convex louver angles (15.5°, 20.0°, 24.0°), louver pitch (0.953 mm, 1.588 mm) and fin pitch (8 fins/in., 10 fins/in., 15 fins/in.) are studied in detail for the Reynolds number ranging from 100 to 1100. It was proven that, for equal louver pitch, both the average Nusselt number and pressure drop coefficient are increased as the louver angle is increased; while for equal louver angles, they are decreased as the louver pitch is increased. Hsieh and Jang (2006) proposed continuously increased or decreased louver angle models and carried out a 3-D numerical analysis on heat transfer and fluid flow. Their results showed that continuously variable louver angle types employed in heat exchangers could effectively enhance the heat transfer performance. They also revealed that the maximum area reduction could reach up to 25.5% compared with a plain fin surface. Jang and Tsai (2011) utilized the simplified conjugate-gradient method (SCGM) to search the optimal louver angle of a fin heat exchanger. The area reduction for using louver surface compared to the plain surface was the objective function to be maximized. The maximum area reduction ratios of the louvered fins were 65.3, 66.9, 65.6, 63.7 and 62.2% with Re = 100 ~ 500 and Lp = 1.0 mm. Hsieh and Jang (2012) numerically studied the optimal design of a louver finned-tube heat exchanger applying the Taguchi method. Eighteen kinds of patterns were made by mixed levels on each factor. The optimal design values for each parameter were all reported.

The foregoing literature review reveals that no related 3-D numerical analysis for the optimization of louvered angle and louvered pitch and their coupled effects on the thermal and hydraulic characteristics of a louver finned and tube heat exchanger has been published. This has motivated the present investigation. In the present research, the optimization of louver angle and its pitch is studied and solved numerically using a commercial CFD code ANSYS FLUENT (2009) along with a simplified conjugate-gradient method. To achieve optimization goals, the area reduction ratio is the objective function to be maximized. The influence of louver pitch (Lp = 2.0 ~ 3.2 mm) and louver angle (θ = 15°< θ < 40°) on the heat transfer performance and friction loss at different Reynolds numbers are discussed in detail. The optimal design values for two operating parameters at different Reynolds number are also presented.

MATHEMATICAL ANALYSIS

Governing equation

Figure 1 describes the physical model and relevant geometric dimensions of the louver finned and tube heat exchanger. The unit is a mini-meter. The louver angle (θ = 15° ~ 40°) and louver pitch (Lp = 2.0 ~ 3.2 mm) as shown in Figure 2 are the main operating parameters in the present study. The fluid is considered 3-D incompressible turbullent flow with constant properties, and the flow is assumed to be steady with no viscous dissipation. Equations for continuity, momentum (Reynolds averaged Navier-Stokes equations), energy, turbulent kinetic energy, k, and the dissipation rate, ε, can be expressed in tensor form as follows:

$$\frac{\partial \overline{u_i}}{\partial x_i} = 0 \tag{1}$$

$$\frac{\partial}{\partial x_j}\rho\left(\overline{u_i u_j}\right) = -\frac{\partial \overline{P}}{\partial x_i} + \frac{\partial}{\partial x_j}\left[\mu_{eff}\left(\frac{\partial \overline{u_i}}{\partial x_j} + \frac{\partial \overline{u_j}}{\partial x_i}\right) - \rho\left(\overline{u_i' u_j'}\right)\right] \tag{2}$$

$$\frac{\partial}{\partial x_j}\rho C\left(\overline{u_j T}\right) = \overline{u_j}\frac{\partial \overline{P}}{\partial x_j} + \overline{u_j'}\frac{\partial \overline{P'}}{\partial x_j} + \frac{\partial}{\partial x_j}\left[k\frac{\partial \overline{T}}{\partial x_j} - \rho C\overline{u_j' T'}\right] \tag{3}$$

$$\frac{\partial}{\partial x_i}\left(\rho \overline{u_i} k\right) = -\frac{\partial}{\partial x_i}\left(\frac{\mu_{eff}}{\sigma_k}\frac{\partial k}{\partial x_i}\right) + \rho(\Pr - \varepsilon) \tag{4}$$

$$\frac{\partial}{\partial x_i}\left(\rho \overline{u_i} \varepsilon\right) = -\frac{\partial}{\partial x_i}\left(\frac{\mu_{eff}}{\sigma_\varepsilon}\frac{\partial \varepsilon}{\partial x_i}\right) + \rho\frac{\varepsilon}{k}\left[\left(c_1 + c_3\frac{\Pr}{\varepsilon}\right)\Pr - c_2\varepsilon\right] \tag{5}$$

where $\Pr = (\mu_t/\rho)[2(\partial u_i/\partial x_i)^2 - 2(\nabla u_i)^2/3]$, $\mu_{eff} = \mu + \mu_t$, $\mu_t = \rho c_\mu\left(k^2/\varepsilon\right)$, $c_\mu = 0.09$, $c_1 = 0.15$, $c_2 = 1.90$, $c_3 = 0.25$, $\sigma_k = 0.75$ *and* $\sigma_\varepsilon = 1.15$

Equation 2 contains Reynolds stresses that are modeled by Chen's extended k-ε turbulence model (Chen and Kim, 1987; Wang and Chen, 1993), where k is the turbulent kinetic energy, and ε is the dissipation rate. In Chen's model, the production time scale as well as the dissipation time scale is used in closing the ε equation. This extra production time scale is claimed to allow the energy transfer mechanism of turbulence to respond to the mean strain rate more effectively. This results in an extra constant in the ε equation. As to the velocity distribution in the near-wall region ($y^+ \leq 11.63$), the following law of the wall (Liakopoulos, 1984) is applied:

$$u^+ = \ln\left[\frac{(y^+ + 11)^{4.02}}{({y^+}^2 - 7.37y^+ + 83.3)^{0.79}}\right] + 5.63\tan^{-1}(0.12y^+ - 0.441) - 3.81 \tag{6}$$

Where

$$y^+ \equiv \frac{\rho u_\tau y}{\mu} \quad \text{and} \quad u_\tau = \sqrt{\frac{\tau_w}{\rho}} \tag{7}$$

Parameter definition of performance factor

The local pressure drop can be expressed in terms of the dimensionless pressure coefficient C_p defined as:

$$Cp = \frac{p - p_{in}}{\frac{1}{2}\rho {U_{in}}^2} \tag{8}$$

where P_{in} is the pressure at inlet and U_{in} is the inlet velocity. The local heat transfer coefficient *h* is defined as:

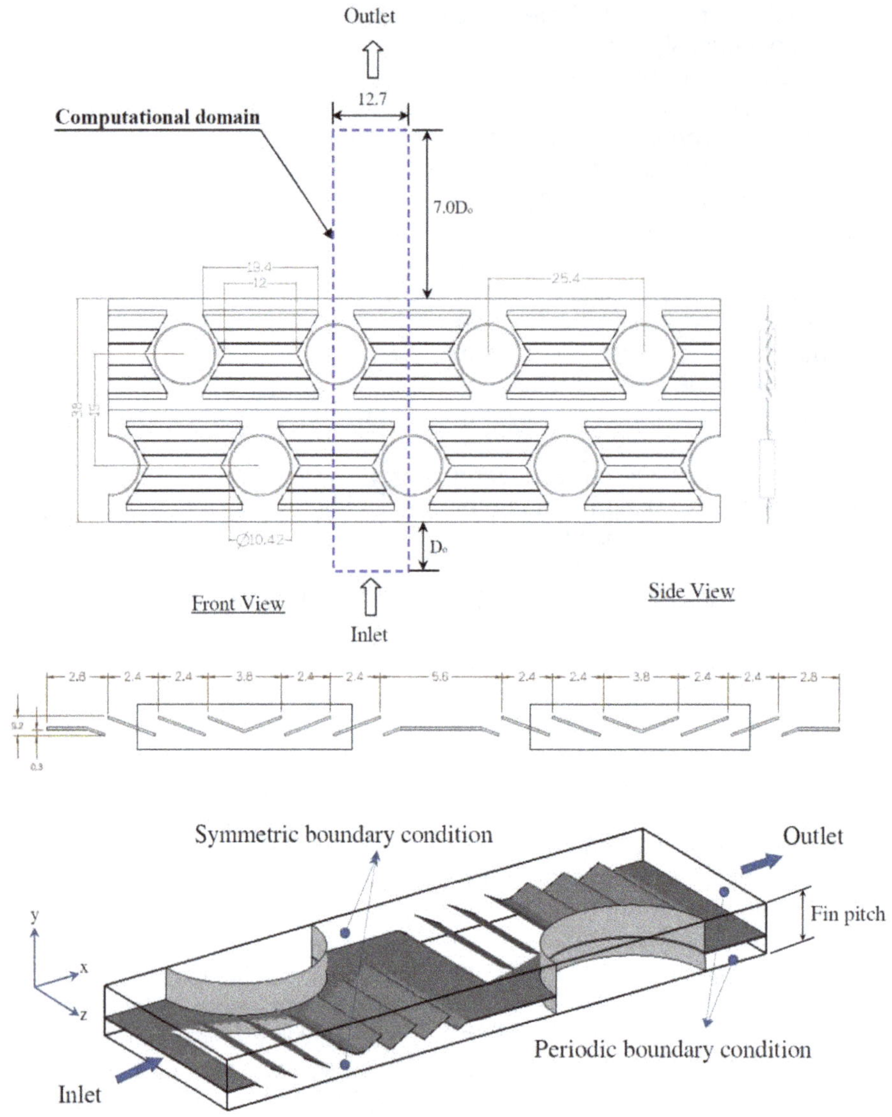

Figure 1. The physical model and computational domain (fin thickness, t =0.115 mm).

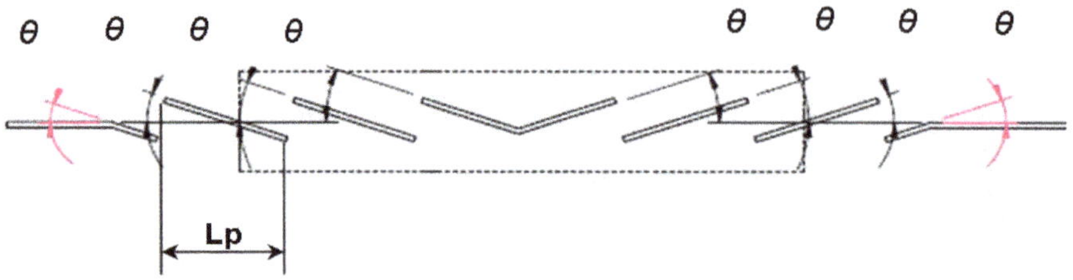

Louver angle θ	15°	20°	25°	30°	35°	40°	
Louver pitch Lp	2.0mm	2.2mm	2.4mm	2.6mm	2.8mm	3.0mm	3.2mm

Figure 2. The louver pattern for different louver angles and louver pitches.

$$h = \frac{q''}{T_w - T_b} \tag{9}$$

where q'' is the local heat flux. T_b is the local bulk mean temperature. T_w is the wall temperature. The local heat transfer coefficient can be expressed in the dimensionless form by the Nusselt number *Nu*, defined as:

$$Nu = \frac{h \cdot D_o}{k} = \frac{\partial \left[\frac{\Theta}{\Theta_b} \right]_{wall} \cdot D_o}{\partial \mathrm{n}} \tag{10}$$

where $\Theta_b = (T_b - T_{in})/(T_w - T_{in})$ is the local dimensionless bulk mean temperature and *n* is the dimensionless unit vector normal to the wall and D_o is the outside diameter of tube. The average Nusselt number $\overline{Nu}$ can be obtained by

$$\overline{\mathrm{Nu}} = \frac{\int \mathrm{Nu}\, \mathrm{dA_s}}{\int \mathrm{dA_s}} \tag{11}$$

where dA_s is the infinitesimal area of the wall surface. The friction factor f and Colburn factor j are defined as:

$$f = \frac{p - p_{in}}{\frac{1}{2}\rho U_{in}^{2}} \times \frac{D_o}{4L} \tag{12}$$

$$j = \frac{\overline{\mathrm{Nu}}}{\mathrm{Re}_D \mathrm{Pr}^{1/3}} \tag{13}$$

where P_{in} is the pressure at the inlet, L is the flow length, Re_D is the Reynolds number defined as $\mathrm{Re}_D = \mathrm{U_{max}} \mathrm{D_o}/\nu$, U_{max} is the air velocity at minimum free flow area, Pr is the Prandtal number defined as $\mathrm{Pr} = \nu/\alpha$, α is the thermal diffusivity, and ν is the kinematic viscosity.

Boundary condition

Since the governing equations are elliptic, it is necessary to impose boundary conditions at all of the boundaries in the computational domain. The upstream boundary is established at a distance of one tube diameter in front of the leading edge of the fin. At this boundary, the flow velocity U_{in} is assumed to be uniform, and the temperature T_{in} is taken to be 300K. At the downstream end of the computational domain, located seven times the tube diameter from the last downstream row tube, the streamwise gradients (Neumann boundary conditions) for all the variables are set to zero. At the solid surfaces, no-slip conditions and constant wall temperature T_W (353K) are specified. On the symmetry planes (two X-Y planes), normal gradients are set to zero. On the upper and lower X-Z planes, periodic boundary conditions are imposed. Additionally, at the solid-fluid interface,

$$\mathrm{T_s} = T_f \; ; \; -k_s \cdot \partial T_s / \partial n = -k_f \cdot \partial T_f / \partial n \tag{14}$$

Performance evaluation criteria (PEC)

Many performance evaluation criteria (PEC) have been developed for evaluating the performance of heat exchangers. The VG-1 (variable geometry) performance criteria, as described by Webb (1994), represents the possibility of surface area reduction by using enhanced surfaces having fixed heat duty, temperature difference and pumping power.

$$\frac{hA}{h_o A_o} = \frac{j}{j_o} \frac{A}{A_o} \frac{G}{G_o} \tag{15}$$

where the subscripts of 'o' refer to the reference plate fin, and G is the mass velocity. The pumping power is calculated as:

$$\omega = \left(f \frac{A}{A_m} \frac{G^2}{2\rho} \right) \left(\frac{G A_m}{\rho} \right) \tag{16}$$

where A_m is the flow area at minimum cross section. The pumping power ratio relative to the reference plane fin can be obtained by:

$$\frac{\omega}{\omega_o} = \frac{f}{f_o} \frac{A}{A_o} \left(\frac{G}{G_o} \right)^3 \tag{17}$$

and by the elimination of the term

$$\frac{hA / h_o A_o}{(\omega / \omega_o)^{1/3} (A / A_o)^{2/3}} = \frac{j / j_o}{(f / f_o)^{1/3}} \tag{18}$$

Under the pumping power constraint of case VG-1, that is ($\omega/\omega_o = 1$), we may obtain the area reduction ratio relative to the reference plane fin as:

$$\frac{A}{A_o} = \left(\frac{f}{f_o} \right)^{1/2} \left(\frac{j_o}{j} \right)^{3/2} \tag{19}$$

NUMERICAL METHOD AND OPTIMIZATION

In this study, the governing equations are solved numerically using a control volume based finite difference formulation, ANSYS FLUENT

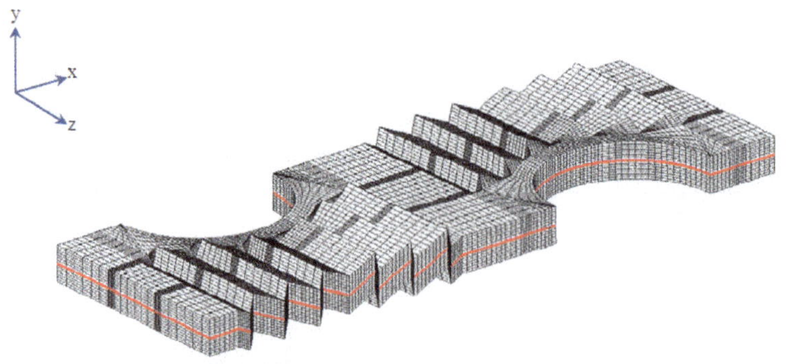

Figure 3. Computational grid system.

(2009). The numerical methodology is briefly described here. Finite difference approximations are employed to discretize the transport equations on non-staggered grid mesh systems. A third-order upwind TVD (total variation diminishing) scheme is used to model the convective terms of governing equations. Second-order central difference schemes are used for the viscous and source terms. A pressure based predictor/multi-corrector solution procedure is employed to enhance velocity–pressure coupling and continuity-satisfied flow filed. A grid system of 288 × 19 × 31 grid points was adopted typically in the computation domain as shown in Figure 3. However, a careful check for the grid-independence of the numerical solutions has been made to ensure the accuracy and validity of the numerical results. For this purpose, three grid systems, 335 × 23 × 37, 288 × 19 × 31 and 241 × 14 × 23, were tested. It was found that for U_{in} = 3.0 m/s, the relative errors in the local pressure and temperature between the solutions of 335 × 23 × 37, 288 × 19 × 31 were less than 3%. The convergence criterion is satisfied when the residuals of all variables are less than 1.0×10^{-7}. Computations were performed on a Pentium 4 3.0G personal computer and typical CPU times were 5000–6000 s.

In the present study, the simplified conjugate-gradient method (Jang and Tsai, 2011) is combined with a finite differential method code (ANSYS FLUENT, 2009) as an optimizer to search the optimum louver angle (θ) and louver pitch (Lp). The objective functions $J\ (x_1,x_2)$ are defined as the maximum area reduction ratio relative to the palin fin surface $(1\text{-}A/A_o)$.

Above all, the SCGM method evaluates the gradient of the objective function, and then it sets up a new conjugate direction for the updated design variables with the help of a direct numerical sensitivity analysis. The initial guess for the value of each search variable is made, and in the successive steps, the conjugate-gradient coefficients and the search directions are evaluated to estimate the new search variables. The solutions obtained from the finite difference method are then used to calculate the value of the objective function, which is further transmitted back to the optimizer for the purpose of calculating the consecutive searching directions. The procedure for applying this method is described in the following:

(1) Generate an initial guess for two design variables (x_1,x_2) –louver angle (θ) and and louver pitch (Lp).
(2) Adopt the finite difference method to predict the velocity field (U) and temperature fields (T) associated with the latest θ and Lp, and then calculate the objective function $J\ (x_1,x_2)$.
(3) When the value of $J(x_1,x_2)$ reaches a maximum, the optimization process is terminated. Otherwise, proceed to step 4.
(4) Determine the gradient functions, $(\partial J/\partial x_1)^{(k)}$ and $(\partial J/\partial x_2)^{(k)}$, by applying a small perturbation $(\Delta x_1,\Delta x_2)$ to each value of x_1 and x_2, and calculate the corresponding change in objective function (ΔJ). Then, the gradient function with respect to each value of the design variables (x_1,x_2) can be calculated by the direct numerical differentiation as

$$\frac{\partial J_1^{(k)}}{\partial x_1} = \frac{J_1^{(k)} - J^{(k)}}{\Delta x_1} \text{ and } \frac{\partial J_2^{(k)}}{\partial x_2} = \frac{J_2^{(k)} - J^{(k)}}{\Delta x_2} \tag{20}$$

(5) Calculate the conjugate-gradient coefficients $\gamma^{(k)}$, and the search directions, $\xi_1^{(k+1)}$ and $\xi_2^{(k+1)}$, for each search variable. For the first step with k = 1, $\gamma^{(1)} = 0$.

$$\gamma^{(k)} = \frac{\sum_{n=1}^{2}\left(\frac{\partial J_n^{(k)}}{\partial x_n}\right)^2}{\sum_{n=1}^{2}\left(\frac{\partial J_n^{(k-1)}}{\partial x_n}\right)^2} \tag{21}$$

$$\xi_1^{(k)} = \frac{\partial J_1^{(k)}}{\partial x_1} + \gamma^{(k)}\xi_1^{(k-1)} \text{ and } \xi_2^{(k)} = \frac{\partial J_2^{(k)}}{\partial x_2} + \gamma^{(k)}\xi_2^{(k-1)} \tag{22}$$

(6) Assign values to the coefficients of descent direction (*β*) for all values of the design variables (x_1, x_2). Specifically, those values are chosen by a trial-and-error process. In general, the coefficients of descent direction (*β*) are within a range of 0.2 ~ 0.01.
(7) Update the design variables with

$$x_1^{(k+1)} = x_1^{(k)} + \beta\xi_1^{(k)} \text{ and } x_2^{(k+1)} = x_2^{(k)} + \beta\xi_2^{(k)} \tag{23}$$

A flowchart of the SCGM optimization process is plotted in Figure 4.

RESULTS AND DISCUSSION

The present study mainly evaluated the influences of louver angle (θ) and louver pitch (Lp) on the local and overall flow and heat transfer characteristics of louver finned and tube heat exchangers. Furthermore, optimization analyses to θ and Lp were utilized in order to search the optimum combination of (θ, Lp) and maximum objective function $(1 - A/A_0)$. The relevant numerical results were achieved in the range of $589 < Re_D < 3533$ ($0.5\text{m/s} < U_{in} < 3.0\text{m/s}$), 15° < θ < 40°, and 2.0 mm < Lp < 3.2 mm. In order to validate the reliability of the numerical

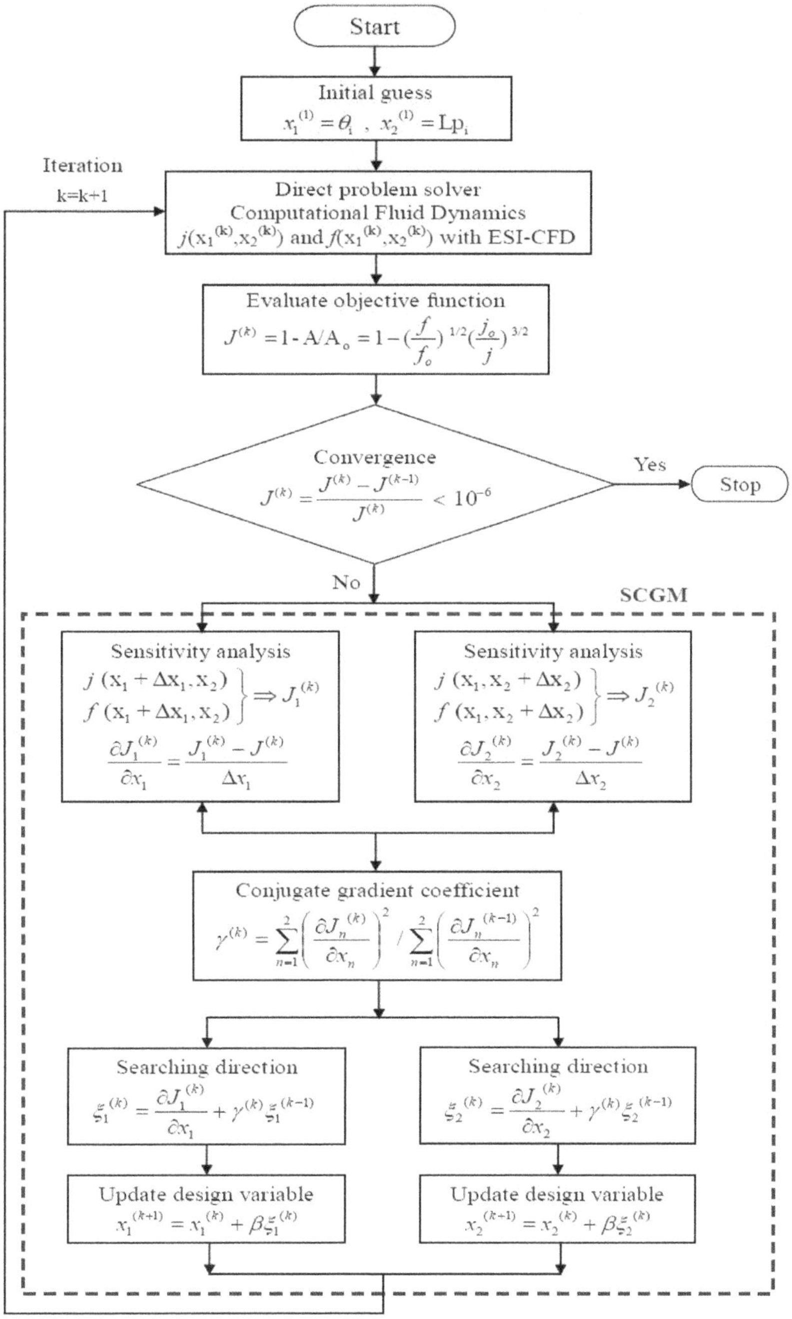

Figure 4. Flowchart for the optimization method.

simulation procedure, numerical simulations were carried out at the same operating conditions as the experimental louver finned-tube heat exchangers with two rows (Wang et al., 1998). Figure 5 shows the comparisons of j and f factors between the simulated results and the experimental results. The present results showed good agreements within a maximum of 10% discrepancy.

The flow and thermal field of a louver finned and tube heat exchanger is very complicated. Figure 6a and b show the streamline velocity and temperature distribution, respectively, for louver finned-tube with U_{in} =3.0 m/s, θ=15° and Lp=2.4 mm. The flow entering a louvered fin array quickly becomes louver directed. Then the flow passing the round cylinder (the first row of tubes) divides into opposite paths of equal velocity and path length over the cylinder surface. Apparently, the streamlines near the tube side wall are very dense and flow velocity accelerates quickly. The reason is that, the geometric shape of the channel near the tube side wall is convergent and divergent. The vortices appear at the downstream behind

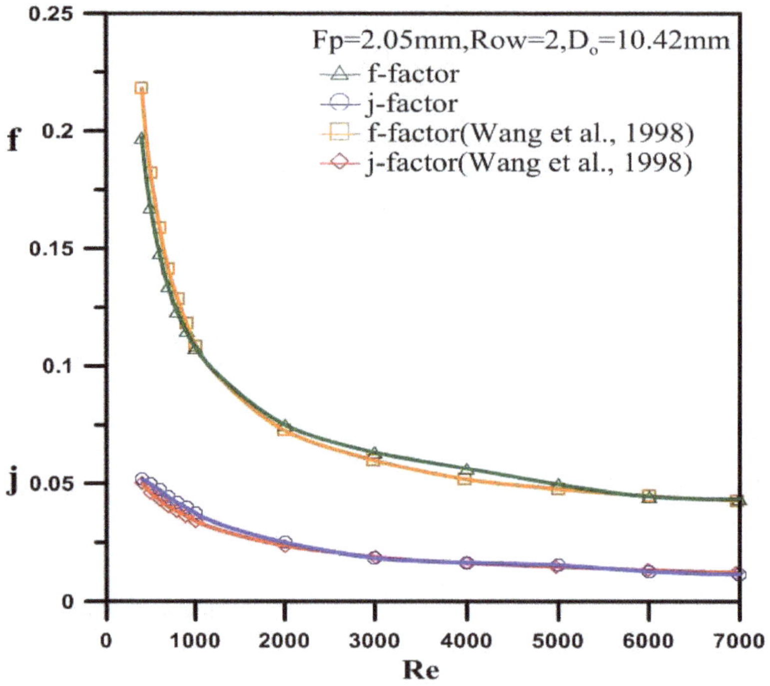

Figure 5. Comparison of the j and f factors for the present study and previous literature.

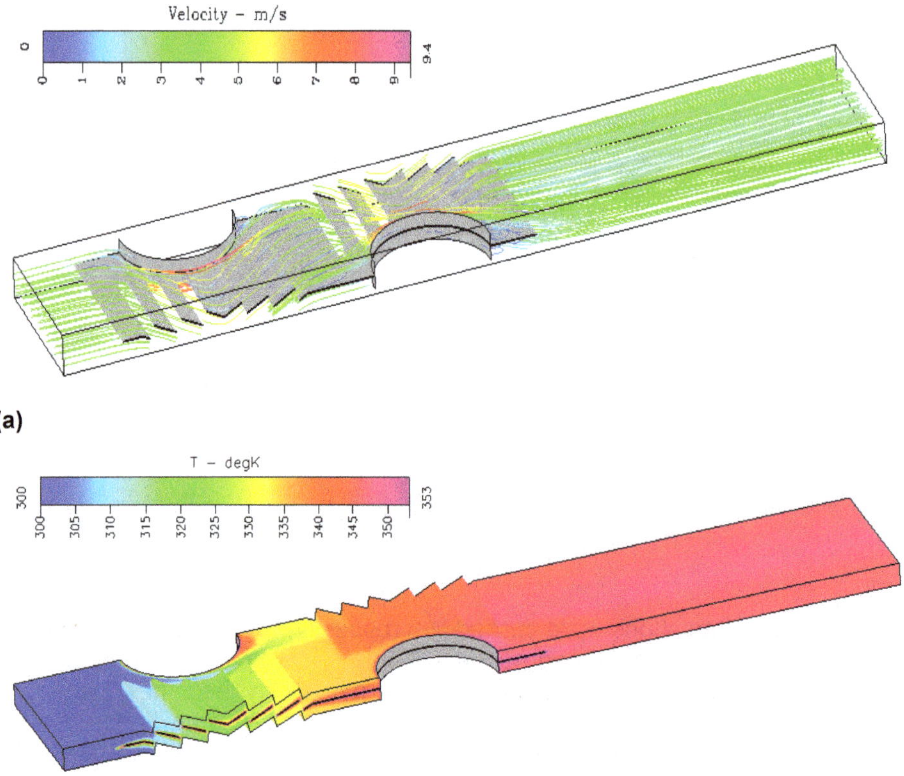

(a)

(b)

Figure 6. (a) Streamline velocity and (b) temperature distribution with Re=3533(U_{in}=3.0m/s) for θ = 15° and Lp=2.4 mm.

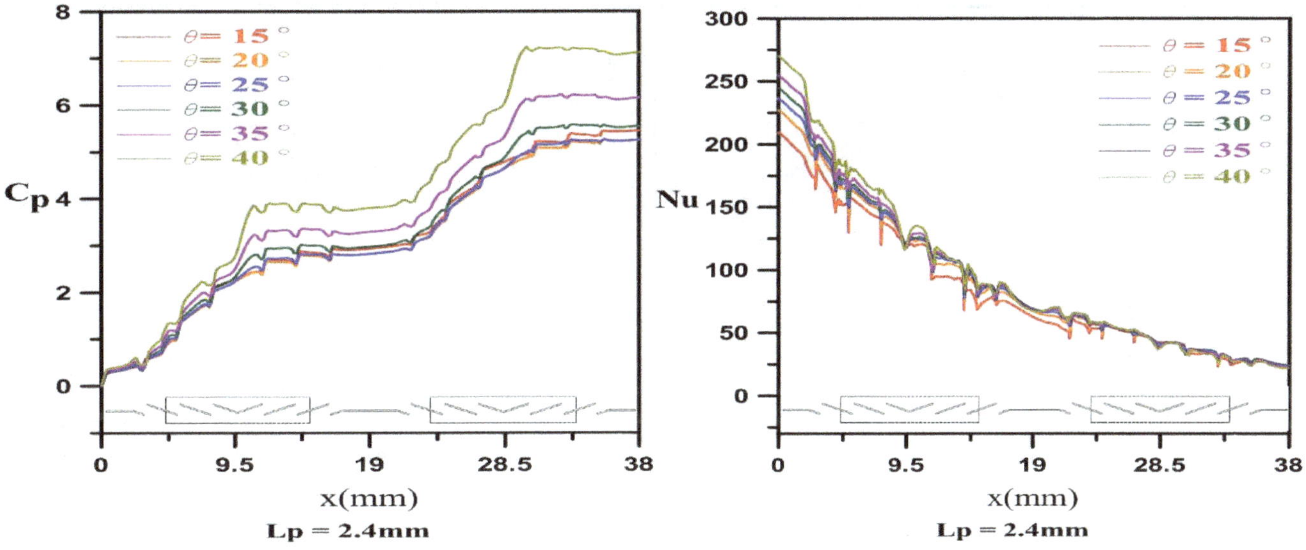

Figure 7. The variation of (a) C_p and (b) Nu along downstream direction for different louver angle with Re=3533(U_{in}=3.0m/s).

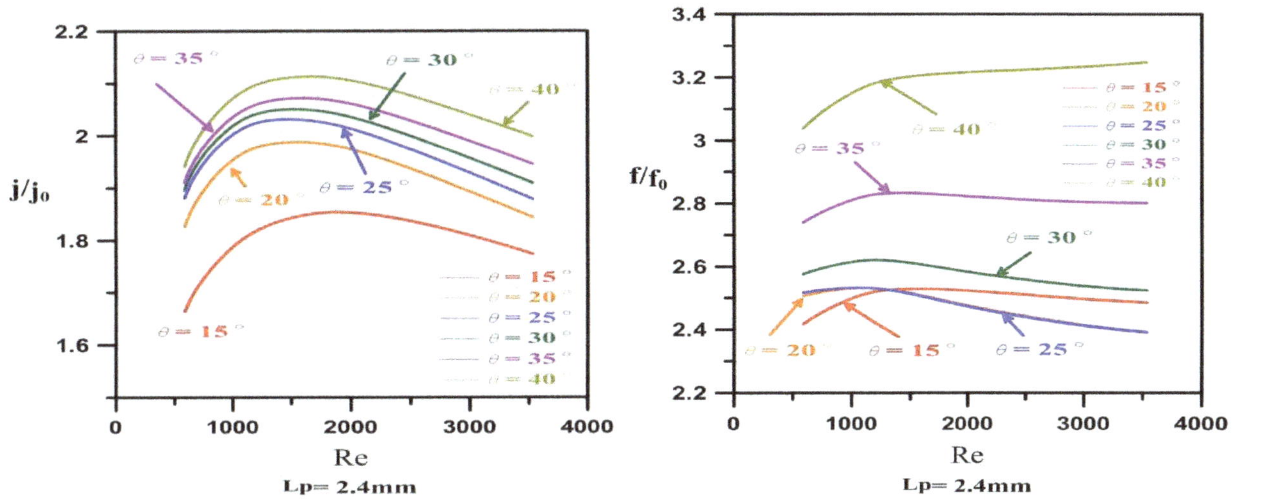

Figure 8. The (a) j/j_o and (b) f/f_o versus Reynolds number for different louver angle (θ) with louver pitch (Lp=2.4 mm).

the tube cylinder. The temperature gradients near the wall are quite large, which indicates a corresponding enhanced heat transfer.

Figure 7a and b present the variations of the local pressure drop coefficient (C_p) and Nusselt number (Nu), respectively, along the downstream direction with inlet frontal velocity (U_{in} = 3.0 m/s) and louver pitch (Lp =2.4 mm) for six different louver angles (θ = 15, 20, 25, 30, 35 and 40°). One can see that the there is a local maximum of Nu at the upstream inlet.

To evaluate how much performance is improved, j/j_o and f/f_o are used to interpret the data, where j/j_o and f/f_0 are the Colburn factor ratio and friction factor ratio between louver and without louver, respectively. Figures 8a and b illustrate the variations of j/j_o and f/f_o, versus Re_D, respectively, for six different louver angles (15, 20, 25, 30, 35 and 40°) with louver pitch (Lp=2.4 mm). The maximum heat transfer improvement interpreted by j/j_o are 1.853, 1.985, 2.026, 2.047, 2.071 and 2.113, and the corresponding friction factor ratio f/f_o are 2.528, 2.494, 2.492, 2.597, 2.829 and 3.211, respectively.

Figure 9a and b illustrate the variations of j/j_o and f/f_o versus Re, respectively, for seven different louver pitch (2.0, 2.2, 2.4, 2.6, 2.8, 3.0, and 3.2 mm) with louver angle

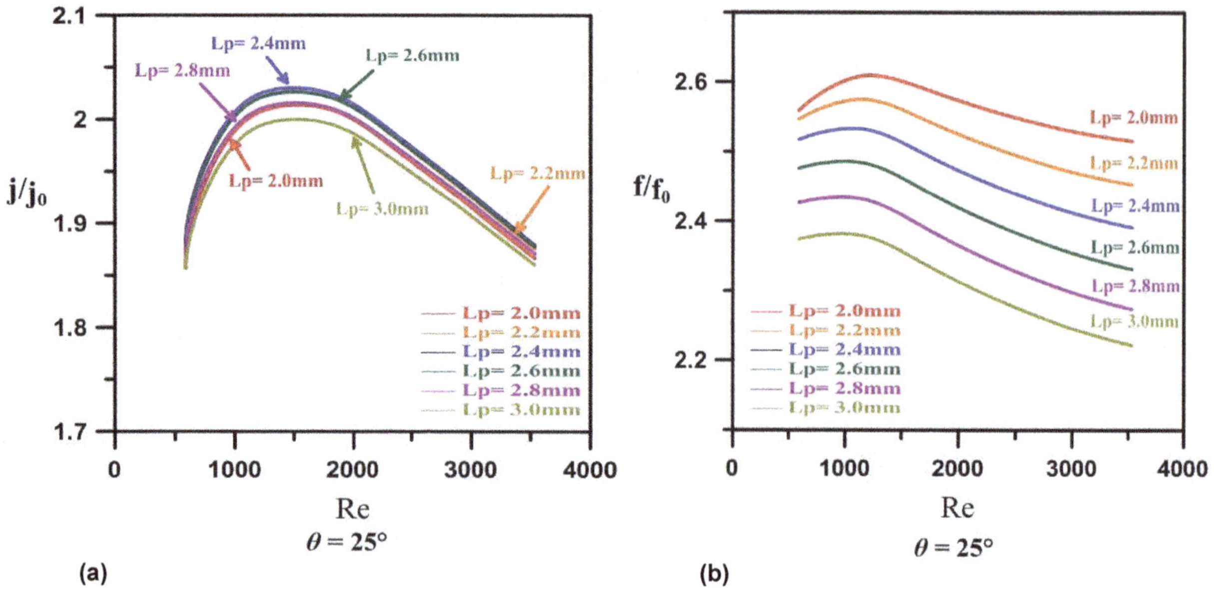

Figure 9. The (a) j/j_o and (b) f/f_o Reynolds number for different louver pitch (Lp) with louver angle (θ=25°).

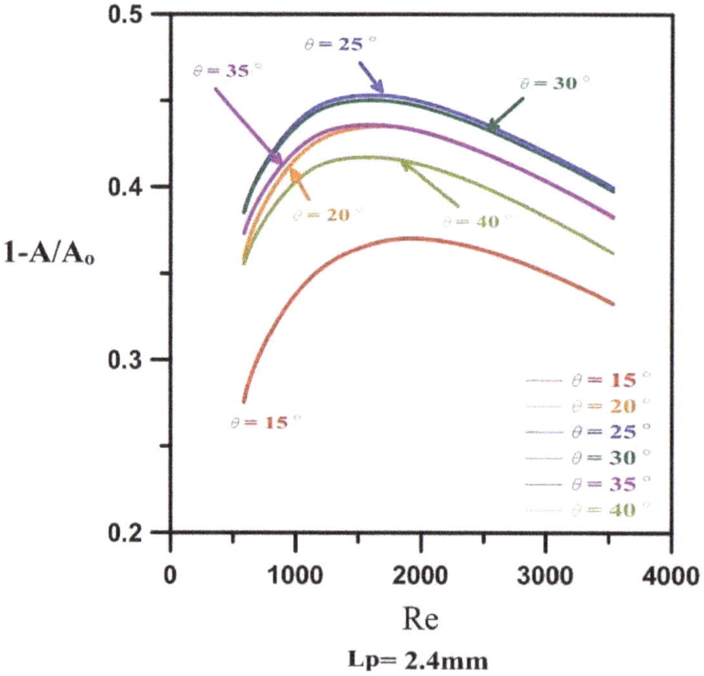

Figure 10. The area reduction versus Reynolds number for different louver angle (θ) with louver pitch (Lp=2.4 mm).

(θ=25°). The maximum heat transfer improvement interpreted by j/j_o are 2.011, 2.023, 2.026, 2.022, 2.012, 1.997 and 1.976, and the corresponding friction factor ratio f/f_o are 2.587, 2.541, 2.492, 2.439, 2.385, 2.333 and 2.286, respectively. The present results indicated that the variable louver angle and pitch patterns applied in heat exchangers could effectively enhance the heat transfer performance.

The possible area reduction $1-A/A_o$ (where A and A_o denote the surface areas for variable louver θ ranging from 15 to 40° and conventional plain fins, respectively) with Lp = 2.4 mm is presented in Figure 10. One can see

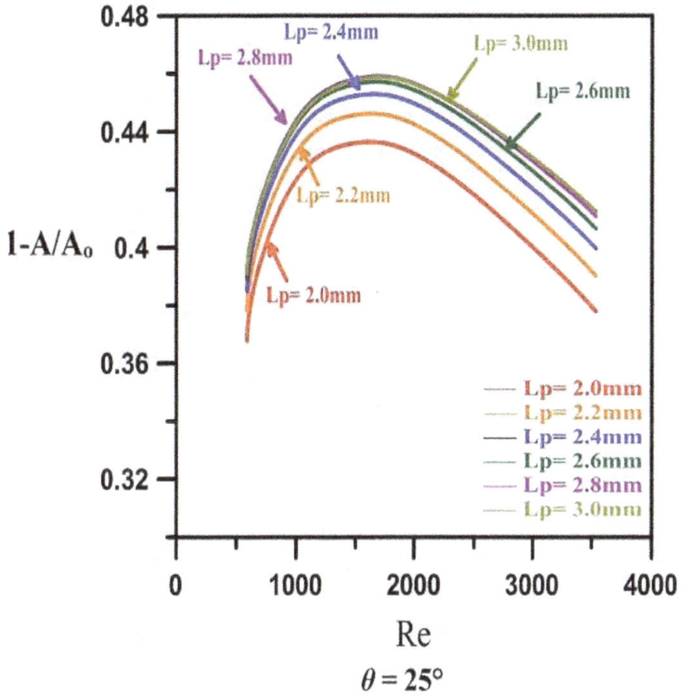

Figure 11. The area reduction versus Reynolds number for different louver pitch (Lp) with louver angle (θ=25°).

that the greatest area reduction ratio is as much as 37.0, 43.5, 45.3, 45.0, 43.5 and 41.6% with specific values of Re_D = 589, 1178, 1766, 2355, 2944 and 3533, respectively, it gives the greatest area reduction at Re = 1766 and θ =25°. Figure 11 presents the area reduction ratio for Lp ranging from 2.0 to 3.2 mm with θ = 25°, the greatest area reduction ratio is as much as 43.6, 44.6, 45.3, 45.7, 45.9, 45.9 and 45.6% with specific values of Re = 589, 1178, 1766, 2355, 2944 and 3533, respectively, it gives the greatest area reduction at Re = 1766 and Lp=3.0 mm.

Figure 12 displays the iteration process used to search the optimum louver angle (θ) and louver pitch (Lp) combination for the maximization of objective function (that is, area reduction ratio, $1\text{-}A/A_o$) at Re_D = 1766 (U_{in} = 1.5 m/s). The constant area reduction ratio contours are plotted as a function of θ and Lp, where the dark red area represents the maximum area reduction ratio. It is seen that, with the initial values (θi = 15° ,Lpi = 3.0 mm) and (θi = 40°, Lpi = 3.0 mm), by using the simple conjugated gradient method (SCGM), the optimal θ and Lp combination is obtained (θ = 24.09°, Lp = 2.91 mm) for around 19 and 18 iterations, respectively. The area reduction ratio is 45.9%. Thus, the current optimization method provides a tremendous savings in regard to computational time for the present physcial model. The searched optimum combination of θ and Lp with specific values of Re_D = 589, 1178, 1766, 2355, 2944 and 3533 (U_{in}=0.5 to 3.0 m/s) are tabulated in Table 1. It is seen that, an area reduction ratio of 39 to 46% is achieved across the range of Re_D.

Conclusion

Three dimenional turbulent fluid flow and heat transfer in two row fin-and-tube heat exchanger with and without louver fins are studied numerically. The optimization of the louvered angle (θ) and louvered pitch (Lp) is executed by using a simplified conjugate-gradient method. A searched procedure for the optimum louver angle (θ) and louver pitch (Lp), ranging from 15°< θ < 40° and 2.0 mm < Lp < 3.2 mm, respectively, is executed. The searched optimum objective function associated with an optimal combination of θ and Lp for different Re_D are obtained for less than 30 iterations. This demonstrates that the current optimization method provides a tremendous savings in regard to computational time for the present physcial model. In addition, the results showed that the maximum area reduction ratios may reach 39~46% combined with the optimal design of (θ, Lp) at U_{in}= 0.5~3.0 m/s.

ACKNOWLEDGEMENT

Financial support for this work was provided by the National Science Council of Taiwan, under contract NSC NSC 101-2221-E-006 -109 -MY2.

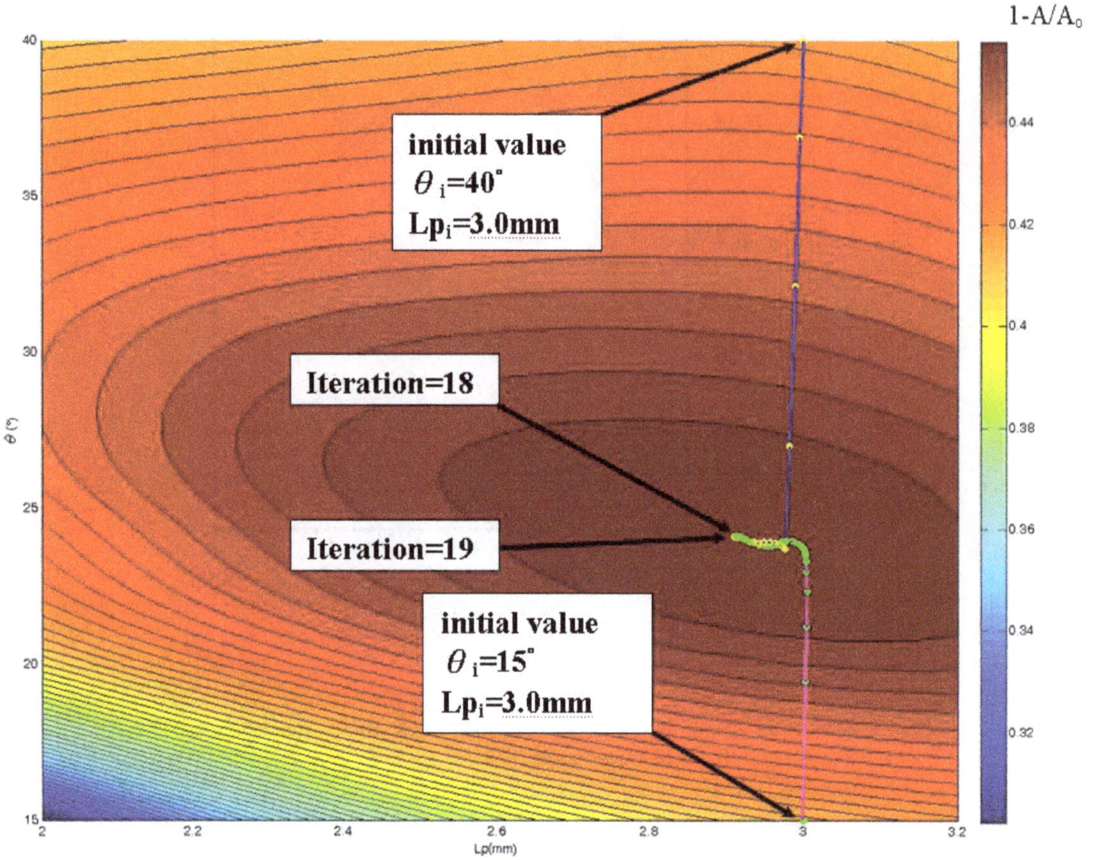

Figure 12. Iteration process to search the optimum combination of θ and Lp (U_{in}=1.5m/s).

Table 1. The searched optimum combination of θ and Lp for different Re_D.

Re	U_{in}(m/s)	Initial value		θ	Lp(mm)	j/j_o	f/f_o	$1-A/A_o$ (%)	Iteration numbers
		$θ_i$	Lp_i						
589	0.5	15.0	3.0	25.11	2.93	1.862	2.393	39.1	29
1178	1.0	15.0	3.0	24.23	2.88	2.003	2.411	45.2	25
1766	1.5	15.0	3.0	24.09	2.91	2.004	2.353	45.9	19
2355	2.0	15.0	3.0	24.10	2.97	1.963	2.291	45.0	18
2944	2.5	15.0	3.0	24.47	2.98	1.912	2.251	43.3	13
3533	3.0	15.0	3.0	24.80	2.99	1.859	2.220	41.2	14

Nomenclature: **A,** total surface area (m^2); **C,** fluid heat capacity (J /kg°C); **Cp,** pressure drop coefficient; D_o, outside diameter of tube (m); ***f,*** friction factor; **h,** heat transfer coefficient ($W/m^{2}°C$); **j,** Colburn factor; **k,** thermal conductivity (W/m°C); **Lp,** louver pitch (m); **Nu,** local Nusselt number, hD_o/k; $\overline{Nu}$, average Nusselt number; **P,** pressure (Pa); **Pr,** Prandtl number, ν/α; **q,** heat flux (W/m^2); **Re_D,** Reynolds numbers, $U_{max}D_o/ν$; **T,** temperature (°C); **T_w,** wall temperature (°C); **T_{in},** inlet temperature (°C); **T_b,** bulk mean temperature (°C); **U_{in},** frontal velocity (m/s); **U_{max},** air velocity at minimum flow area (m/s); **x,y,z,** coordinates; **α,** thermal diffusivity (m^2/s); **θ,** louver angle (degree); **ν,** kinematic viscosity (m^2/s); **ρ,** density of fluid (kg/m^3); **μ,** dynamic viscosity (kg/ms).

REFERENCES

Achaichia A, Cowell TA (1988). Heat transfer and pressure drop characteristics of flat tube and louvered plate fin surfaces, Experimental Thermal Fluid Sci. 1:147-157.

ANSYS FLUENT (2009). A Release 12.0, Documentation for ANSYS Workbench. ANSYS Ltd.

Chen YS, Kim SW (1987). Computation of Turbulent Flows Using an Extended K - ε turbulence Closure Model NASA CR-179204.

Davenport CJ (1983). Correlations for heat transfer and flow friction characteristics of louvred fin, AIChE Symposium Series. 19-27.
Hiramatsu M, Ishimaru T, Matsuzaki K (1990). Research on fins for air conditioning heat exchangers. JSME Int. J. Ser. II, 33:749-756.
Hsieh CT, Jang JY (2006). 3-D thermal-hydraulic analysis for louver fin heat exchangers with variable louver angle. Appl. Thermal Eng. 26:1629–1639.
Hsieh CT, Jang JY (2012). Parametric study and optimization of louver finned-tube heat exchangers by Taguchi method, Appl.Thermal Eng. 42:101-110.
Ikuta S, Sasaki Y, Tanaka K, Takagi M, Himeno R (1990). Numerical analysis of heat transfer around louver assemblies. Int. Congress & Exposition Tech., Detroit, Michigan, USA.
Jang JY, Shieh KP, Ay H (2001). 3-D thermal-hydraulic analysis in convex louver finned -tube heat exchangers, ASHRAE Annual Meeting, Cincinnati, OH, USA, June 22-27, pp. 501-509.
Jang JY, Tsai YC (2011). Optimum Louver Angle Design for a Louvered Fin Heat Exchanger, Int. J. Phys. Sci. 6:6422-6438.
Kays WM, London AL (1950). Heat transfer and flow friction characteristics of some compact heat exchanger surfaces-part I: test system and procedure, Trans. ASME 72:1075-1085.
Liakopoulos A (1984). Explicit representation of the complete velocity profile in a turbulent boundary layer, AIAA J. 22:844-846.
Sahnoun A, Webb RL (1992). Prediction of heat transfer and friction for louver fin geometry, ASME Journal of Heat Transfer 114:893-900.
Suga K, Aoki H, Shinagawa T (1990). Numerical analysis on two dimensional flow and heat transfer of louvered fins using overlaid grids. JSME Int. J. Ser. II, 33:122-127.
Suga K, Aoki H (1991). Numerical study on heat transfer and pressure drop in multilouvered fins. In Proceed. of ASME/JSME Thermal Eng. Joint Conf. 4:361-368.
Sunden B, Svantesson J (1992). Correlation of j- and f- Factors for multilouvered heat transfer surfaces. In proceedings of the 3rd UK National Heat Transfer Conf. 805-811.
Wang CC, Chi KY, Chang YJ (1998). An experimental study of heat transfer and friction characteristics of typical louver finned-tube heat exchangers, Int. J. Heat Mass Transfer. 41:817-822.
Wang TS, Chen YS (1993). Unified Navier-Stokes flowfield and performance analysis of liquid rocket engines, AIAA J. 9:678-685.
Webb RL (1994). Principles of Enhanced Heat Transfer, New York, John Wiley & Sons.

11

Magnetic field effects on entropy generation in heat and mass transfer in porous cavity

Nawaf H. Saeid

Department of Mechanical, Materials and Manufacturing Engineering, The University of Nottingham Malaysia Campus, 43500 Semenyih, Selangor, Malaysia.

The entropy generation for natural convection heat and mass transfer in a two dimensional porous cavity subjected to a magnetic field is selected for numerical investigation. The Darcy model is used in the mathematical formulation of the fluid flow in porous media. The mathematical model is derived in dimensionless form and the governing equations are solved using the finite volume method. The governing parameters arise in the mathematical model are the Rayleigh number, Lewis number, buoyancy ratio and Hartmann number. The results are presented as average Nusselt number ($\overline{Nu}$), Sherwood numbers ($\overline{Sh}$) and dimensionless form of local entropy generation rate (N_s) for different values of the governing parameters. The numerical results show that increasing the magnetic field parameter (Hartmann number) leads to deterioration of the flow circulation strength in the cavity and this leads to a decrease in the rates of the heat and mass transfer as well as the rate of entropy generation. The results show a stagnate fluid everywhere in the cavity when the buoyancy forces generated due to temperature and concentration differences are in the same order and opposite directions. In this case, the values of $\overline{Nu}$, $\overline{Sh}$ and N_s are the minimum. Increasing or decreasing the value of the buoyancy ratio parameter leads to enhance the fluid circulation and hence increase the values of $\overline{Nu}$, $\overline{Sh}$ and N_s. The average Sherwood number can be increased with increasing Lewis number. It is observed that the strength of the fluid circulation in the cavity is reduced by increasing the Lewis number. This leads to the decrease in the average Nusselt number and the entropy generation by increasing Lewis number.

Key words: Natural convection, mass transfer, entropy generation, magnetic field, porous media.

INTRODUCTION

The natural convection in porous media has been studied and analysed widely in recent years. This interest was estimated due to many applications in, for example, packed sphere beds, high performance insulation for buildings, electronic packages, chemical catalytic reactors, to name a few. Representative studies in this area may be found in the books by Kaviany (1999), Nield and Bejan (2006) and Ingham and Pop (2001).

Double-diffusive convection in porous media concerns the processes of combined (simultaneous) heat and mass transfer which are driven by buoyancy forces. The buoyancy force is not only affected by the difference of temperature, but also it is affected by the difference of concentration in the fluid. A detailed review of double-diffusive natural convection in porous media can be found in research done by Mojtabi and Mojtabi (2005). Natural convection in a cavity saturated with porous media in the presence of magnetic field is relatively a new topic and needs more investigation. The effects of the magnetic field applied on electrically conducting fluids have been reported for various systems involving convection heat transfer. A detailed review of magnetic convection incrystal growth, including the effect of magnetic control can be found in the book by Ozoe (2005). There are many industrial applications of the electrically conducting fluids in the presence of a magnetic field such as crystal growth, electronic packages, metallurgical applications involving continuous casting and solidification of metal alloys and others. In such cases, the fluid experiences a Lorentz force, which tends to aid/oppose the fluid flow and hence increase/reduce the flow velocities. Oreper and Szekely (1983) showed that the strength of the magnetic field is one of the key factors in controlling the quality of the crystal. A numerical study is conducted by Rudraiah et al. (1995) to investigate the effect of magnetic field on the flow driven by the combined mechanism of buoyancy and thermocapillarity in a rectangular open cavity filled with a low Prandtl number fluid. The detailed flow structure and the associated heat transfer characteristics inside the cavity are presented. An investigation is conducted by Bian et al. (1996) to study the effect of an electromagnetic field on natural convection in inclined rectangular porous cavity saturated with an electrically conducting fluid. Bian et al. (1996) presented a linear stability analysis to determine the effect of the magnetic field on the onset of convection in a horizontal layer heated from below.

The irreversibility phenomena which are expressed by entropy generation are of important interest during the design of any thermodynamic system. Many studies concerning entropy generation in natural convection in porous media have been carried out, for example, Baytas (2000), Mahmud and Fraser (2004), Hooman et al. (2007), Famouri and Hooman (2008), Kaluri and Basak (2011), Mchirgui et al. (2012).

Baytas (2000) considered the entropy generation in natural convection in a tilted saturated porous cavity. The analysis of the study (Baytas, 2000) shows that, the calculations of local entropy generation maps are feasible and can be used for the selection of a suitable angle of inclination. The problem of entropy generation in a fluid saturated porous cavity for laminar magneto hydrodynamic natural convection heat transfer is analyzed by Mahmud and Fraser (2004). They studied the effect of Rayleigh and Hartmann numbers on average Nusselt number, entropy generation number, and Bejan number. Hooman et al. (2007) investigated analytically the entropy generation optimization of forced convection in porous-saturated ducts of rectangular cross-section. In this study, the authors indicated that it is possible to compare, evaluate, and optimize alternative rectangular duct design options in terms of heat transfer, pressure drop, and entropy generation. The entropy generation in natural convection by heated partition in a cavity, with adiabatic horizontal and isothermally cooled vertical walls is considered for numerical investigation by Famouri and Hooman (2008). Their results indicated that the fluid friction term has nearly no contribution to entropy production and the heat transfer irreversibility increases monotonically with the Nusselt number and the dimensionless temperature difference. A numerical study is carried out by Hooman et al. (2008) to investigate the entropy generation due to forced convection in a parallel plate channel filled by a saturated porous medium. The authors concluded that the dimensionless degree of irreversibility increases with increase in the porous media shape factor and the Brinkman number, and a decrease in the dimensionless heat flux or temperature difference. Kaluri and Basak (2011) considered the investigations on the entropy generation during natural convection in porous square cavities with distributed heat sources. Parametric study is carried out by Kaluri and Basak (2011) for four different configurations of discretely heated cavities based on the location of the heat sources on the walls of the cavities. Mchirgui et al. (2012) studied numerically the entropy generation in double-diffusive convection through a square porous cavity using Darcy–Brinkman formulation. The cavity is saturated with a binary perfect gas mixture and subjected to horizontal thermal and concentration gradients. Their results indicated that the entropy generation considerably depends on the Darcy number and the porosity induces the increase of entropy generation, especially at higher values of Rayleigh number.

It is noted that, the entropy generation during the double diffusive convection in enclosed cavities submitted to a magnetic field has not received much attention. The aim of this paper is to study numerically the problem of entropy generation in heat and mass transfer in square porous cavity filled with electrically conducting fluid and subjected to a magnetic field. A schematic diagram of the porous cavity and coordinate system is shown in Figure 1. Horizontal temperature and concentration differences are specified between the vertical walls and zero mass and heat fluxes are imposed at the horizontal walls.

Mathematical formulation

The mathematical model in the present problem is formulated based on the following assumption:

1. The convective fluid and the porous media are in local thermal equilibrium,
2. The properties of the fluid and the porous media are constants,
3. The mass flux produced by temperature gradients (Soret effect) and the heat flux produced by a concentration gradients (Dufour effect) are neglected,

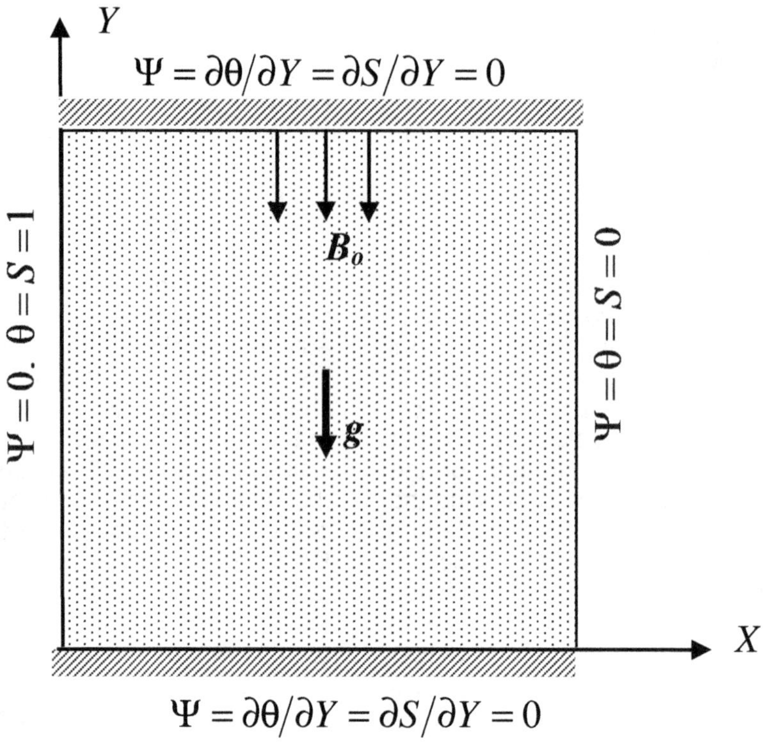

Figure 1. Schematic diagram of the physical model and coordinate system.

4. The viscous drag and inertia terms of the momentum equations are negligible, which are valid assumptions for low Darcy and particle Reynolds numbers,
5. The Darcy law is applicable under these assumptions, the conservation equations for steady flow can be written as:

$$\nabla.\vec{\mathbf{V}} = 0 \tag{1}$$

$$\vec{\mathbf{V}} = \frac{K}{\mu}\left(-\nabla p + \rho\vec{\mathbf{g}} + \vec{\mathbf{I}} \times \vec{\mathbf{B}}\right) \tag{2}$$

$$\nabla.\vec{\mathbf{I}} = 0; \quad \vec{\mathbf{I}} = \sigma\left(-\nabla\varphi + \vec{\mathbf{V}} \times \vec{\mathbf{B}}\right) \tag{3}$$

$$\nabla.(\vec{\mathbf{V}}T - \alpha\nabla T) = 0 \tag{4}$$

$$\nabla.(\vec{\mathbf{V}}C - D\nabla C) = 0 \tag{5}$$

where $\vec{\mathbf{V}}$ (ms^{-1}) is the velocity vector, K (m^2) is the permeability of the porous medium, μ ($kg.m^{-1}.s^{-1}$) is the dynamic viscosity, p (Nm^{-2}) is the pressure, ρ ($kg.m^{-3}$) is the density, $\vec{\mathbf{g}}$ (ms^{-2}) is the acceleration vector, $\vec{\mathbf{B}}$ (Wbm^{-2} or Tesla) is the external magnetic field vector, $\vec{\mathbf{I}}$ (A) is the electric current vector, σ (Ω^{-1} m^{-1}) is the fluid electrical conductivity, φ (V), is the electric potential, T(K) is the fluid temperature, C ($mol.m^{-3}$) is the concentration, α and D (m^2s^{-1}) are diffusivity of heat and constituent through the fluid saturated porous matrix respectively. Garandet et al. (1992) proposed an analytical solution of the equations of magnetohydrodynamics that can be used to model the effect of a transverse magnetic field on buoyancy driven convection in a two-dimensional cavity. According to Garandet et al. (1992), Equation (3) can be reduced to $\nabla^2\varphi = 0$. The unique solution is $\nabla\varphi = 0$ since there is always an electrically insulating boundary around enclosure, where the gradients normal to the walls are zeros ($\partial\varphi/\partial n = 0$). It follows that the electric field vanishes everywhere as discussed by Alchaar et al. (1995).

The solution that saturates the porous matrix is modelled as a Boussinesq incompressible fluid whose density variation can be expressed using the Oberbeck–Boussinesq approximation:

$$\rho \cong \rho_o\{1 - \beta_T(T - T_o) - \beta_c(C - C_o)\} \tag{6}$$

Where β_T and β_c are the thermal and concentration expansion coefficients. Subscript *o* stands for a reference state. For two-dimensional flow the pressure p in Equations (2) can be eliminated by cross differentiation and a single momentum equation can be derived. The governing equations may be written in dimensionless form using the following non-dimensional variables:

$$X,Y=(x,y)/L;\ U,V=(u,v)L/\alpha;$$
$$\theta=(T-T_c)/\Delta T;\ S=(C-C_c)/\Delta C \tag{7}$$

Where,
$\Delta T=(T_h-T_c)$ and $\Delta C=(C_h-C_c)$. The subscripts h and c stand for high and low values respectively. The dimensionless forms of the governing Equations (1) to (5) become:

$$\frac{\partial^2\Psi}{\partial X^2}+\left(1-Ha^2\right)\frac{\partial^2\Psi}{\partial Y^2}=-Ra\left(\frac{\partial\theta}{\partial X}+N\frac{\partial S}{\partial X}\right) \tag{8}$$

$$\frac{\partial\Psi}{\partial Y}\frac{\partial\theta}{\partial X}-\frac{\partial\Psi}{\partial X}\frac{\partial\theta}{\partial Y}=\frac{\partial^2\theta}{\partial X^2}+\frac{\partial^2\theta}{\partial Y^2} \tag{9}$$

$$\frac{\partial\Psi}{\partial Y}\frac{\partial S}{\partial X}-\frac{\partial\Psi}{\partial X}\frac{\partial S}{\partial Y}=\frac{1}{Le}\left(\frac{\partial^2 S}{\partial X^2}+\frac{\partial^2 S}{\partial Y^2}\right) \tag{10}$$

Where the stream function defined as $U=\partial\Psi/\partial Y$ and $V=-\partial\Psi/\partial X$, and the governing parameters are Rayleigh number, Buoyancy ratio, Lewis number and Hartmann number defined, respectively as:

$$Ra=\frac{\rho Kg\beta_T\Delta TL}{\alpha\mu};\ N=\frac{\beta_C}{\beta_T}\frac{\Delta C}{\Delta T};\ Le=\frac{\alpha}{D};$$
$$Ha=B_0\sqrt{\sigma K/\mu} \tag{11}$$

where B_o is the magnitude of $\vec{\mathbf{B}}$. The governing equations in the present problem are subjected to the following boundary conditions:

at $X=0$, $\quad \Psi=0,\quad \theta=1,\quad S=1$ (12a)

at $X=1$, $\quad \Psi=0,\quad \theta=0,\quad S=0$ (12b)

at $Y=0$, $\quad \Psi=0,\quad \frac{\partial\theta}{\partial Y}=0,\quad \frac{\partial S}{\partial Y}=0$ (12c)

at $Y=1$, $\quad \Psi=0,\quad \frac{\partial\theta}{\partial Y}=0,\quad \frac{\partial S}{\partial Y}=0$ (12d)

The results is presented in terms of the average Nusselt number $\overline{Nu}$ and the average Sherwood number $\overline{Sh}$ on the vertical walls, which are defined as follows:

$$\overline{Nu}=\int_0^1\left(\frac{\partial\theta}{\partial X}\right)_{X=0,1}dY;\ \overline{Sh}=\int_0^1\left(\frac{\partial S}{\partial X}\right)_{X=0,1}dY \tag{13}$$

Entropy generation

In the present convection heat and mass transfer problems, fluid friction, heat and mass transfer and the coupling between heat and mass transfer in addition to the magnetic field contribute to the rate of entropy generation. Accordingly, the volumetric entropy generation is the sum of the irreversibilities generation due to temperature gradients, viscous dissipation, magnetic field effect and concentration gradients. The irreversibility of the clear fluid is neglected based on the Darcy model and the viscous dissipation can be represented by velocity square term and the effective properties of the porous media (Baytas, 2000). Hence the volumetric entropy generation in heat and mass convection through a porous medium with the effect of magnetic field can be calculated by the following equation (Woods, 1975):

$$\dot{s}_{gen}=\frac{k}{T_o^2}(\nabla T)^2+\frac{\mu}{KT_o}\left(u^2+v^2\right)+\frac{\sigma B^2}{T_o}u^2+\frac{RD}{C_o}(\nabla C)^2+\frac{RD}{T_o}(\nabla T.\nabla C) \tag{14}$$

Dimensionless form of Equation (14) can be obtained by utilising the dimensionless variable defined in (7) as:

$$\dot{S}_{GEN}=\left\{\left(\frac{\partial\theta}{\partial X}\right)^2+\left(\frac{\partial\theta}{\partial Y}\right)^2\right\}+N_\mu\left\{\left(\frac{\partial\Psi}{\partial Y}\right)^2+\left(\frac{\partial\Psi}{\partial X}\right)^2+Ha^2\left(\frac{\partial\Psi}{\partial Y}\right)^2\right\}$$
$$+N_C\left\{\left(\frac{\partial S}{\partial X}\right)^2+\left(\frac{\partial S}{\partial Y}\right)^2\right\}+N_{TC}\left\{\left(\frac{\partial\theta}{\partial X}\right)\left(\frac{\partial S}{\partial X}\right)+\left(\frac{\partial\theta}{\partial Y}\right)\left(\frac{\partial S}{\partial Y}\right)\right\} \tag{15}$$

where N_μ, N_C and N_{TC} are irreversibility distribution ratios related to velocity gradients, concentration gradients and mixed product of concentration and thermal gradients, respectively. They are defined as follows:

$$N_\mu=\frac{\mu T_o}{k}\left\{\frac{\alpha^2}{K(\Delta T)^2}\right\};N_C=\frac{RDT_o^2}{kC_o}\left(\frac{\Delta C}{\Delta T}\right)^2;$$
$$N_{TC}=\frac{RDT_o}{k}\left(\frac{\Delta C}{\Delta T}\right) \tag{16}$$

Table 1. Comparison of $\overline{Nu}$ with Mahmud and Fraser (2004) results (*Ra* = 250 and *N* = 0).

Ha	0	2	4	6	10
Present results	5.883	3.195	1.516	1.140	1.021
Mahmud and Fraser (2004)	5.90	3.15	1.50	1.15	1.05

The local volumetric entropy generation would be integrated over the whole cavity to obtain the entropy generation number for the whole cavity volume as:

$$N_S = \int_0^1 \int_0^1 \dot{S}_{GEN} \ dXdY \quad (17)$$

NUMERICAL METHOD

The dimensionless governing Equations (8) to (10) subjected to the boundary conditions (12) are integrated numerically using the finite volume method (Patankar, 1980; Versteeg and Malalasekera, 2007). The power law scheme proposed by Patankar (1980) is used for the convection terms formulation of the energy and mass transfer equations. The resulting algebraic equations were solved by line-by-line using the Tri-Diagonal Matrix Algorithm iteration. The iteration process is terminated under the following condition:

$$\sum_{i,j} \left|\phi_{i,j}^{n} - \phi_{i,j}^{n-1}\right| \Big/ \sum_{i,j} \left|\phi_{i,j}^{n}\right| \le 10^{-7} \quad (18)$$

where ϕ is the general dependent variable which can stands for either θ, S or Ψ and *n* denotes the iteration step. The developed code is an extension of the code verified and validated in previous studies (Saeid and Mohamad, 2005; Saeid and Pop, 2004). The local volumetric entropy generation is calculated after calculating the temperature, concentration and stream function in each control volume in the mesh. Finally the entropy generation number for the cavity volume is calculated using Equation (17) via the numerical integration.

The present numerical results are compared with the results obtained by Mahmud and Fraser (2004) for the effect of magnetic field on the convective heat transfer in porous cavity. The results presented in Table 1 reflect the accuracy of the present results using uniform mesh of 40×40 square control volumes.

RESULTS AND DISCUSSION

The effect of the magnetic field on the entropy generation is represented by the second bracket in Equation (15). The irreversibility distribution ratios related to velocity gradients (N_μ) will be considered as a governing parameter in the range $0.001 \le N_\mu \le 1$. In order to reduce the governing parameters, the effect of the irreversibility distribution ratios related to concentration gradients (N_C) and mixed product of concentration and thermal gradients (N_{TC}) are fixed. The values of irreversibility coefficients $N_C = 0.5$ and $N_{TC} = 0.01$ are considered as recommended by Mchirgui et al. (2012) for the case of a square porous cavity filled with a binary perfect gas mixture without the effects of the magnetic field.

The results are generated to show the effect of the following governing parameters $0 \le Ha \le 10$, $-5 \le N \le 5$, $0.001 \le N_\mu \le 1$, $1 \le Ra \le 1000$ and $0.1 \le Le \le 10$ on the average Nusselt number, the average Sherwood number and the entropy generation number.

It is important to note that the dimensionless temperature and dimensionless concentration distributions are identical for the case of *Le* =1, which leads to $\overline{Nu} = \overline{Sh}$.

The classical natural convection in porous cavity without mass transfer and zero magnetic field is considered first as a reference case. The variation of the average Nusselt number ($\overline{Nu}$) and the local entropy generation number (N_s) with Rayleigh number (*Ra*) are shown in Figure 2 with fixed values of *N* = 0, *Le* = 1 and *Ha* = 0. Figure 2a shows the classical variation of $\overline{Nu}$ with Ra. It is important to note that the irreversibility distribution ratio (N_μ) and the Rayleigh number (Ra) are related to each other as they are functions of the effective properties of the porous media. Hence the dependence of the heat transfer on the properties of the fluid and the porous media is justified through the dependence of $\overline{Nu}$ on Ra. Mathematically; the effect of the effective properties on the thermal field can be represented by *Ra*, *N*, *Le* and *Ha* as can be seen in the dimensionless governing Equations (8) to (10). The irreversibility distribution ratio (N_μ) is not appeared in these equations and, therefore, it affects the thermal field implicitly through the effective properties of the porous media.

The effect of the irreversibility distribution ratio (N_μ) on the entropy generation number (N_s) variation with Rayleigh number (*Ra*) is shown in Figure 2b. The results show that increasing N_μ leads to higher entropy generation in the cavity. From the definition of N_μ, in

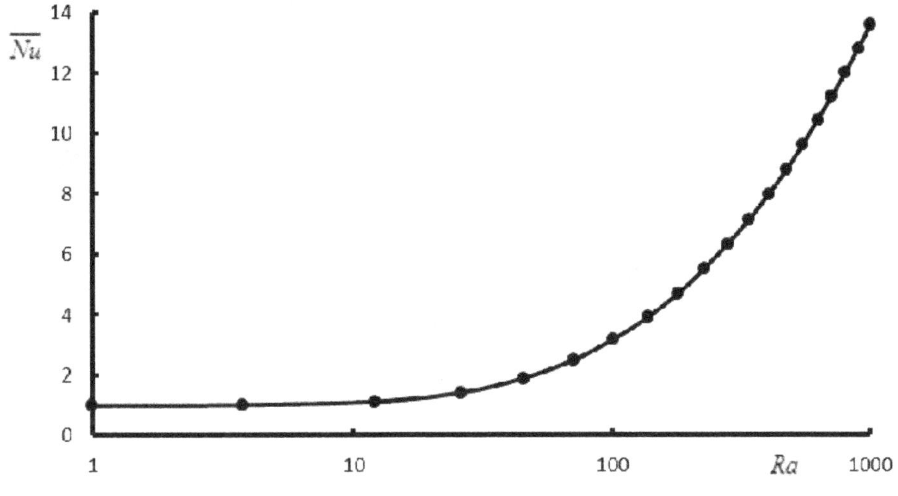

(a)

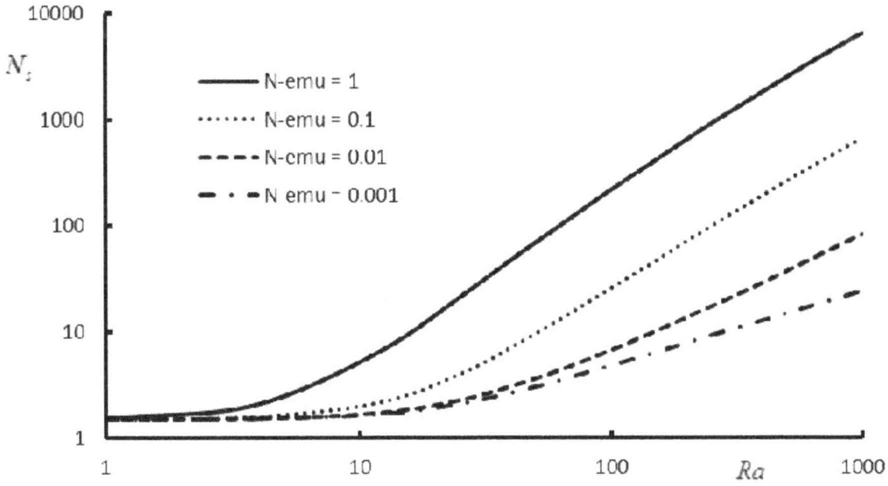

(b)

Figure 2. Variation of $\overline{Nu}$ and N_s with *Ra* at constant *N* = 0, *Ha* = 0 and *Le* = 1.

order to reduce the rate of entropy generation, it is necessary to reduce the reference temperature and reduce the viscosity and the thermal diffusivity of the fluid. The rate of entropy generation can be minimized for the flow through a high permeable porous medium.

The variation $\overline{Nu}$ and N_s with buoyancy ratio (*N*) are shown in Figure 3 with fixed values of N_μ = 0.1, *Le* = 1 and *Ha* = 0. The results depicted in Figure 3 show minimum values of $\overline{Nu}$ and N_s are observed at *N* = -1 for all the values of Ra. Negative values of *N* mean that the buoyancy forces generated due to temperature and concentration differences are in opposite directions. For *Le* = 1, *Ha* = 0 and *N* = -1, Equation (8) reduced to $\nabla^2\Psi = 0$ with $\Psi = 0$ at the importable walls, leads to a stagnate fluid with $\Psi = 0$ everywhere in the cavity. In this case, the heat and mass transfer are pure diffusion processes, in which the values of $\overline{Nu}$ and N_s are the minimum. Increasing or decreasing the value of the buoyancy ratio parameter (*N*) leads to enhance the fluid circulation due to the increase of the resultant buoyancy force in the cavity. This leads to the increase the values of $\overline{Nu}$ and N_s as shown in Figure 3.

The effect of the magnetic field on the variation of $\overline{Nu}$ and N_s is show in Figure 4 for different values of *Ra* and fixed values of N_μ = 0.1, *Le* = 1 and *N* = 1. Maximum rates of heat transfer as well as entropy generation are observed at Ha = 0. Increasing the

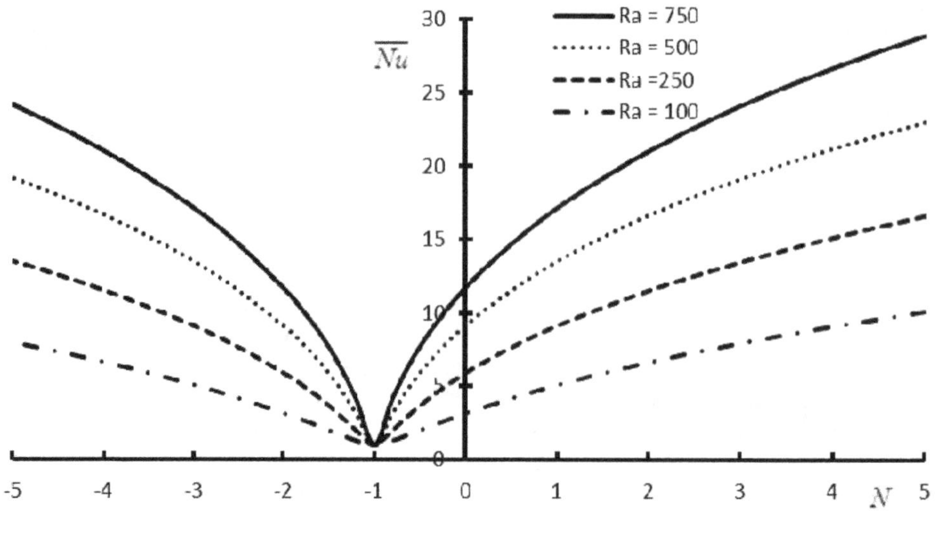

(a)

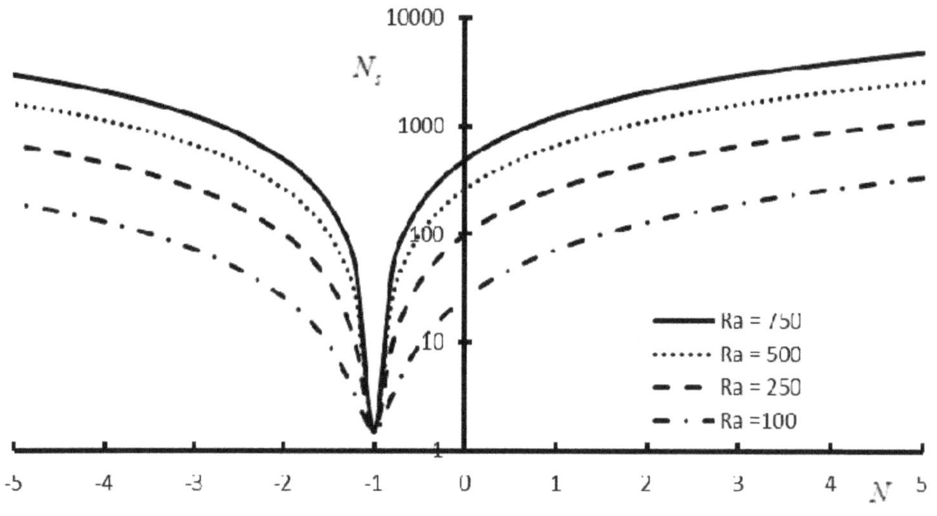

(b)

Figure 3. Variation of $\overline{Nu}$ and N_s with N at constant N_μ = 0.1, Ha = 0 and Le = 1.

magnetic field leads to increase the Lorentz forces and slowdown the fluid flow which leads to decrease $\overline{Nu}$ rapidly as shown in Figure 5a. For high values of *Ha* (*Ha* > 10), the effect of *Ra* on the heat and mass transfer process is vanished and the values of $\overline{Nu}$ approaching unity. The heat and mass transfer processes are associated with the entropy generation and again the rate of entropy generation is maximum when the rate of heat and mass transfer is maximum as shown in Figure 4b.

Finally, the effect of Lewis number (*Le*) of the entropy generation for natural convection heat and mass transfer is studied for different values of Hartmann number (*Ha*). The results presented in Figure 5 show the variations of $\overline{Nu}$, $\overline{Sh}$ and N_s with *Le* at constant N_μ = 0.1, N = 1 and *Ra* = 100. At low values of *Le* (0 < *Le* < 1), the values of $\overline{Nu}$ are almost constant at fixed values of other parameters while $\overline{Sh}$ is increasing slightly with the increase of *Le* as shown in Figures 5 (a) and (b). In this range of *Le*, the heat and mass transfer processes are

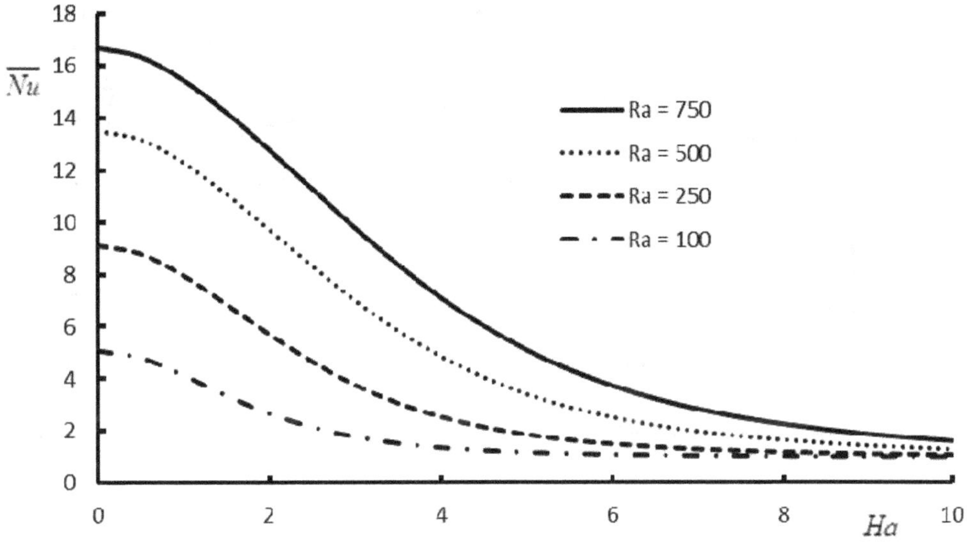

(a)

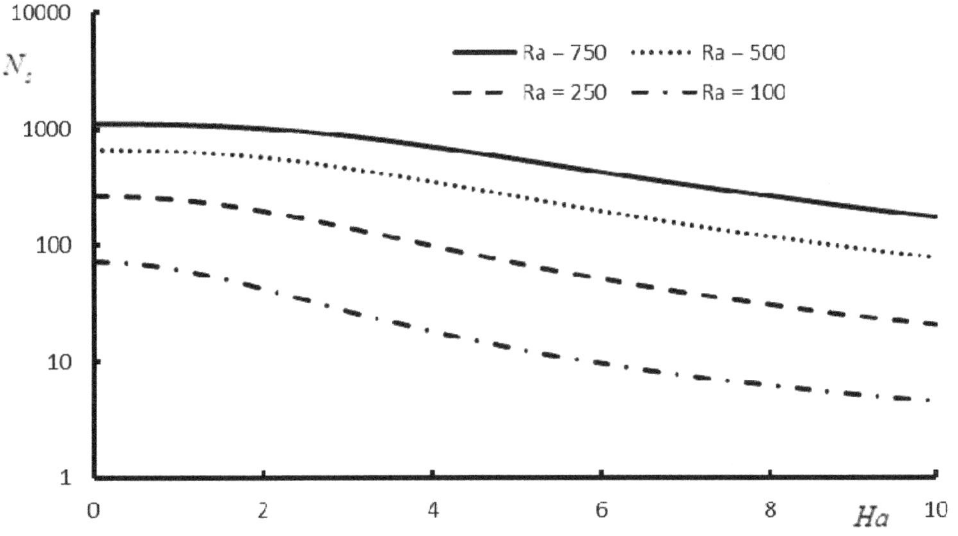

(b)

Figure 4. Variation of $\overline{Nu}$ and N_s with *Ha* at constant N_μ = 0.1, *N* = 1 and *Le* = 1.

carried out by convection and diffusion and the increase f $\overline{Sh}$ is due to the magnifying the convection term (Equation 10).

On the other hand, the heat transfer process is not affected much by increasing the Lewis number. This is due to the fact that there is no clear effect of the concentration variation on the buoyancy forces or the circulation of the fluid in the cavity at low values of *Le* and *Ra*. The strength of the circulation of the fluid in the cavity can be measured by maximum absolute stream function in the cavity. This value is found to be $|\Psi|_{max}$ = 9.1069 for the case with *Le* = 0.1 and the following parameters: N_μ = 0.1, *N* = 1, *Ha* = 0 and *Ra* = 100. This leads to a fixed rate of entropy generation, as shown in Figure 5(c), since the mass transfer effect on the entropy generation due to the irreversible generated by viscosity effects. The contour plots for isotherms and iso-concentrations shown

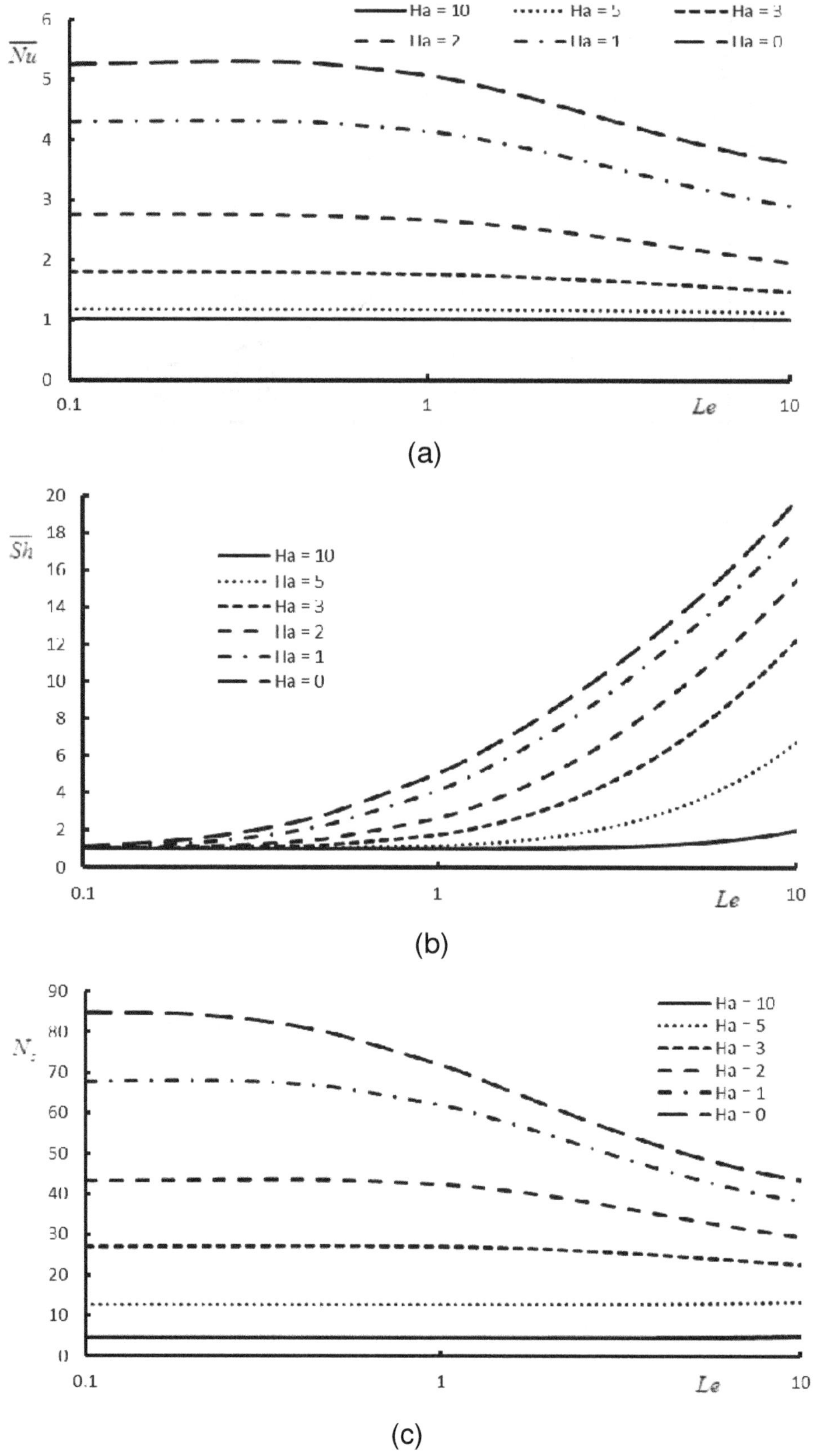

Figure 5. Variation of $\overline{Nu}$, $\overline{Sh}$ and N_s with *Le* at constant N_μ = 0.1, *N* = 1 and *Ra* = 100.

in Figure 6(a) indicate that the mass transfer process is almost pure diffusion at Le = 0.1 as the contour lines are almost vertical lines. Increasing the Lewis number enhances the convection mode in the mass transfer process. This became obvious by comparing the contour plots for isotherms and iso-concentrations presented in Figures 6(a), (b) and (c) for Le = 0.1, Le = 5 and Le =10 respectively. Note that the contour plots for isotherms and iso-concentrations are the same for the case of Le = 1 and almost similar to the isotherms shown in Figure 6 and therefore not shown.

It was observed that the strength of the fluid circulation in the cavity was reduced by increasing the Lewis number. The strengths of the fluid circulation are represented respectively by $|\Psi|_{max}$ = 9.1069, $|\Psi|_{max}$ = 5.1475 and $|\Psi|_{max}$ = 4.7883 for the cases of Le = 0.1, Le = 5 and Le = 10 with fixed following parameters: N_{μ} = 0.1, N = 1, Ha = 0 and Ra = 100. Figure 6 shows that the isotherms are not affected much with increasing Le and other fixed parameters compare to the contours of the iso-concentrations. This leads to increase in the average Sherwood number with the increase of Le as shown in Figure 5(b) while the average Nusselt number and the entropy generation are decreasing slightly by increasing Le as shown in Figures 5(a) and (c). Figure 5 shows also that the heat, mass and entropy generation are reduced by increasing the magnetic field and, at Ha = 10, the variations of $\overline{Nu}$, $\overline{Sh}$ and N_s with Le are almost constant.

Conclusion

The entropy generation for the combined natural convection heat and mass transfer in a porous cavity subjected to a magnetic field is considered for investigation in the present study. In order to reduce the rate of entropy generation, it is necessary to reduce the reference temperature and reduce the viscosity and the thermal diffusivity of the working fluid. The rate of entropy generation can be minimized for the flow through a high permeable porous medium. The numerical results for Le = 1, Ha = 0 and N = -1, show a stagnate fluid everywhere in the cavity since the buoyancy forces generated due to temperature and concentration differences are in the same order and opposite directions. In this case, the heat and mass transfer is a pure diffusion process, in which the values of $\overline{Nu}$ and N_s are the minimum. Increasing or decreasing the value of the buoyancy ratio parameter (N) leads to enhance the fluid circulation due to the increase of the resultant buoyancy force in the cavity. This leads to the increase the values of $\overline{Nu}$ and N_s. The average Sherwood number is increasing with the increase of Lewis number due to the magnifying the convection mode in the mass transfer process. It is observed that the strength of the fluid circulation in the cavity is reduced by increasing the Lewis number. This leads to the decrease slightly in the average Nusselt number and the entropy generation are decreasing by increasing Le. The numerical results show that increasing the magnetic field parameter (Hartmann number) leads to reduce the flow circulation strength in the cavity and this leads to decrease the rate of heat and mass transfer as well as the rate of entropy generation.

Nomenclature

$\vec{\mathbf{B}}$ External magnetic field vector, (Wbm^{-2} or Tesla)
C Molar concentration, ($mol.m^{-3}$)
D Effective mass diffusivity through the fluid saturated porous matrix. (m^2s^{-1})
$\vec{\mathbf{g}}$ Acceleration vector, (ms^{-2})
Ha Hartmann number, Equation (11).
$\vec{\mathbf{I}}$ Electric current vector, (A)
K Permeability of the porous medium (m^2)
L Cavity length (m)
Le Lewis number, Equation (11).
N Buoyancy ratio parameter, Equation (11)
N_{μ} Irreversibility distribution ratio related to velocity gradients, Equation (16)
N_S Dimensionless entropy generation, Equation (17)
$\overline{Nu}$ Average Nusselt number, Equation (13)
p pressure, (Nm^{-2})
R Universal gas constant ($Jkg^{-1}\ K^{-1}$)
Ra Rayleigh number, Equation (11)
S Dimensionless molar concentration, Equation (7)
$\dot{s}_{gen}$ Entropy generation rate per unit volume ($Wm^{-3}K^{-1}$)
$\dot{S}_{GEN}$ Dimensionless entropy generation rate per unit volume.
$\overline{Sh}$ Average Sherwood numbers, Equation (13)
T fluid temperature (K)
u, v Velocity components in x, y directions (m/s)
U, V Dimensionless velocity components in X, Y directions, Equation (7)
$\vec{\mathbf{V}}$ Velocity vector, (ms^{-1})
x, y Cartesian coordinates (m)
X, Y Dimensionless coordinates, Equation (7)

Greek symbols

α Effective thermal diffusivity through the fluid saturated porous matrix. (m^2s^{-1})
β_T Thermal expansion coefficients (K^{-1})
β_c Solute expansion coefficient ($m^3\ mol^{-1}$)

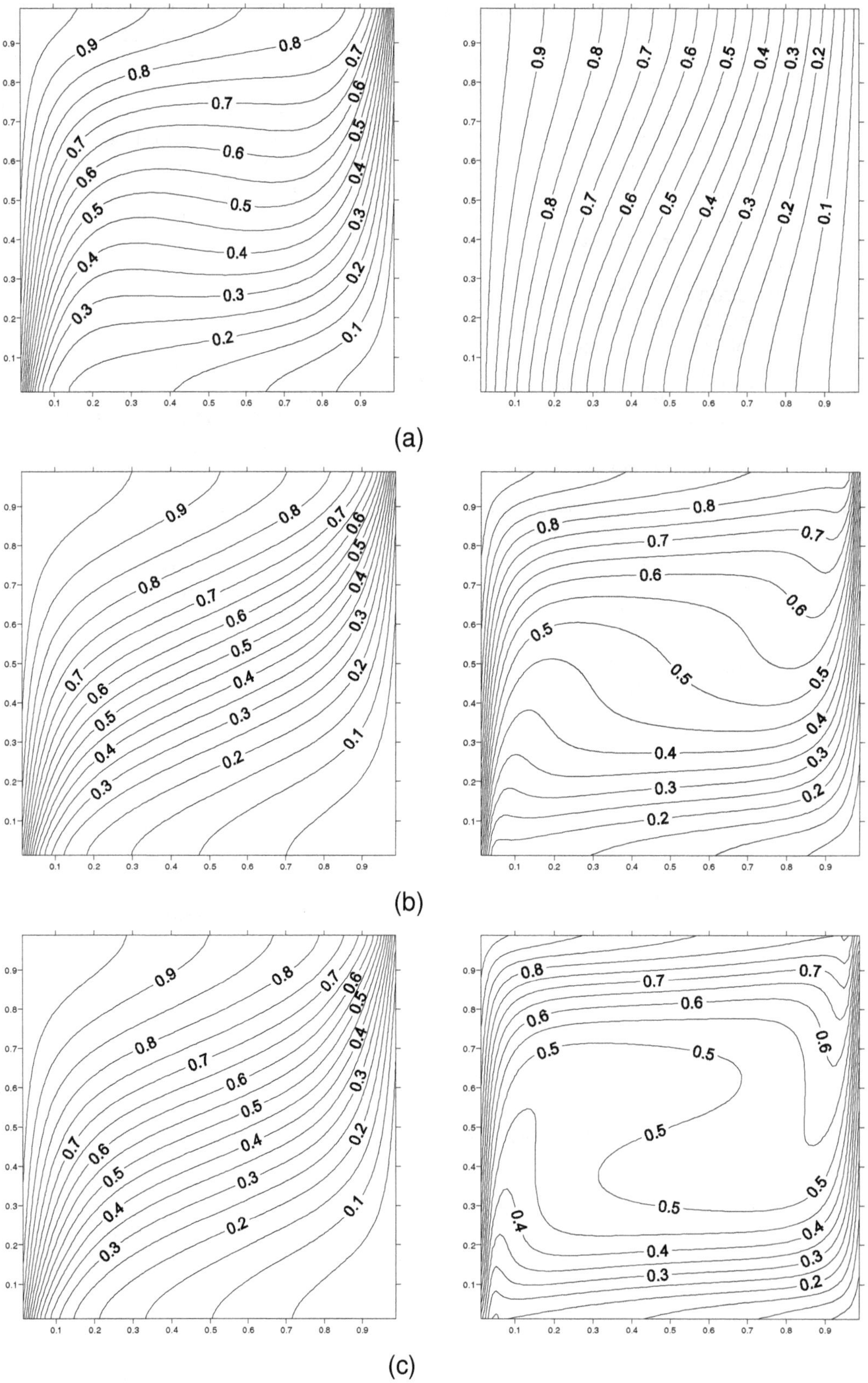

Figure 6. Isotherms (left) Iso-concentrations (right) with N_{μ} = 0.1, *N* = 1, *Ha* = 0 and *Ra* = 100. (a) *Le* = 0.1 ($|\Psi|_{max}$ = 9.1069). (b) *Le* =5 ($|\Psi|_{max}$ =5.1475). (c) *Le* =10 ($|\Psi|_{max}$ = 4.7883).

θ Dimensionless temperature, Equation (7)
Ψ Dimensionless stream function.
μ Dynamic viscosity, ($kg.m^{-1}.s^{-1}$)
ρ Density, ($kg.m^{-3}$)
σ Fluid electrical conductivity, ($\Omega^{-1}\ m^{-1}$)
φ Electric potential, (V).

REFERENCES

Alchaar S, Vasseur P, Bilgen E (1995). The effect of a magnetic field on natural convection in a shallow cavity heated from below. Chem. Eng. Comm. 134:195-209.

Baytas AC (2000). Entropy generation for natural convection in an inclined porous cavity. Int. J. Heat Mass Transf. 43:2089-2099.

Bian W, Vasseur P, Bilgen E, Meng F (1996). Effect of an electromagnetic field on natural convection in an inclined porous layer. Int. J. Heat Fluid Flow 17:36-44.

Famouri M, Hooman K (2008). Entropy generation for natural convection by heated partitions in a cavity. Int. Comm. Heat Mass Transf. 35:492-502.

Garandet JP, Alboussiere T, Moreau R (1992). Buoyancy driven convection in a rectangular enclosure with a transverse magnetic field. Int. J. Heat Mass Transf. 35:741-748.

Hooman K, Gurgenci H, Merrikh AA (2007). Heat transfer and entropy generation optimization of forced convection in porous-saturated ducts of rectangular cross-section. Int. J. Heat Mass Transf. 50:2051-2059.

Hooman K, Hooman F, Mohebpour SR (2008). Entropy generation for forced convection in a porous channel with isoflux or isothermal walls. Int. J. Exergy 5:78-96.

Kaluri RS Basak T (2011). Role of entropy generation on thermal management during natural convection in porous square cavities with distributed heat sources. Chem. Eng. Sci. 66:2124-2140.

Kaviany M (1999). Principles of Heat Transfer in Porous Media. Springer, New York. http://books.google.com.ng/books/about/Principles_of_Heat_Transfer_in_Porous_Me.html?id=pX61QAOKKqsC&redir_esc=y.

Mahmud S, Fraser RA (2004). Magnetohydrodynamic free convection and entropy generation in a square porous cavity. Int. J. Heat Mass Transf. 47:3245-3256.

Mchirgui A, Hidouri N, Magherbi M, Brahim AB (2012). Entropy generation in double-diffusive convection in a square porous cavity using Darcy–Brinkman formulation. Transp. Porous Med. 93:223-240.

Mojtabi A, Mojtabi MC (2005). Double-Diffusive Convection in Porous Media; in K. Vafai (Ed.), Handbook of Porous Media. 2nd edition, Taylor & Francis Group, New York. http://www.crcnetbase.com/doi/pdf/10.1201/9780415876384.fmatt.

Nield DA, Bejan A (2006). Convection in Porous Media. 3rd ed. Springer, New York. 3rd ed.

Oreper GM, Szekely J (1983). The effect of an externally imposed magnetic field on buoyancy driven flow in a rectangular cavity. J. Cryst. Growth 64:505-515.

Ozoe H (2005). Magnetic Convection. Imperial College Press. pof.aip.org/phfle6/v19/i8/p087104_s1?view=refs.

Patankar SV (1980). Numerical Heat Transfer and Fluid Flow. McGraw-Hill, New York. www.ewp.rpi.edu/hartford/~ernesto/F2012/.../Patankar-NHTFF-1980.pdf.

Pop I, Ingham DB (2001). Convective Heat Transfer: Mathematical and Computational Modelling of Viscous Fluids and Porous Media. Pergamon, Oxford.

Rudraiah N, Venkatachalappa M, Subbaraya CK (1995). Combined surface tension and buoyancy-driven convection in a rectangular open cavity in the presence of a magnetic field. Int. J. Nonlinear Mech. 30:759-770.

Saeid NH, Mohamad AA (2005). Natural convection in a porous cavity with spatial sidewall temperature variation. Int. J. Num. Meth. Heat Fluid Flow 15:555-566.

Saeid NH, Pop I (2004). Transient free convection in a square cavity filled with a porous medium. Int. J. Heat Mass Transf. 47:1917-1924.

Versteeg K, Malalasekera W (2007). An Introduction to Computational Fluid Dynamics: The Finite Volume Method. Pearson Prentice Hall.

Woods LC (1975). The Thermodynamics of Fluid Systems. Oxford University Press: Oxford, UK.

12

On the hydrogenation-dehydrogenation of graphene-layer-nanostructures: Relevance to the hydrogen on-board storage problem

Yu S. Nechaev[1] and Nejat T. Veziroglu[2]

[1]Bardin Institute for Ferrous Metallurgy, Kurdjumov Institute of Metals Science and Physics, Vtoraya Baumanskaya St., 9/23, Moscow 105005, Russia.
[2]International Association for Hydrogen Energy, 5794 SW 40 St. #303, Miami, FL 33155, USA.

Herein, results of thermodynamic analysis of some theoretical and experimental [thermal desorption (TDS), scanning tunneling microscopy (STM), scanning tunneling spectroscopy (STS), high-resolution electron energy loss spectroscopy/low-energy electron diffraction (HREELS/LEED), photoelectron spectroscopy (PES), angle-resolved photoemission spectroscopy (ARPES), Raman spectroscopy and others] data on "reversible" hydrogenation and dehydrogenation of some graphene-layer-nanostructures are presented. In the framework of the formal kinetics and the approximation of the first order rate reaction, some thermodynamic quantities for the reaction of hydrogen sorption (the reaction rate constant, the reaction activation energy, the per-exponential factor of the reaction rate constant) have been determined. Some models and characteristics of hydrogen chemisorption on graphite (on the basal and edge planes) have been used for interpretation of the obtained quantities, with the aim of revealing the atomic mechanisms of hydrogenation and dehydrogenation of different graphene-layer-systems. The cases of both non-diffusion rate limiting kinetics and diffusion rate limiting kinetics are considered. Some open questions and perspectives remain in solving the actual problem in effective hydrogen on-board storage; using the graphite nanofibers (GNFs) is also considered.

Key words: Epitaxial and membrane graphenes, other graphene-layer-systems, hydrogenation-dehydrogenation, thermodynamic characteristics, atomic mechanisms, the hydrogen on-board efficient storage problem.

INTRODUCTION

As noted in a number of articles 2007 through 2014, hydrogenation of graphene-layers-systems, as a prototype of covalent chemical functionality and an effective tool to open the band gap of graphene, is of both fundamental and applied importance (Geim and Novoselov, 2007; Palerno, 2013).

It is relevant to the current problems of thermodynamic stability and thermodynamic characteristics of the hydrogenated graphene-layers-systems (Sofo et al., 2007; Openov and Podlivaev, 2010; Han et al., 2012), and also to the current problem of hydrogen on-board storage (Akiba, 2011; Zuettel, 2011; DOE targets, 2012).

In the case of epitaxial graphene on substrates, such as SiO_2 and others, hydrogenation occurs only on the top basal plane of graphene, and it is not accompanied with a strong (diamond-like) distortion of the graphene network, but only with some ripples. The first experimental indication of such a specific single-side hydrogenation came from Elias et al. (2009). The authors mentioned a possible contradiction with the theoretical results of Sofo et al. (2007), which had down-played the possibility of a single side hydrogenation. They proposed an important facilitating role of the material ripples for hydrogenation of graphene on SiO_2, and believed that such a single-side hydrogenated epitaxial graphene can be a disordered material, similar to graphene oxide, rather than a new graphene-based crystal - the experimental graphane produced by them (on the free-standing graphene membrane).

On the other hand, it is expedient to note that changes in Raman spectra of graphene caused by hydrogenation were rather similar (with respect to locations of D, G, D′, 2D and (D+D′) peaks) both for the epitaxial graphene on SiO_2 and for the free-standing graphene membrane (Elias et al., 2009).

As it is supposed by many scientists, such a single side hydrogenation of epitaxial graphene occurs, because the diffusion of hydrogen along the graphene-SiO_2 interface is negligible, and perfect graphene is impermeable to any atom and molecule (Jiang et al., 2009). But, firstly, these two aspects are of the kinetic character, and therefore they cannot influence the thermodynamic predictions (Sofo et al., 2007; Boukhvalov et al., 2008; Zhou et al., 2009). Secondly, as shown in the present analytical study, the above noted two aspects have not been studied in an enough degree.

As shown in Elias et al. (2009), when a hydrogenated graphene membrane had no free boundaries (a rigidly fixed membrane) in the expanded regions of it, the lattice was stretched isotropically by nearly 10%, with respect to the pristine graphene. This amount of stretching (10%) is close to the limit of possible elastic deformations in graphene (Nechaev and Veziroglu, 2013), and indeed it has been observed that some of their membranes rupture during hydrogenation. It was believed (Elias et al., 2009) that the stretched regions were likely to remain non-hydrogenated. They also found that instead of exhibiting random stretching, hydrogenated graphene membranes normally split into domain-like regions of the size of the order of 1 μm, and that the annealing of such membranes led to complete recovery of the periodicity in both stretched and compressed domains (Elias et al., 2009).

It can be supposed that the rigidly fixed graphene membranes are related, in some degree, to the epitaxial graphenes. Those may be rigidly fixed by the cohesive interaction with the substrates.

As was noted in Xiang et al. (2010), the double-side hydrogenation of graphene is now well understood, at least from a theoretical point of view. For example, Sofo et al. (2007) predicted theoretically a new insulating material of CH composition called graphane (double-side hydrogenated graphene), in which each hydrogen atom adsorbs on top of a carbon atom from both sides, so that the hydrogen atoms adsorbed in different carbon sublattices are on different sides of the monolayer plane (Sofo et al., 2007). The formation of graphane was attributed to the efficient strain relaxation for sp^3 hybridization, accompanied by a strong (diamond-like) distortion of the graphene network (Sofo et al., 2007; Xiang et al., 2009). In contrast to graphene (a zero-gap semiconductor), graphane is an insulator with an energy gap of $E_g \approx$ 5.4 eV (Openov and Podlivaev, 2010; Lebegue et al., 2009).

Only if hydrogen atoms adsorbed on one side of graphene (in graphane) are retained, we obtain graphone of C_2H composition, which is a magnetic semiconductor with $E_g \approx$ 0.5 eV, and a Curie temperature of $T_c \approx$ 300 to 400K (Zhou et al., 2009).

As was noted in Openov and Podlivaev (2012), neither graphone nor graphane are suitable for real practical applications, since the former has a low value of E_g, and undergoes a rapid disordering because of hydrogen migration to neighboring vacant sites even at a low temperature, and the latter cannot be prepared on a solid substrate (Podlivaev and Openov, 2011).

It is also expedient to refer to a theoretical single-side hydrogenated graphene (SSHG) of CH composition (that is, an alternative to graphane (Sofo et al. (2007)), in which hydrogen atoms are adsorbed only on one side (Pujari et al., 2011; Dzhurakhalov and Peeters, 2011). In contrast to graphone, they are adsorbed on all carbon atoms rather than on every second carbon atom. The value of E_g in SSHG is sufficiently high (1.6 eV lower than in graphane), and it can be prepared on a solid substrate in principle. But, this quasi-two-dimensional carbon-hydrogen theoretical system is shown to have a relatively low thermal stability, which makes it difficult to use SSGG in practice (Openov and Podlivaev, 2012; Pujari et al., 2011).

As was noted in Pujari et al. (2011), it may be inappropriate to call the covalently bonded SSHG system sp^3 hybridized, since the characteristic bond angle of 109.5° is not present anywhere that is, there is no diamond-like strong distortion of the graphene network, rather than in graphane. Generally in the case of a few hydrogen atoms interacting with graphene or even for graphane, the underlining carbon atoms are displaced from their locations. For instance, there may be the diamond-like local distortion of the graphene network, showing the signature of sp^3 bonded system. However, in SSHGraphene all the carbon atoms remain in one plane, making it difficult to call it sp^3 hybridized. Obviously, this is some specific sp^3- like hybridization.

The results of Nechaev (2010), and also Table 1A and B in the present paper, of thermodynamic analysis of a number of experimental data point that some specific

Table 1A. Theoretical, experimental and analytical values of some related quantities.

Material	Value/quantity				
	$\Delta H_{(C\text{-}H)}$ (eV)	$\Delta H_{(bind.)}$, eV	$\Delta H_{(C\text{-}C)}$, (eV)	$\Delta H_{(des.)}$ (eV) $\Delta H_{(ads.)}$ (eV)	$K_{0(des.)}$, s^{-1} ($L \approx (D_{0app.}/K_{0(des.)})^{1/2}$)
Graphane CH (Sofo et al. , 2007)	2.5 ± 0.1(analysis)	6.56 (theory)	2.7 (analysis)		
Graphane CH (Dzhurakhalov and Peeters, 2011)	1.50 (theory)	5.03 (theory)	2.35 (analysis)		
Graphane CH (Openov and Podlivaev, 2010)	2.46 ± 0.17 (analysis)			2.46 ± 0.17 (theory)	2.0 × 10^{15} (analysis)
Free-standing graphene-like membrane (Elias et al., 2009)	There are no experimental values in the work			if 2.5 ± 0.1 if 2.6 ± 0.1 (1.0 ± 0.2) (analysis)	then 7 × 10^{12} then 5 × 10^{13} ($K_{0(ads.)} \approx K_{0(des.)}$)
Hydrogenated epitaxial graphene (Elias et al., 2009)	There are no experimental values in the work			then 1.84 then 1.94 if 0.3 if 0.6 if 0.9 (0.3 ± 0.2) (analysis)	if 7 × 10^{12} if 5 × 10^{13} then 0.2 then 80 then 3.5 × 10^4 ($K_{0(ads.)} \approx K_{0(des.)}$) ($L$~$d_{sample}$)
Hydrogenated epitaxial* graphene, TDS-peak #1 (Elias et al., 2009)				0.6 ± 0.3 (as processes ~ I-II,~ model "G", Figure 4) (analysis)	2 × 10^7 (or 2 × 10^3-2 × 10^{11}) (L~ d_{sample}) (analysis)
Hydrogenated epitaxial* graphene, TDS-peak #2 (Elias et al., 2009)				0.6 ± 0.3 (as for processes ~I-II, ~model "G", Figure 4) (analysis)	1 × 10^6 (or 4 × 10^2 - 2 × 10^9) (L~d_{sample}) (analysis)
Hydrogenated epitaxial* graphene, TDS-peak #3 (Elias et al., 2009)				0.23 ± 0.05 (as process ~I,~ models "F","G", Figure 4) (analysis)	2.4 (or 0.8-7) (L~d_{sample}) (analysis)
Rigidly fixed hydrogenated graphene membrane (Elias et al., 2009)	There are no experimental values in the work			There are no experimental values in the work	There are no experimental values in the work
Graphene (Dzhurakhalov and Peeters, 2011)		7.40 (theory)	4.93 (analysis)		
Graphite (Nechaev and Veziroglu, 2013)		7.41 ± 0.05 (analysis)	4.94 ± 0.03 (analysis)		
Diamond (Nechaev and Veziroglu, 2013)		7.38 ± 0.04 (analysis)	3.69 ± 0.02 (analysis)		

Table 1B. Theoretical, experimental and analytical values of some related quantities.

Material	Value/quantity			
	$\Delta H_{(C-H)}$, eV	$\Delta H_{(C-C)}$, eV	$\Delta H_{(des.)}$, eV	$K_{0(des.)}$, s^{-1}
Hydrofullerene $C_{60}H_{36}$ (Pimenova et al., 2002)	2.64 ± 0.01 (experiment)			
Hydrogenated carbon nanotubes C_2H(Bauschlicher and So, 2002)	2.5 ± 0.2 (theory)			
Hydrogenated isotropic graphite, graphite nanofibers and nanostructured graphite(Nechaev, 2010)	2.50 ± 0.03 (analysis, process III, model "F*")	4.94 ± 0.03 (analysis)	2.6 ± 0.03 (analysis, process III)	There are empirical values in the work (analysis of experiment)
Hydrogenated isotropic graphite, graphite nano-fibers, nanostructured graphite, defected carbon nanotubes (Nechaev, 2010)	2.90 ± 0.05 [analysis, process II, models "H","G" (Figure 4)]		1.24 ± 0.03 (analysis, process II)	There are empirical values in thework (analysis of experiment)
Hydrogenated isotropic graphite, carbon nanotubes (Nechaev, 2010)	2.40 ± 0.05 [analysis, process I, models "F", "G" (Figure 4)]		0.21 ± 0.02 (analysis, process I)	There are empirical values in the work (analysis of experiment)
Hydrogenated isotropic and pyrolytic and nanostructured graphite (Nechaev, 2010)	3.77 ± 0.05 [analysis, process IV, models "C", "D" (Figure 4)]		3.8 ± 0.5 (analysis, process IV)	There are empirical values in the work (analysis of experiment)

local sp^3- like hybridization, without the diamond-like strong distortion of the graphene network, may be manifested itself in the cases of hydrogen atoms dissolved between graphene layers in isotropic graphite, graphite nanofibers (GNFs) and nanostructured graphite, where obviously there is a situation similar (in a definite degree) to one of the rigidly fixed graphene membranes. As far as we know, it has not been taken into account in many recent theoretical studies.

In this connection, it is expedient to note that there are a number of theoretical works showing that hydrogen chemisorption corrugates the graphene sheet in fullerene, carbon nanotubes, graphite and graphene, and transforms them from a semimetal into a semiconductor (Sofo et al., 2007; Elias et al., 2009). This can even induce magnetic moments (Yazyev and Helm, 2007; Lehtinen et al., 2004; Boukhvalov et al., 2008).

Previous theoretical studies suggest that single-side hydrogenation of ideal graphene would be thermodynamically unstable (Boukhvalov et al., 2008; Zhou et al., 2009). Thus, it remains a puzzle why the single-side hydrogenation of epitaxial graphenes is possible and even reversible, and why the hydrogenated species are stable at room temperatures (Elias et al., 2009; Sessi et al., 2009). This puzzling situation is also considered in the present analytical study.

Xiang et al. (2010) noted that their test calculations show that the barrier for the penetration of a hydrogen atom through the six-member ring of graphene is larger than 2.0 eV. Thus, they believe that it is almost impossible for a hydrogen atom to pass through the six-member ring of graphene at room temperature (from a

Figure 1. Structure of the theoretical graphane in chair configuration. The carbon atoms are shown in gray and the hydrogen atoms in white. The figure shows the diamond-like distorted hexagonal network with carbon in sp^3 hybridization (Sofo et al., 2007).

private communication with Xiang et al. (2009).

In the present analytical study, a real possibility of the penetration is considered when a hydrogen atom can pass through the graphene network at room temperature. This is the case of existing relevant defects in graphene, that is, grain boundaries, their triple junctions (nodes) and/or vacancies (Brito et al., 2011; Zhang et al., 2014; Banhart et al., 2011; Yazyev and Louie, 2010; Kim et al., 2011; Koepke et al., 2013; Zhang and Zhao, 2013; Yakobson and Ding, 2011; Cockayne et al., 2011; Zhang et al., 2012; Eckmann et al., 2012). The present study is related to revealing the atomic mechanisms of reversible hydrogenation of epitaxial graphenes, compared with membrane graphenes.

In the next parts of this paper, results of thermodynamic analysis, comparison and interpretation of some theoretical and experimental data are presented, which are related to better understanding and/or solving of the open questions mentioned above. It is related to a further development and modification of our previous analytical results (2010-2014), particularly published in the openaccess journals. Therefore, in the present paper, the related figures 1- 25 from our "open" publication (Nechaev and Veziroglu, 2013) are referred.

CONSIDERATION OF SOME ENERGETIC CHARACTERISTICS OF THEORETICAL GRAPHANES

In the work of Sofo et al. (2007), the stability of graphane, a fully saturated extended two-dimensional hydrocarbon derived from a single grapheme sheet with formula CH, has been predicted on the basis of the first principles and total-energy calculations. All of the carbon atoms are in sp^3 hybridization forming a hexagonal network (a strongly diamond-like distorted graphene network) and the hydrogen atoms are bonded to carbon on both sides of the plane in an alternative manner. It has been found that graphane can have two favorable conformations: a chair-like (diamond-like, Figure 1) conformer and a boat-like (zigzag-like) conformer (Sofo et al., 2007).

The diamond-like conformer (Figure 1) is more stable than the zigzag-like one. This was concluded from the results of the calculations of binding energy ($\Delta H_{bind.(graphane)}$) (that is, the difference between the total energy of the isolated atoms and the total energy of the compounds), and the standard energy of formation ($\Delta H^0_{f298(graphane)}$) of the compounds ($CH_{(graphane)}$) from crystalline graphite ($C_{(graphite)}$) and gaseous molecular hydrogen ($H_{2(gas)}$) at the standard pressure and temperature conditions (Sofo et al., 2007; Dzhurakhalov and Peeters, 2011).

For the diamond-like graphane, the former quantity is $\Delta H_{bind.(graphane)}$ = 6.56 eV/atom, and the latter one is $\Delta H_1 = \Delta H^0_{f298(graphane)}$ = - 0.15 eV/atom. The latter quantity corresponds to the following reaction:

$$C_{(graphite)} + ½H_{2(gas)} \rightarrow CH_{(graphane)}, \qquad (\Delta H_1) \qquad (1)$$

Where ΔH_1 is the standard energy (enthalpy) change for this reaction.

By using the theoretical quantity of $\Delta H^0_{f298(graphane)}$, one can evaluate, using the framework of the thermodynamic method of cyclic processes (Karapet'yants and

Karapet'yants, 1968; Bazarov, 1976), a value of the energy of formation (ΔH_2) of graphane ($CH_{(graphane)}$) from graphene ($C_{(graphene)}$) and gaseous atomic hydrogen ($H_{(gas)}$). For this, it is necessary to take into consideration the following three additional reactions:

$$C_{(graphene)} + H_{(gas)} \rightarrow CH_{(graphane)}, (\Delta H_2) \quad (2)$$

$$C_{(graphene)} \rightarrow C_{(graphite)}, \quad (\Delta H_3) \quad (3)$$

$$H_{(gas)} \rightarrow ½ H_{2(gas)}, \quad (H_4) \quad (4)$$

where ΔH_2, ΔH_3 and ΔH_4 are the standard energy (enthalpy) changes.

Reaction 2 can be presented as a sum of Reactions 1, 3 and 4 using the framework of the thermodynamic method of cyclic processes (Bazarov, 1976):

$$\Delta H_2 = (\Delta H_3 + \Delta H_4 + \Delta H_1). \quad (5)$$

Substituting in Equation 5 the known experimental values (Karapet'yants and Karapet'yants, 1968; Dzhurakhalov and Peeters, 2011) of ΔH_4 = -2.26 eV/atom and ΔH_3 = -0.05 eV/atom, and also the theoretical value (Sofo et al., 2007) of ΔH_1 = -0.15 эВ/atom, one can obtain a desired value of ΔH_2 = -2.5 ± 0.1 eV/atom. The quantity of $-\Delta H_2$ characterizes the breakdown energy ofC-H sp^3 bond in graphane (Figure 1), relevant to the breaking away of one hydrogen atom from the material, which is $\Delta H_{(C\text{-}H)graphane} = -\Delta H_2$ = 2.5 ± 0.1 eV (Table 1A).

In evaluating the above mentioned value of ΔH_3, one can use the experimental data (Karapet'yants and Karapet'yants, 1968) on the graphite sublimation energy at 298K ($\Delta H_{subl.(graphite)}$ = 7.41 ± 0.05 eV/atom), and the theoretical data (Dzhurakhalov and Peeters, 2011) on the binding cohesive energy at about 0K for graphene ($\Delta H_{cohes.(graphene)}$ = 7.40 eV/atom). Therefore, neglecting the temperature dependence of these quantities in the interval of 0 to 298K, one obtains the value of $\Delta H_3 \approx$ -0.05 eV/atom.

$\Delta H_{cohes.(graphene)}$ quantity characterizes the breakdown energy of1.5 C-C sp^2 bond in graphene, relevant to the breaking away of one carbon atom from the material. Consequently, one can evaluate the breakdown energy ofC-C sp^2 bonds in graphene, which is $\Delta H_{(C\text{-}C)grapheme}$ = 4.93 eV. This theoretical quantity coincides with the similar empirical quantities obtained in (Nechaev and Veziroglu, 2013) from $\Delta H_{subl.(graphite)}$ forC-C sp^2 bonds in graphene and graphite, which are $\Delta H_{(C\text{-}C)graphene} \approx \Delta H_{(C\text{-}C)graphite}$ = 4.94 ± 0.03 eV. The similar empirical quantity for C-C sp^3 bonds in diamond obtained from the diamond sublimation energy $\Delta H_{subl.(diamond)}$ (Karapet'yants and Karapet'yants, 1968) is $\Delta H_{(C\text{-}C)diamond}$ = 3.69 ± 0.02 eV (Nechaev and Veziroglu, 2013).

It is important to note that chemisorption of hydrogen on graphene was studied (Dzhurakhalov and Peeters, 2011) using atomistic simulations, with a second generation reactive empirical bond order of Brenner interatomic potential. As shown, the cohesive energy of graphane (CH) in the ground state is $\Delta H_{cohes.(graphane)}$ = 5.03 eV/atom (C). This results in the binding energy of hydrogen, which is $\Delta H_{(C\text{-}H)graphane}$ = 1.50 eV/atom (Dzhurakhalov and Peeters, 2011) (Table 1A).

The theoretical $\Delta H_{bind.(graphane)}$ quantity characterizes the breakdown energy of one C-H sp^3 bond and 1.5 C-C sp^3 bonds (Figure 1). Hence, by using the above mentioned values of $\Delta H_{bind.(graphane)}$ and $\Delta H_{(C\text{-}H)graphane}$, one can evaluate the breakdown energy ofC-C sp^3 bonds in the theoretical graphane (Sofo et al., 2007), which is $\Delta H_{(C\text{-}C)graphane}$ = 2.7 eV (Table 1). Also, by using the above noted theoretical values of $\Delta H_{cohes.(graphane)}$ and $\Delta H_{(C\text{-}H)graphane}$, one can evaluate similarly the breakdown energy ofC-C sp^3 bonds in the theoretical graphane (Dzhurakhalov and Peeters, 2011), which is $\Delta H_{(C\text{-}C)graphane}$ = 2.35 eV (Table 1A).

CONSIDERATION AND INTERPRETATION OF THE DATA ON DEHYDROGENATION OF THEORETICAL GRAPHANE, COMPARING WITH THE RELATED EXPERIMENTAL DATA

In Openov and Podlivaev (2010) and Elias et al. (2009) the process of hydrogen thermal desorption (TDS) from graphane has been studied using the method of molecular dynamics. The temperature dependence (for T = 1300 - 3000K) of the time ($t_{0.01}$) of hydrogen desorption onset (that is, the time $t_{0.01}$ of removal ~1% of the initial hydrogen concentration $C_0 \approx 0.5$ (in atomic fractions), $-\Delta C/C_0 \approx 0.01$, $C/C_0 \approx 0.99$) from the $C_{54}H_{7(54+18)}$ clustered with 18 hydrogen passivating atoms at the edges to saturate the dangling bonds of sp^3-hybridized carbon atoms have been calculated. The corresponding activation energy of $\Delta H_{(des.)} = E_a$ = 2.46 ± 0.17 eV and the corresponding (temperature independent) frequency factor A = (2.1 ± 0.5) × 10^{17} s^{-1} have also been calculated. The process of hydrogen desorption at T = 1300 - 3000K has been described in terms of the Arrhenius-type relationship:

$$1/t_{0.01} = A \exp(-E_a / k_B T), \quad (6)$$

where k_B is the Boltzmann constant.

Openov and Podlivaev (2010) predicted that their results would not contradict the experimental data (Elias et al., 2009), according to which the nearly complete desorption of hydrogen ($-\Delta C/C_0$ □ 0.9, $C/C_0 \approx 0.1$) from a free-standing graphane membrane (Figure 2B) was achieved by annealing it in argon at T = 723K for 24 h (that is, $t_{0.9(membr.\ [5])\ 723K}$ = 8.6 × 10^4 s). However, as the analysis presented below shows, this declaration (Openov and Podlivaev, 2010) is not enough adequate.

By using Equation (6), Openov and Podlivaev, 2010) evaluated the quantity of $t_{0.01(graphane[4])}$ for T = 300K

($\sim 1 \cdot 10^{24}$ s)and for T = 600K ($\sim 2 \times 10^{3}$ s). However, they noted that the above two values of $t_{0.01(\text{graphane})}$ should be considered as rough estimates. Indeed, using Equation 6, one can evaluate the value of $t_{0.01(\text{graphane}[4])723\text{K}} \approx 0.7$ s for T = 723K, which is much less (by five orders) than the $t_{0.9(\text{membr.}[5])723\text{K}}$ value in Elias et al. (2009).

In the framework of the formal kinetics approximation in the first order rate reaction (Bazarov, 1976) a characteristic quantity for the reaction of hydrogen desorption is $\tau_{0.63}$ - the time of the removal of ~ 63% of the initial hydrogen concentration C_0 (that is, $-\Delta C/C_0 \approx$ 0.63, $C/C_0 \approx 0.37$) from the hydrogenated graphene. Such a first order rate reaction (desorption) can be described by the following equations (Nechaev, 2010; Nechaev and Veziroglu, 2013; Bazarov, 1976):

$$dC / dt = - KC, \tag{7}$$

$$(C / C_0) = \exp(- Kt) = \exp(- t / \tau_{0.63}), \tag{8}$$

$$K = (1/\tau_{0.63}) = K_0 \exp(-\Delta H_{\text{des.}} / k_B T), \tag{9}$$

Where C is the averaged concentration at the annealing time t, $K = (1/\tau_{0.63})$ is the reaction (desorption) rate constant, $\Delta H_{\text{des.}}$is the reaction (desorption) activation energy, and K_0, the per-exponential (or frequency) factor of the reaction rate constant.

In the case of a diffusion rate limiting kinetics, the quantity of K_0 is related to a solution of the corresponding diffusion problem ($K_0 \approx D_0 / L^2$, where D_0 is the per-exponential factor of the diffusion coefficient, L is the characteristic diffusion length) (Nechaev, 2010; Nechaev and Veziroglu, 2013).

In the case of a non-diffusion rate limiting kinetics, which is obviously related to the situation of Openov and Podlivaev (2010) and Elias et al. (2009), the quantity of K_0 may be the corresponding vibration (for (C-H) bonds) frequency ($K_0 = \nu_{(\text{C-H})}$), the quantity $\Delta H_{(\text{des.})} = \Delta H_{(\text{C-H})}$ (Table 1), and Equation (9) corresponds to Polanyi-Wigner (Nechaev, 2010; Nechaev and Veziroglu, 2013).

By substituting in Equation (8) the quantities of $t = t_{0.01(\text{graphane}[4])723\text{K}}$ and (C/C_0) = 0.99, one can evaluate the desired quantity $\tau_{0.63(\text{graphane}[4])723\text{K}} \approx 70$ s. Analogically, the quantity of $t_{0.9(\text{graphane}[4])723\text{K}} \approx 160$ s can be evaluated, which is less by about three orders - than the experimental value (Elias et al., 2009) of $t_{0.9(\text{membr.}[5])723\text{K}}$. In the same manner, one can evaluate the desired quantity $\tau_{0.63(\text{membr.}[5])723\text{K}} \approx 3.8 \times 10^{4}$ s, which is higher (by about three orders) than $\tau_{0.63(\text{graphane}[4])723\text{K}}$.

By using Equation (9) and supposing that $\Delta H_{\text{des.}} = E_a$ and $K = 1/\tau_{0.63(\text{graphane}[4])723\text{K}}$, one can evaluate the analytical quantity of $K_{0(\text{graphane}[4])} = 2 \times 10^{15}$ s^{-1} for graphane of (Openov and Podlivaev, 2010) (Table 1A).

By substituting in Equation (9) the quantity of $K = K_{(\text{membr.}[5])723\text{K}} = 1/\tau_{0.63(\text{membr.}[5])723\text{K}}$ and supposing that $\Delta H_{\text{des.}(\text{membr.}[5])} \approx \Delta H_{\text{C-H}(\text{graphane}[3,4])} \approx 2.5$ eV (Sofo et al., 2007; Nechaev and Veziroglu, 2013; Openov and Podlivaev, 2010) (Table 1A), one can evaluate the quantity of $K_{0(\text{membr.}[5])} = \nu_{(\text{membr.}[5])} \approx 7 \times 10^{12}$ s^{-1} for the experimental graphane membranes of Elias et al. (2009). The obtained quantity of $\nu_{(\text{membr.}[5])}$ is less by one and a half orders of the vibrational frequency $\nu_{\text{RD}} = 2.5 \times 10^{14}$ s^{-1}, corresponding to the D Raman peak (1342 cm^{-1}) for hydrogenated graphene membrane and epitaxial graphene on SiO_2 (Figure 2). The activation of the D Raman peak in the hydrogenated samples authors (Elias et al., 2009) attribute to breaking of the translation symmetry of C-C sp^2 bonds after formation of C-H sp^3 bonds.

The quantity $\nu_{(\text{membr.}[5])}$ is less by one order of the value (Xie et al., 2011) of the vibration frequency $\nu_{\text{HREELS}} = 8.7 \times 10^{13}$ s^{-1} corresponding to an additional HREELS peak arising from C-H sp^3 hybridization; a stretching appears at 369 meV after a partial hydrogenation of the epitaxial graphene. Xie et al. (2011) suppose that this peak can be assigned to the vertical C-H bonding, giving direct evidence for hydrogen attachment on the epitaxial graphene surface.

Taking into account ν_{RD} and ν_{HREELS} quantities, and substituting in Equation (9) quantities of $K = 1/\tau_{0.63(\text{membr.}[5])723\text{K}}$ and $K_0 \approx K_{0(\text{membr.}[5])} \approx \nu_{\text{HREELS}}$, one can evaluate $\Delta H_{\text{des.}(\text{membr.}[5])} = \Delta H_{\text{C-H}(\text{membr.}[5])} \approx 2.66$ eV (Table 1A). In such approximation, the obtained value of $\Delta H_{\text{C-H}(\text{membr.}[5])}$ coincides (within the errors) with the experimental value (Pimenova et al., 2002) of the breakdown energy of C-H bonds in hydrofullerene $C_{60}H_{36}$ ($\Delta H_{\text{C-H}(\text{C60H36})} = 2.64 \pm 0.01$ eV, Table 1B).

The above analysis of the related data shows that the experimental graphene membranes (hydrogenated up to the near-saturation) can be used. The following thermodesorption characteristics of the empirical character, relevant to Equation (9): $\Delta H_{\text{des.}(\text{membr.}[5])} = \Delta H_{\text{C-H}(\text{membr.}[5])} = 2.6 \pm 0.1$ eV, $K_{0(\text{membr.}[5])} = \nu_{\text{C-H}(\text{membr.}[5])} \approx 5 \times 10^{13}$ s^{-1} (Table 1A). The analysis also shows that this is a case for a non-diffusion rate limiting kinetics, when Equation (9) corresponds to Polanyi-Wigner (Nechaev, 2010; Nechaev and Veziroglu, 2013). Certainly, these tentative results could be directly confirmed and/or modified by receiving and treating within Equations (8) and (9) of the experimental data on $\tau_{0.63}$ at several annealing temperatures.

The above noted fact that the empirical (Elias et al., 2009; Nechaev and Veziroglu, 2013) quantity $\tau_{0.63(\text{membr.}[5])723\text{K}}$ is much larger (by about 3 orders), than the theoretical (Openov and Podlivaev, 2010; Nechaev and Veziroglu, 2013) one ($\tau_{0.63(\text{graphane}[4])723\text{K}}$), is consistent with that mentioned in (Elias et al., 2009). The alternative possibility has been supposed in Elias et al., (2009) that (i) the experimental graphane membrane (a free-standing one) may have "a more complex hydrogen bonding, than the suggested by the theory", and that (ii) graphane (CH) (Sofo et al., 2007) may be until now the theoretical material.

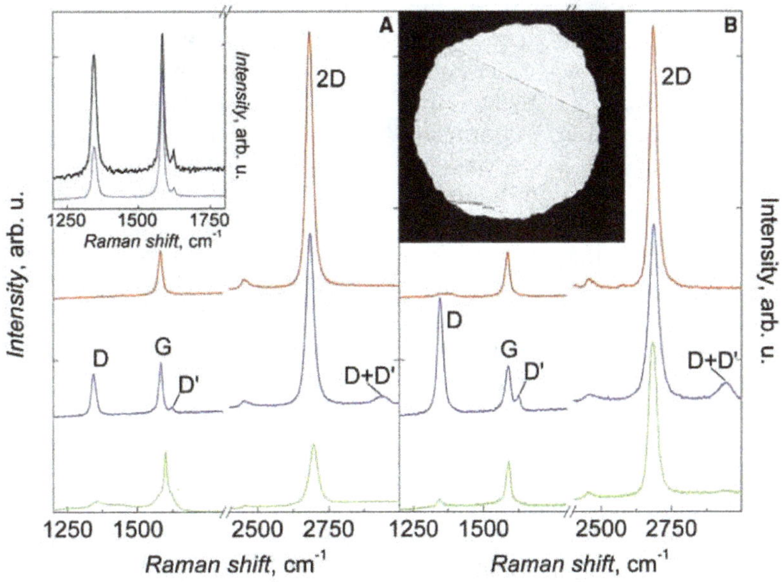

Figure 2. Changes in Raman spectra of graphene caused by hydrogenation (Elias et al., 2009). The spectra are normalized to have a similar integrated intensity of the G peak. (**A**) Graphene on SiO_2. (**B**) Free-standing graphene. Red, blue, and green curves (top to bottom) correspond to pristine, hydrogenated, and annealed samples, respectively. Graphene was hydrogenated for ~2 hours, and the spectra were measured with a Renishaw spectrometer at wavelength 514 nm and low power to avoid damage to the graphene during measurements. (Left inset) Comparison between the evoluation of D and D′ peaks for single- and double-sided exposure to atomic hydrogen. Shown is a partially hydrogenated state achieved after 1 hour of simultaneous exposure of graphene on SiO_2 (blue curve) and of a membrane (black curve). (Right inset) TEM image of one of the membranes that partially covers the aperture 50 µm in diameter.

CONSIDERATION OF THE EXPERIMENTAL DATA ON HYDROGENATION-DEHYDROGENATION OF MONO- AND BI-LAYER EPITAXIAL GRAPHENES, AND COMPARING THE RELATED DATA FOR FREE-STANDING GRAPHENE

Characteristics of hydrogenation-dehydrogenation of mono-layer epitaxial graphenes

In Elias et al. (2009), both the graphene membrane samples considered above, and the epitaxial graphene and bi-graphene samples on substrate SiO_2 were exposed to cold hydrogen DC plasma for 2 h to reach the saturation in the measured characteristics. They used a low-pressure (0.1 mbar) hydrogen-argon mixture of 10% H_2. Raman spectra for hydrogenated and subsequently annealed free-standing graphene membranes (Figure 2B) are rather similar to those for epitaxial graphene samples (Figure 2A), but with some notable differences.

If hydrogenated simultaneously for 1 h, and before reaching the saturation (a partial hydrogenation), the D peak area for a free-standing membrane is two factors greater than the area for graphene on a substrate (Figure 2, the left inset). This indicates the formation of twice as many C-H sp^3 bonds in the membrane. This result also agrees with the general expectation that atomic hydrogen attaches to both sides of the membranes. Moreover, the D peak area became up to about three times greater than the G peak area after prolonged exposures (for 2 h, a near-complete hydrogenation) of the membranes to atomic hydrogen.

The integrated intensity area of the D peak in Figure 2B corresponding to the adsorbed hydrogen saturation concentration in the graphene membranesis larger by a factor of about 3 for the area of the D peak in Figure 2A, corresponding to the hydrogen concentration in the epitaxial graphene samples.

The above noted Raman spectroscopy data (Elias et al., 2009) on dependence of the concentration (C) of adsorbed hydrogen from the hydrogenation time (t) (obviously, at about 300K) can be described with Equation (8) (Xiang et al., 2010; Bazarov, 1976). By using the above noted Raman spectroscopy data (Elias et al., 2009) (Figure 2), one can suppose that the near-saturation ((C/C_0) ≈ 0.95) time ($t_{0.95}$) for the free standing graphene membranes (at ~300K) is about 3 h, and a maximum possible (but not defined experimentally) value of $C_{0(membr.)}$ ≈ 0.5 (atomic fraction, that is, the atomic ratio

(H/C) =1). Hence, using Equation (8)* results in the quantities of $\tau_{0.63(membr.[5])hydr.300K} \approx 1.0$ h, $C_{3h(membr.[5])} \approx 0.475$, $C_{2h(membr.[5])} \approx 0.43$ and $C_{1h(membr.[5])} \approx 0.32$, where, $C_{3h(membr.[5])}$, $C_{2h(membr.[5])}$ and $C_{1h(membr.[5])}$ being the adsorbed hydrogen concentration at the hydrogenation time (t) equal to 3, 2 and 1 h, respectively. It is expedient to note that the quantity of $C_{0(membr.[5])} \approx 0.5$ corresponds to the local concentration of $C_{0(membr.[5]one\_side)} \approx 0.33$ for each of the two sides of a membrane, that is, the local atomic ratio (H/C) = 0.50.

The evaluated value of $\tau_{0.63(membr.[5])hydr.300K}$ (for process of hydrogenation of the free standing graphene membranes (Elias et al., 2009) is much less (by about 26 orders) of the evaluated value of the similar quantity of $\tau_{0.63(membr.[5])dehydr.300K} \approx (0.4 - 2.7) \times 10^{26}$ h (if $\Delta H_{(des.)}$ = (2.49 - 2.61) eV, $K_{0(des.)} = (0.7 -5) \times 10^{13}$ s^{-1}, Table 1A) for process of dehydrogenation of the same free standing graphene membranes (Elias et al., 2009). This shows that the activation energy of the hydrogen adsorption ($\Delta H_{(ads.)}$) for the free standing graphene membranes (Elias et al., 2009) is considerably less than the activation energy of the hydrogen desorption ($\Delta H_{(des.)}$ = (2.5 or 2.6) eV). Hence, by using Equation (9) and supposing that $K_{0(ads.)} \approx K_{0(des.)}$, one can obtain a reasonable value of $\Delta H_{(ads.)membr.[5]} = 1.0 \pm 0.2$ eV (Table 1). The heat of adsorption of atomic hydrogen by the free standing graphene membranes (Elias et al., 2009) may be evaluated as (Nechaev, 2010; Bazarov, 1976): $(\Delta H_{(ads.)membr.[5]} - \Delta H_{(des.)membr.[5]}) = -1.5 \pm 0.2$ eV (an exothermic reaction).

One can also suppose that the near-saturation ($(C/C_0) \approx 0.95$) time ($t_{0.95}$) for the epitaxial graphene samples (at ~300K) is about 2 h. Hence, by using Equation 8 and the above noted data (Elias et al., 2009) on the relative concentrations $[(C_{1h(membr.[5])} / C_{1h(epitax.[5])}) \approx 2$, and $((C_{3h(membr.[5])} / C_{3h(epitax.[5])}) \approx 3]$, one can evaluate the quantities of $\tau_{0.63(epitax.[5])hydr.300K} \approx 0.7$ h and $C_{0(epitax.[5])} \approx 0.16$. Obviously, $C_{0(epitax.[5])}$ is related only for one of the two sides of an epitaxial graphene layer, and the local atomic ration is (H/C) ≈ 0.19. It is considerably less (about 2.6 times) of the above considered local atomic ratio (H/C) = 0.5 for each of two sides the free standing hydrogenated graphene membranes.

The obtained value of $\tau_{0.63(epitax.[5])hydr.300K} \approx 0.7$ h (for process of hydrogenation of the epitaxial graphene samples (Elias et al., 2009) is much less (by about two - seven orders) of the evaluated values of the similar quantity for the process of dehydrogenation of the same epitaxial graphene samples (Elias et al., 2009) ($\tau_{0.63(epitax.[5])dehydr.300K} \approx (1.5 \times 10^2 - 1.0 \times 10^7)$ h, for $\Delta H_{(des.)}$ = (0.3 - 0.9) eV and $K_{0(des.)} = (0.2 - 3.5 \times 10^4)$ s^{-1}, Table 1A). Hence, by using Equation 9 and supposing that $K_{0(ads.)} \approx K_{0(des.)}$ (a rough approximation), one can obtain a reasonable value of $\Delta H_{(ads.)epitax.[5]} \approx 0.3 \pm 0.2$ eV (Table 1A). The heat of adsorption of atomic hydrogen by the free standing graphene membranes (Elias et al., 2009) may be evaluated as (Nechaev, 2010; Bazarov, 1976): $(\Delta H_{(ads.)epitax.[5]} - \Delta H_{(des.)epitax.[5]}) = -0.3 \pm 0.2$ eV (an exothermic reaction).

The smaller values of $C_{0(epitax.[5])} \approx 0.16$ and $(H/C)_{(epitax.[5])} \approx 0.19$ (in comparison with $C_{0(membr.[5]one\_side)} \approx 0.33$ and $(H/C)_{(membr.[5]one\_side)} \approx 0.50$) may point to a partial hydrogenation localized in some defected nanoregions (Brito et al., 2011; Zhang et al., 2014; Banhart et al., 2011; Yazyev and Louie, 2010; Kim et al., 2011; Koepke et al., 2013; Zhang and Zhao, 2013; Yakobson and Ding, 2011; Cockayne et al., 2011; Zhang et al., 2012; Eckmann et al., 2012) for the epitaxial graphene samples (even after their prolonged (3 h) exposures, that is, after reaching their near-saturation. Similar analytical results, relevance to some other epitaxial graphenes are also presented.

Characteristics of dehydrogenation of mono-layer epitaxial graphenes

According to a private communication from D.C. Elias,a near-complete desorption of hydrogen ($-\Delta C/C_0 \approx 0.95$) from a hydrogenated epitaxial graphene on a substrate SiO_2 (Figure 2A) has been achieved by annealing it in 90% Ar/10% H_2 mixture at T = 573K for 2 h (that is, $t_{0.95(epitax.[5])573K} = 7.2 \times 10^3$ s). Hence, by using Equation 8, one can evaluate the value of $\tau_{0.63(epitax.[5])573K} = 2.4 \times 10^3$ s for the epitaxial graphene (Elias et al., 2009), which is about six orders less than the evaluated value of $\tau_{0.63(membr.[5])573K} = 1.5 \times 10^9$ s for the free-standing membranes (Elias et al., 2009).

The changes in Raman spectra of graphene (Elias et al., 2009) caused by hydrogenation were rather similar in respect to locations of D, G, D′, 2D and (D+D′) peaks, both for the epitaxial graphene on SiO_2 and for the free-standing graphene membrane (Figure 2). Hence, one can suppose that $K_{0(epitax.[5])} = \nu_{C\text{-}H(epitax.[5])} \approx K_{0(membr.[5])} = \nu_{C\text{-}H(membr.[5])} \approx (0.7$ or $5) \times 10^{13}$ s^{-1} (Table 1A). Then, by substituting in Equation 9 the values of $K = K_{(epitax.[5])573K} = 1/\tau_{0.63(epitax.[5])573K}$ and $K_0 \approx K_{0(epitax.[5])} \approx K_{0(membr.[5])}$, one can evaluate $\Delta H_{des.(epitax.[5])} = \Delta H_{C\text{-}H(epitax.[5])} \approx$ (1.84 or 1.94) eV (Table 1A). Here, the case is supposed of a non-diffusion-rate-limiting kinetics, when Equation 9 corresponds to thePolanyi-Wigner one (Nechaev, 2010). Certainly, these tentative thermodynamic characteristics of the hydrogenated epitaxial graphene on a substrate SiO_2 could be directly confirmed and/or modified by further experimental data on $\tau_{0.63(epitax.)}$ at various annealing temperatures.

It is easy to show that: 1) these analytical results (for the epitaxial graphene (Elias et al., 2009) are not consistent with the presented below analytical results for the mass spectrometry data (Figure 3, TDS peaks ## 1-3, Table 1A) on TDS of hydrogen from a specially prepared single-side (obviously, epitaxial*) graphane (Elias et al., 2009); and 2) they cannot be described in the framework of the theoretical models and characteristics of thermal

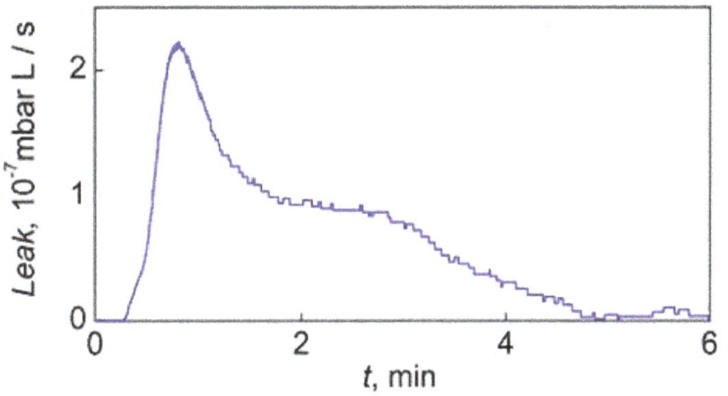

Figure 3. Desorption of hydrogen from single-side graphane (Elias et al., 2009). The measurments were done by using a leak detector tuned to sense molecular hydrogen. The sample was heated to 573 K (the heater was switched on at t = 10 s). Control samples (exposed to pure argon plasma) exhibited much weaker and featureless response (< $5 \cdot 10^{-8}$ mbar L/s), which is attributed to desorption of water at heated surfaces and subtracted from the shown data (water molecules are ionized in the mass-spectrometer, which also gives rise to a small hydrogen signal).

stability of SSHG (Openov and Podlivaev, 2012) or graphone (Podlivaev and Openov, 2011).

According to further consideration presented below (both here and subsequently), the epitaxial graphene case (Elias et al., 2009) may be related to a hydrogen desorption case of a diffusion rate limiting kinetics, when $K_0 \neq \nu$, and Equation (9) does not correspond to the Polanyi-Wigner one (Nechaev, 2010).

By using the method of Nechaev, (2010) of treatment from the TDS spectra, relevant to the mass spectrometry data (Elias et al., 2009) (Figure 3) on TDS of hydrogen from the specially prepared single-side (epitaxial*) graphane (under heating from room temperature to 573K for 6 min), one can obtain the following tentative results:

(1) The total integrated area of the TDS spectra corresponds to ~10^{-8} g of desorbed hydrogen that may correlate with the graphene layer mass (unfortunately, it's not considered in Elias et al. (2009), particularly, for evaluation of the C_0 quantities);
(2) The TDS spectra can be approximated by three thermodesorption (TDS) peaks (# # 1-3);
(3) TDS peak # 1 (~30 % of the total area, $T_{max\#1} \approx 370$ K) can be characterized by the activation energy of $\Delta H_{(des.)} = E_{TDS\text{-}peak\ \#\ 1} = 0.6 \pm 0.3$ eV and by the per-exponential factor of the reaction rate constant $K_{0(TDS\text{-}peak\ \#1)} \approx 2 \cdot 10^7$ s^{-1};
(4) TDS peak # 2 (~15% of the total area, $T_{max\#2} \approx 445$K) can be characterized by the activation energy $\Delta H_{(des.)} = E_{TDS\text{-}peak\ \#2} = 0.6 \pm 0.3$ eV, and by the per-exponential factor of the reaction rate constant $K_{0(TDS\text{-}peak\ \#2)} \approx 1 \times 10^6$ s^{-1};
(5) TDS peak # 3 (~55% of the total area, $T_{max\#3} \approx 540$K) can be characterized by the activation energy $\Delta H_{(des.)} = E_{TDS\text{-}peak\ \#3} = 0.23 \pm 0.05$ eV and by the per-exponential factor of the reaction rate constant $K_{0(TDS\text{-}peak\ \#3)} \approx 2.4$ s^{-1}.

These analytical results (on quantities of $\Delta H_{(des.)}$ and K_0) show that all three of the above noted TDS processes ($\#1_{TDS}$, $\#2_{TDS}$ and $\#3_{TDS}$) can not been described in the framework of the Polanyi-Wigner equation (Nechaev, 2010; Nechaev and Veziroglu, 2013) (due to the obtained low values of the $K_{0(des.)}$ and $\Delta H_{(des.)}$ quantities, in comparison with the $\nu_{(C\text{-}H)}$ and $\Delta H_{(C\text{-}H)}$ ones).

As shown below, these results may be related to a hydrogen desorption case of a diffusion-rate-limiting kinetics (Nechaev, 2010; Nechaev and Veziroglu, 2013), when in Equation (9) the value of $K_0 \approx D_{0app.} / L^2$ and the value of $\Delta H_{des.} = Q_{app.}$, where D_{0app} is the per-exponent factor of the apparent diffusion coefficient $D_{app.} = D_{0app.} \exp(-Q_{app.}/k_B T)$, $Q_{app.}$ is the apparent diffusion activation energy, and L is the characteristic diffusion size (length), which (as shown below) may correlate with the sample diameter (Elias et al., 2009) ($L \sim d_{sample} \approx 4 \times 10^{-3}$ cm, Figure 2, Right inset).

TDS process (or peak) $\#3_{TDS}$ (Figure 3, Table 1A) may be related to the diffusion-rate-limiting TDS process (or peak) I in (Nechaev, 2010), for which the apparent diffusion activation energy is $Q_{app.I} \approx 0.2$ eV $\approx E_{TDS\text{-}peak\#3}$ and $D_{0app.I} \approx 3 \times 10^{-3}$ cm^2/s, and which is related to chemisorption models "F" and/or "G" (Figure 4).

By supposing of $L \sim d_{sample}$, that is, of the order of diameter of the epitaxial graphene specimens (Elias et al., 2009), one can evaluate the quantity of $D_{0app.(TDS\text{-}peak\#3)} \approx L^2 \cdot K_{0(TDS\text{-}peak\#3)} \approx 4 \times 10^{-5}$ cm (or within the errors limit, it is of (1.3 - 11) $\times 10^{-5}$ cm, for $E_{TDS\text{-}peak\ \#3}$ values 0.18 - 0.28 eV, Table 1A). The obtained values of

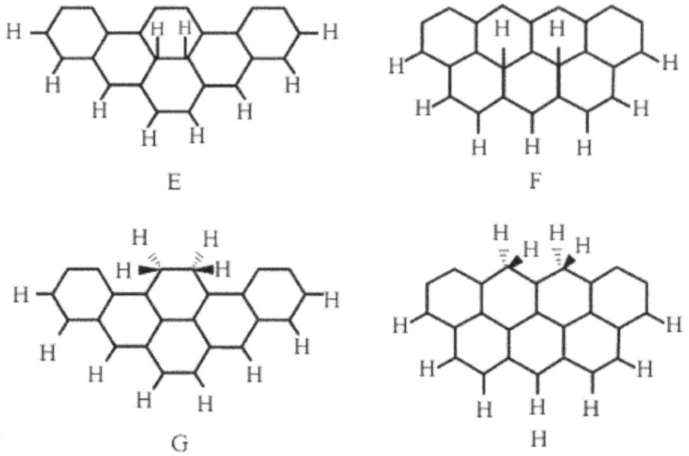

Figure 4. Schematics of some theoretical models (*ab initio* molecular orbital calculations (Yang and Yang, 2002) of chemisorption of atomic hydrogen on graphite on the basal and edge planes.

$D_{0app.(TDS\text{-}peak\#3)}$ satisfactory (within one-two orders, that may be within the errors limit) correlate with the $D_{0app.I}$ quantity. Thus, the above analysis shows that for TDS process (or peak) # 3_{TDS} (Elias et al., 2009), the quantity of L may be of the order of diameter (d_{sample}) of the epitaxial* graphene samples.

Within approach (Nechaev, 2010), model "F" (Figure 4) is related to a "dissociative-associative" chemisorption of molecular hydrogen on free surfaces of graphene layers of the epitaxial samples (Elias et al., 2009). Model "G" (Figure 4) is related, within (Nechaev, 2010) approach, to a "dissociative-associative" chemisorption of molecular hydrogen on definite defects in graphene layers of the epitaxial samples (Elias et al., 2009), for instance, vacancies, grain boundaries (domains) and/or triple junctions (nodes) of the grain-boundary network (Brito et al., 2011; Zhang et al., 2014; Banhart et al., 2011; Yazyev and Louie, 2010; Kim et al., 2011; Koepke et al., 2013; Zhang and Zhao, 2013; Yakobson and Ding, 2011; Cockayne et al., 2011; Zhang et al., 2012; Eckmann et al., 2012), where the dangling carbon bonds can occur.

TDS processes (or peaks) $\#1_{TDS}$ and $\#2_{TDS}$ (Elias et al., 2009) (Table 1A) may be (in some extent) related to the diffusion-rate-limiting TDS processes (or peaks) I and II in (Nechaev, 2010).

Process II is characterized by the apparent diffusion activation energy $Q_{app.II} \approx 1.2$ eV (that is considerably higher of quantities of $E_{TDS\text{-}peak\#1}$ and $E_{TDS\text{-}peak\#2}$) and$D_{0app.II} \approx 1.8 \cdot 10^3$ cm^2/s. It is related to chemisorption model "H" (Figure 4). Within approach (Nechaev, 2010), model "H" is related (as and model "G") to a "dissociative - associative" chemisorption of molecular hydrogen on definite defects in graphene layers of the epitaxial samples (Elias et al., 2009), for instance, vacancies, grain boundaries (domains) and/or triple junctions (nodes) of the grain-boundary network noted above, where the dangling carbon bonds can occur.

By supposing the possible values of $E_{TDS\text{-}peaks\#\#1,2} = 0.3$, 0.6 or 0.9 eV, one can evaluate the quantities of $K_{0(TDS\text{-}peak\#1)}$ and $K_{0(TDS\text{-}peak\#2)}$ (Table 1A). Hence, by supposing of $L \sim d_{sample}$, one can evaluate the quantities of $D_{0app.(TDS\text{-}peak\#1)}$ and $D_{0app.(TDS\text{-}peak\#2)}$, some of them correlatewith the $D_{0app.I}$ quantity or with $D_{0app.II}$ quantity. It shows that for TDS processes (or peaks) $\#1_{TDS}$ and $\#2_{TDS}$ (Elias et al., 2009), the quantity of L may be of the order of diameter of the epitaxial* graphene samples.

For the epitaxial graphene (Elias et al., 2009) case, supposing the values of $\Delta H_{des.(epitax.[5])} \approx 0.3$, 0.6 or 0.9 eV results in relevant values of $K_{0(epitax.[5])}$ (Table 1A). Hence, by supposing of $L \sim d_{sample}$, one can evaluate the quantities of $D_{0app.(epitax.[5])}$, some of them correlate with the $D_{0app.I}$ quantity or with $D_{0app.II}$ quantity. It shows that for these two processes, the quantity of L also may be of the order of diameter of the epitaxial graphene samples (Elias et al., 2009).

It is important to note that chemisorption of atomic hydrogen with free-standing graphane-like membranes (Elias et al., 2009) and with the theoretical graphanes may be related to model "F*" considered in (Nechaev, 2010). Unlike model "F" (Figure 4), where two hydrogen atoms are adsorbed by two alternated carbon atoms in a graphene-like network, in model "F*" a single hydrogen atom is adsorbed by one of the carbon atoms (in the graphene-like network) possessing of 3 unoccupied (by hydrogen) nearest carbons. Model "F*" is characterized (Nechaev, 2010) by the quantity of $\Delta H_{(C\text{-}H)\text{"}F^{*}\text{"}} \approx 2.5$ eV, which coincides (within the errors) with the similar quantities ($\Delta H_{(C\text{-}H)}$) for graphanes (Table 1A). As also shown in the previous paper parts, the dehydrogenation processes in graphanes (Elias et al., 2009; Openov and Podlivaev, 2010) may be the case of a non-diffusion rate limiting kinetics, for which the quantity of K_0 is the corresponding vibration frequency ($K_0 = \nu$), and Equation (9) is correspond to the Polanyi-Wigner one.

On the other hand, model "F*" is manifested in the diffusion-rate-limiting TDS process (or peak) III in (Nechaev, 2010) (Table 1B), for which the apparent diffusion activation energy is $Q_{app.III} \approx 2.6$ eV $\approx \Delta H_{(C\text{-}H)\text{"F*"}}$ and $D_{0app.III} \approx 3 \times 10^{-3}$ cm^2/s. Process III is relevant to a dissociative chemisorption of molecular hydrogen between graphene-like layers in graphite materials (isotropic graphite and nanostructured one) and nanomaterials – GNFs (Nechaev, 2010) (Table 1B).

It is expedient also to note about models "C" and "D", those manifested in the diffusion-rate-limiting TDS process (or peak) IV in (Nechaev, 2010) (Table 1B), for which the apparent diffusion activation energy is $Q_{app.IV} \approx 3.8$ eV $\approx \Delta H_{(C\text{-}H)\text{"C","D"}}$ and $D_{0app.IV} \approx 6 \times 10^2$ cm^2/s. Process IV is relevant to a dissociative chemisorption of molecular hydrogen in defected regions in graphite materials (isotropic graphite, pyrolytic graphane and nanostructured one) (Nechaev, 2010) (Table 1B).

But such processes (III and IV) have not manifested, when the TDS annealing of the hydrogenated epitaxial graphene samples (Elias et al., 2009) (Figure 3), unlike some hydrogen sorption processes in epitaxial graphenes and graphite samples considered in some next parts of this paper.

An interpretation of characteristics of hydrogenation-dehydrogenation of mono-layer epitaxial graphenes

The above obtained values (Table 1A and B) of characteristics of dehydrogenation of mono-layer epitaxial graphene samples (Elias et al., 2009) can be presented as follows: $\Delta H_{des.} \sim Q_{app.I}$ or $\sim Q_{app.II}$ (Nechaev, 2010), $K_{0(des.)} \sim (D_{0app.I} / L^2)$ or $\sim (D_{0app.II} / L^2)$ (Nechaev, 2010), $L \sim d_{sample}$, that is, being of the order of diameter of the epitaxial graphene samples. And it is related to the chemisorption models "F", "G" and/or "H" (Figure 4).

These characteristics unambiguously point that in the epitaxial graphene samples (Elias et al., 2009), there are the rate-limiting processes (types of I and/or II (Nechaev, 2010) of diffusion of hydrogen, mainly, from chemisorption "centers" [of "F", "G" and/or "H" types (Figure 4)] localized on the internal graphene surfaces (and/or in the graphene/substrate interfaces) to the frontier edges of the samples. It corresponds to the characteristic diffusion length ($L \sim d_{sample}$) of the order of diameter of the epitaxial graphene samples, which, obviously, cannot be manifested for a case of hydrogen desorption processes from the external graphene surfaces. Such interpretation is direct opposite, relevance to the interpretation of Elias et al. (2009) and a number of others, those probably believe in occurrence of hydrogen desorption processes, mainly, from the external epitaxial graphene surfaces. Such different (in some sense, extraordinary) interpretation is consisted with the above analytical data (Table 1A) on activation energies of hydrogen adsorption for the epitaxial graphene samples ($\Delta H_{(ads.)epitax.[5]} \approx 0.3 \pm 0.2$ eV), which is much less than the similar one for the free standing graphene membranes (Elias et al., 2009) ($\Delta H_{(ads.)membr.[5]} = 1.0 \pm 0.2$ eV). It may be understood for the case of chemisorotion [of "F", "G" and/or "H" types (Figure 4)] on the internal graphene surfaces [neighboring to the substrate (SiO_2) surfaces], which obviously proceeds without the diamond-like strong distortion of the graphene network, unlike graphene (Sofo et al., 2007).

Such an extraordinary interpretation is also consisted with the above analytical results about the smaller values of $C_{0(epitax.[5])} \approx 0.16$ and $(H/C)_{(epitax.[5])} \approx 0.19$, in comparison with $C_{0(membr.[5]one\_side)} \approx 0.33$ and $(H/C)_{(membr.[5]one\_side)} \approx 0.50$. It may point to an "internal" (in the above considered sense) local hydrogenation in the epitaxial graphene layers. It may be, for instance, an "internal" hydrogenation localized, mainly, in some defected nanoregions (Brito et al., 2011; Zhang et al., 2014; Banhart et al., 2011; Yazyev and Louie, 2010; Kim et al., 2011; Koepke et al., 2013; Zhang and Zhao, 2013; Yakobson and Ding, 2011; Cockayne et al., 2011; Zhang et al., 2012; Eckmann et al., 2012), where their near-saturation may be reached after prolonged (3 h) exposures.

On the basis of the above analytical results, one can suppose that a negligible hydrogen adsorption by the external graphene surfaces (in the epitaxial samples of Elias et al., 2009) is exhibited. Such situation may be due to a much higher rigidity of the epitaxial graphenes (in comparison with the free standing graphene membranes), that may suppress the diamond-like strong distortion of the graphene network attributed for graphene of Sofo et al. (2007). It may result (for the epitaxial graphenes of Elias et al. (2009) in disappearance of the hydrogen chemisorption with characteristics of $\Delta H_{(ads.)membr.[5]}$ and $\Delta H_{(des.)membr.[5]}$ (Table 1A) manifested in the case of the free standing graphene membranes of Elias et al. (2009). And the hydrogen chemisorption with characteristics of $\Delta H_{(ads.)epitax.[5]}$ and ($\Delta H_{(des.)epitax.[5]}$ (Table 1A) by the external graphene surfaces, in the epitaxial samples of Elias et al. (2009), is not observed, may be, due to a very fast desorption kinetics, unlike the kinetics in the case of the internal graphene surfaces.

Certainly, such an extraordinary interpretation also needs in a reasonable explanation of results (Figure 2) the fact that the changes in Raman spectra of graphene of Elias et al. (2009) caused by hydrogenation were rather similar with respect to locations of D, G, D′, 2D and (D+D′) peaks, both for the epitaxial graphene on SiO_2 and for the free-standing graphene membrane.

An interpretation of the data on hydrogenation of bi-layer epitaxial graphenes

In Elias et al. (2009), the same hydrogenation procedures

of the 2 h long expositions have been applied also for bi-layer epitaxial graphene on SiO_2/Si wafer. Bi-layer samples showed little change in their charge carrier mobility and a small D Raman peak, compared to the single-layer epitaxial graphene on SiO_2/Si wafer exposed to the same hydrogenation procedures. Elias et al. (2009) believe that higher rigidity of bi-layers suppressed their rippling, thus reducing the probability of hydrogen adsorption.

But such an interpretation (Elias et al., 2009) does not seem adequate, in order to take into account the above, and below (next parts of this paper) the presented consideration and interpretation of a number of data.

By using the above extraordinary interpretation, and results on characteristics ($Q_{app.III} \approx 2.6$ eV, $D_{0app.III} \approx 3 \times 10^{-3}$ cm^2/s (Table 1B) of a rather slow diffusion of atomic hydrogen between neighboring graphene-like layers in graphitic materials and nanostructures (process III, model "F*" (Nechaev, 2010), one can suppose a negligible diffusion penetration of atomic hydrogen between the two graphene layers in the bi-layer epitaxial samples of Elias et al. (2009) (during the hydrogenation procedures of the 2 h long expositions, obviously, at $T \approx 300$K). Indeed, by using values of $Q_{app.III}$ and$D_{0app.III}$, one can estimate the characteristic diffusion size (length) $L \sim 7 \times 10^{-22}$ cm, which points to absence of such diffusion penetration.

In the next next parts of this study, a further consideration of some other known experimental data on hydrogenation and thermal stability characteristics of mono-layer, bi-layer and three-layer epitaxial graphene systems is given, where (as shown) an important role plays some defects found in graphene networks (Brito et al., 2011; Zhang et al., 2014; Banhart et al., 2011; Yazyev and Louie, 2010; Kim et al., 2011; Koepke et al., 2013; Zhang and Zhao, 2013; Yakobson and Ding, 2011; Cockayne et al., 2011; Zhang et al., 2012; Eckmann et al., 2012), relevant to the probability of hydrogen adsorption and the permeability of graphene networks for atomic hydrogen.

Consideration and interpretation of the Raman spectroscopy data on hydrogenation-dehydrogenation of graphene flakes, the scanning tunneling microscopy/ scanning tunnelingspectroscopy (STM/STS) data on hydrogenation-dehydrogenation of epitaxial graphene and graphite (HOPG) surfaces and the high-resolution electron energy loss spectroscopy/low-energy electron diffraction (HREELS/LEED) data on dehydrogenation of epitaxial graphene on SiC substrate

In Wojtaszek et al. (2011), it is reported that the hydrogenation of single and bilayer graphene flakes by an argon-hydrogen plasma produced a reactive ion etching (RIE) system. They analyzed two cases: One where the graphene flakes were electrically insulated from the chamber electrodes by the SiO_2 substrate, and the other where the flakes were in electrical contact with the source electrode (a graphene device). Electronic transport measurements in combination with Raman spectroscopy were used to link the electric mean free path to the optically extracted defect concentration, which is related to the defect distance ($L_{def.}$). This showed that under the chosen plasma conditions, the process does not introduce considerable damage to the graphene sheet, and that a rather partial hydrogenation ($C_H \leq$ 0.05%) occurs primarily due to the hydrogen ions from the plasma, and not due to fragmentation of water adsorbates on the graphene surface by highly accelerated plasma electrons. To quantify the level of hydrogenation, they used the integrated intensity ratio (I_D/I_G) of Raman bands. The hydrogen coverage (C_H) determined from the defect distance ($L_{def.}$) did not exceed ~ 0.05%.

In Nechaev and Veziroglu (2013), the data (Wojtaszek et al., 2011) (Figure 5) has been treated and analyzed. The obtained analytical results (Table 2) on characteristics of hydrogenation-dehydrogenation of graphene flakes (Wojtaszek et al., 2011) may be interpreted within the models used for interpretation of the similar characteristics for the epitaxial graphenes of Elias et al. (2009) (Table 1A), which are also presented (for comparing) in Table 2.

By taking into account the fact that the RIE exposure regime (Wojtaszek et al., 2011) is characterized by a form of $(I_D/I_G) \sim L_{def.}^{-2}$ (for $(I_D/I_G) < 2.5$), $L_{def.} \approx 11 - 17$ nm and the hydrogen concentration $C_H \leq 5 \times 10^4$, one can suppose that the hydrogen adsorption centers in the single graphene flakes (on the SiO_2 substrate) are related in some point, nanodefects (that is, vacancies and/or triple junctions (nodes) of the grain-boundary network) of diameter $d_{def.} \approx$ const. In such a model, the quantity C_H can be described satisfactory as:

$$C_H \approx n_H \, (d_{def.})^2 \, / \, (L_{def.})^2, \tag{10}$$

Where $n_H \approx$ const. is the number of hydrogen atoms adsorbed by a center; $C_H \sim (I_D/I_G) \sim L_{def.}^{-2}$.

It was also found (Wojtaszek et al., 2011) that after the Ar/H_2 plasma exposure, the (I_D/I_G) ratio for bi-layer graphene device is larger than that of the single graphene device. As noted in (Wojtaszek et al. (2011), this observation is in contradiction to the Raman ratios after exposure of graphene to atomic hydrogen and when other defects are introduced. Such a situation may have place in Elias et al. (2009) for bi-layer epitaxial graphene on SiO_2/Si wafer.

In Castellanos-Gomez (2012) and Wojtaszek et al. (2012), the effect of hydrogenation on topography and electronic properties of graphene grown by CVD on top of a nickel surface and HOPG surfaces were studied by scanning tunneling microscopy (STM) and scanning

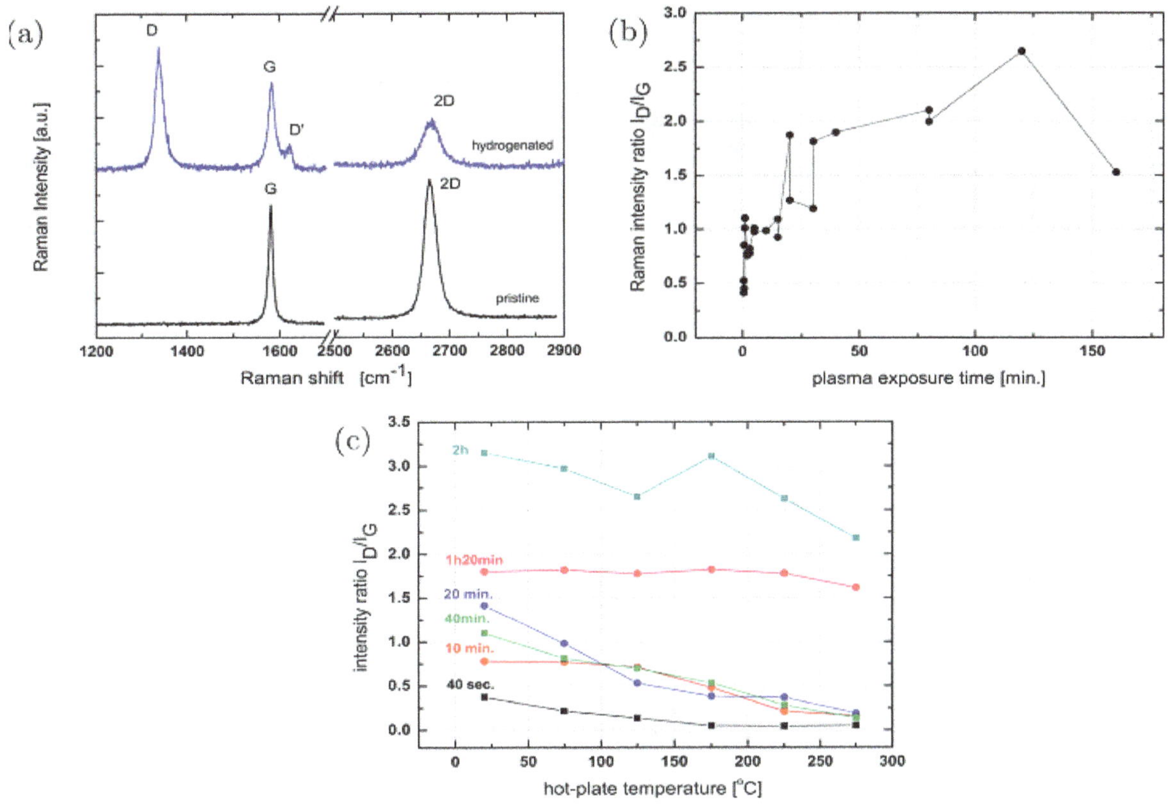

Figure 5. (a) Raman spectrum of pristine single layer graphene – SLG (black) and after 20 min of exposure to the Ar/H_2 plasma (blue) (Wojtaszek et al., 2011). Exposure induces additional Raman bands: a D band around 1340 cm^{-1} and a weaker D′ band around 1620 cm^{-1}. The increase of FWHM of original graphene bands (G, 2D) is apparent. (b) Integrated intensity ratio between the D and G bands (I_D/I_G) of SLG after different Ar/H_2 plasma exposure times. The scattering of the data for different samples is attributed to the floating potential of the graphene flake during exposure. (c) The change of the I_D/I_G ratio of exposed flakes under annealing on hot-plate for 1 min. The plasma exposure time for each flake is indicated next to the corresponding I_D/I_G values. In flakes exposed for less than 1 h the D band could be almost fully suppressed (I_D/I_G < 0.2), which confirms the hydrogen-type origin of defects. In longer exposed samples (80 min and 2 h), annealing does not significantly reduce I_D/I_G, which suggests a different nature of defects, e.g., vacancies.

tunneling spectroscopy (STS). The surfaces were chemically modified using 40 min Ar/H_2 plasma (with 3 W power) treatment (Figure 6) average an energy band gap of 0.4 eV around the Fermi level. Although the plasma treatment modifies the surface topography in an irreversible way, the change in the electronic properties can be reversed by moderate thermal annealing (for 10 min at 553K), and the samples can be hydrogenated again to yield a similar, but slightly reduced, semiconducting behavior after the second hydrogenation. The data (Figure 6) show that the time of desorption from both the epitaxial graphene/Ni samples and HOPG samples of about 90 to 99% of hydrogen under 553K annealing is $t_{0.9(des.)553K}$ (or $t_{0.99(des.)553K}$) ≈ 6 × 10^2 s. Hence, by using Equation (8), one can evaluate the quantity $\tau_{0.63(des.)553K[52]}$≈ 260 (or 130) s, which is close (within the errors) to the similar quantity of $\tau_{0.63(des.)553K[51]}$≈ 70 s for the epitaxial graphene flakes (Wojtaszek et al., 2011) (Table 2).

The data (Figure 6) also show that the time of adsorption (for both the epitaxial graphene/Ni samples and HOPG samples) of about 90 to 99% of the saturation hydrogen amount (under charging at about 300K) is $t_{0.9(ads.)300K}$ (or $t_{0.99(ads.)300K}$) ≈ 2.4 × 10^3 s. Hence, by using Equation (8)*, one can evaluate the quantity $\tau_{0.63(ads.)300K[52]}$≈ (1.1 or 0.5) × 10^2 s, which coincides (within the errors) with the similar quantity of $\tau_{0.63(ads.)300K[51]}$≈ 9 × 10^2 s for the epitaxial graphene flakes (Wojtaszek et al., 2011) (Table 2).

The data (Figure 6) also show that the time of adsorption (for both the epitaxial graphene/Ni samples and HOPG samples) of about 90 - 99% of the saturation hydrogen amount (under charging at about 300K) is $t_{0.9(ads.)300K}$ (or $t_{0.99(ads.)300K}$) ≈ 2.4 × 10^3 s. Hence, by using Equation (8)*, one can evaluate the quantity $\tau_{0.63(ads.)300K[52]}$≈ (1.1 or 0.5) × 10^2 s, which coincides (within the errors) with the similar quantity of $\tau_{0.63(ads.)300K[51]}$≈ 9 × 10^2 s for the epitaxial graphene

Table 2. Analytical values of some related quantities.

Material	Value/Quantity		
	$\Delta H_{(des.)}$, eV {$\Delta H_{(ads.)}$, eV}	$K_{0(des.)}$, s^{-1} {$L \approx (D_{0app.III}/K_{0(des.)})^{1/2}$}	$\tau_{0.63(des.)553K}$, s {$\tau_{0.63(ads.)300K}$, s}
Graphene flakes/SiO_2 (Wojtaszek et al., 2011)	0.11 ± 0.07 (as process ~ I, ~ models "F", "G", Figure 4) {0.1 ± 0.1}	0.15 (for 0.11 eV) {$L \sim d_{sample}$}	0.7×10^2 {0.9×10^3}
Graphene/Ni HOPG (Castellanos-Gomez et al., 2012)			1.3×10^2 - 2.6×10^2 {0.5×10^3 - 1.0×10^3} 1.3×10^2 - 2.6×10^2 {0.5×10^3 - 1.0×10^3}
SiC-D/QFMLG-H (Bocquet et al., 2012)	0.7 ± 0.2 (as processes ~ I - II, ~ model "G", Figure 4)	9×10^2 (for 0.7 eV) {$L \sim d_{sample}$}	2.7×10^3
SiC-D/QFMLG (Bocquet et al., 2012)	2.0 ± 0.6 2.6 (as process ~ III, ~model "F*")	1×10^6 (for 2.0 eV) 6×10^8 (for 2.6 eV) {$L \approx 22$ nm}	1.7×10^{12} 8×10^{14}
Graphene/SiO_2 (Elias et al., 2009) (Table 1A)	If 0.3 if 0.6 if 0.9 (as processes ~ I-II, ~model "G", Figure 4) {0.3 ± 0.2}	then 0.2 then 0.8×10^2 then 3.5×10^4 {$L \sim d_{sample}$}	0.3×10^2 3.7×10^3 4.6×10^3 {2.5×10^3}
Graphene*/SiO_2 (TDS-peak #3) (Elias et al., 2009) (Table 1A)	0.23 ± 0.05 (as process ~ I, ~ models "F", "G", Figure 4)	2.4(for 0.23 eV) {$L \sim d_{sample}$}	0.5×10^2
Graphene*/SiO_2 (TDS-peak #2) (Elias et al., 2009) (Table 1A)	0.6 ± 0.3 (as processes ~ I - II, ~ model "G", Figure 4)	1×10^6 (for 0.6 eV) {$L \sim d_{sample}$}	0.3
Graphene*/SiO_2 (TDS-peak #1) (Elias et al., 2009) (Table 1A)	0.6 ± 0.3 (as processes ~ I - II, ~ model "G", Figure 4)	2×10^7 (for 0.6 eV) {$L \sim d_{sample}$}	1.5×10^{-2}

flakes (Wojtaszek et al., 2011) considered previously (Table 2).

These analytical results on characteristics of hydrogenation-dehydrogenation of epitaxial graphene and graphite surfaces (Castellanos-Gomez et al., 2012; Wojtaszek et al., 2012) (also as the results forgraphene flakes (Wojtaszek et al., 2011) presented previously) may be interpreted within the models used for interpretation of the similar characteristics for the epitaxial graphenes (Elias et al., 2009) (Tables 1 and 2).

As noted in Castellanos-Gomes et al. (2012) and Arramel et al. (2012), before the plasma treatment, the CVD graphene exhibits a Moiré pattern superimposed to the honeycomb lattice of graphene (Figure 6d). This is due to the lattice parameter mismatch between the graphene and the nickel surfaces, and thus the characteristics of the most of the epitaxial graphene samples. On the other hand, as is also noted in Castellanos-Gomes et al. (2012) and Arramel et al., 2012), for the hydrogenated CVD graphene, the expected

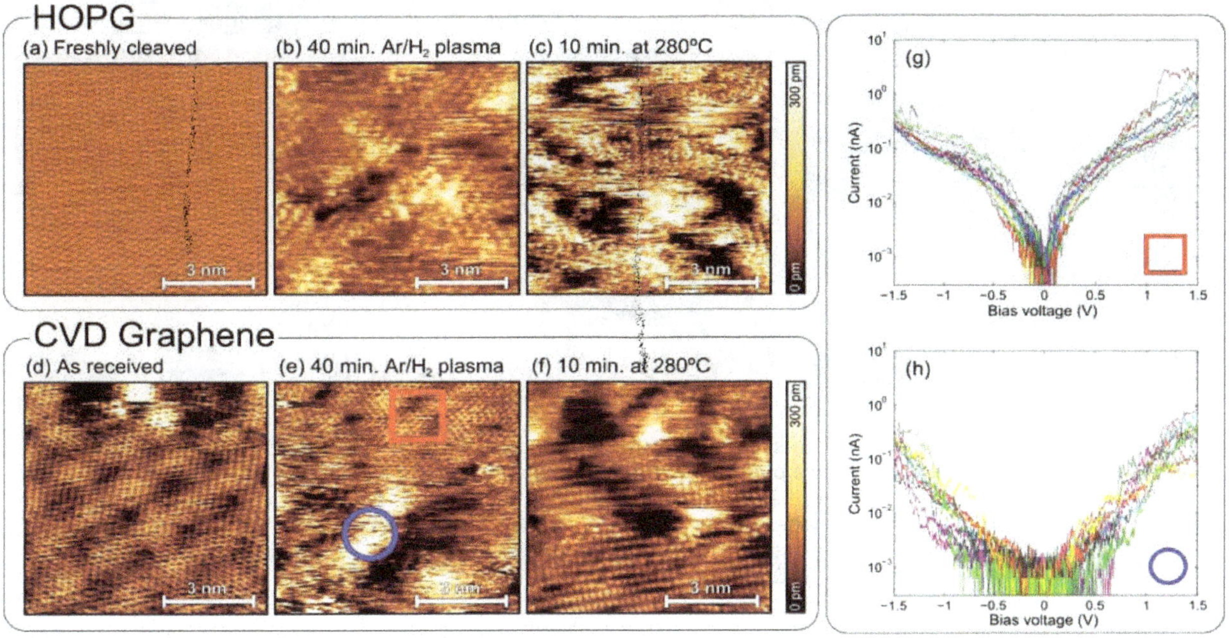

Figure 6. (a-f) Topography images acquired in the constant-current STM mode (Castellanos-Gomez, Wojtaszek et al., 2012): (a-c) HOPG, d-f) graphene grown by CVD on top of a nickel surface at different steps of the hydrogenation/dehydrogenation process. a,d) Topography of the surface before the hydrogen plasma treatment. For the HOPG, the typical triangular lattice can be resolved all over the surface. For the CVD graphene, a Moiré pattern, due to the lattice mismatch between the graphene and the nickel lattices, superimposed onto the honeycomb lattice is observed. b,e) After 40 min of Ar/H_2 plasma treatment, the roughness of the surfaces increases. The surfaces are covered with bright spots where the atomic resolution is lost or strongly distorted. c,f) graphene surface after 10 min of moderate annealing; the topography of both the HOPG and CVD graphene surfaces does not fully recover its original crystallinity. g) Current-voltage traces measured for a CVD graphene sample in several regions with pristine atomic resolution, such as the one marked with the red square in (e). h) The same as (g) but measured in several bright regions, such as the one marked with the blue circle in (e), where the atomic resolution is distorted.

structural changes are twofold. First, the chemisorption of hydrogen atoms will change the sp^2 hybridization of carbon atoms to tetragonal sp^3 hybridization, modifying the surface geometry. Second, the impact of heavy Ar ions, present in the plasma, could also modify the surface by inducing geometrical displacement of carbon atoms (rippling graphene surface) or creating vacancies and other defects (for instance, grain or domain boundaries (Brito et al., 2011; Zhang et al., 2014; Banhart et al., 2011; Yazyev and Louie, 2010; Kim et al., 2011; Koepke et al., 2013; Zhang and Zhao, 2013; Yakobson and Ding, 2011; Cockayne et al., 2011; Zhang et al., 2012; Eckmann et al., 2012). Figure 6e shows the topography image of the surface CVD graphene after the extended (40 min) plasma treatment. The nano-order-corrugation increases after the treatment, and there are brighter nano-regions (of about 1 nm in height and several nm in diameter) in which the atomic resolution is lost or strongly distorted. It was also found (Castellanos-Gomez, Wojtaszek et al., 2012; Castellanos-Gomes, Arramel et al., 2012) that these bright nano-regions present a semiconducting behavior, while the rest of the surface remains conducting (Figure 6g to h).

It is reasonable to assume that most of the chemisorbed hydrogen is localized into these bright nano-regions, which have a blister-like form. Moreover, it is also reasonable to assume that the monolayer (single) graphene flakes on the Ni substrate are permeable to atomic hydrogen only in these defected nano-regions. This problem has been formulated in Introduction. A similar model may be valid and relevant for the HOPG samples (Figure 6a to c).

It has been found out that when graphene is deposited on a SiO_2 surface (Figures 7 and 8) the charged impurities presented in the graphene/substrate interface produce strong inhomogeneities of the electronic properties of graphene.On the other hand, it has also been shown how homogeneous graphene grown by CVD can be altered by chemical modification of its surface by the chemisoption of hydrogen. It strongly depresses the local conductance at low biases, indicating the opening of a band gap in graphene (Castellanos-Gomes, Arramel et al., 2012; Castellanos-Gomez, Smit et al., 2012).

The charge inhomogeneities (defects) of epitaxial hydrogenated graphene/SiO_2 samples do not show long range ordering, and the mean spacing between them is

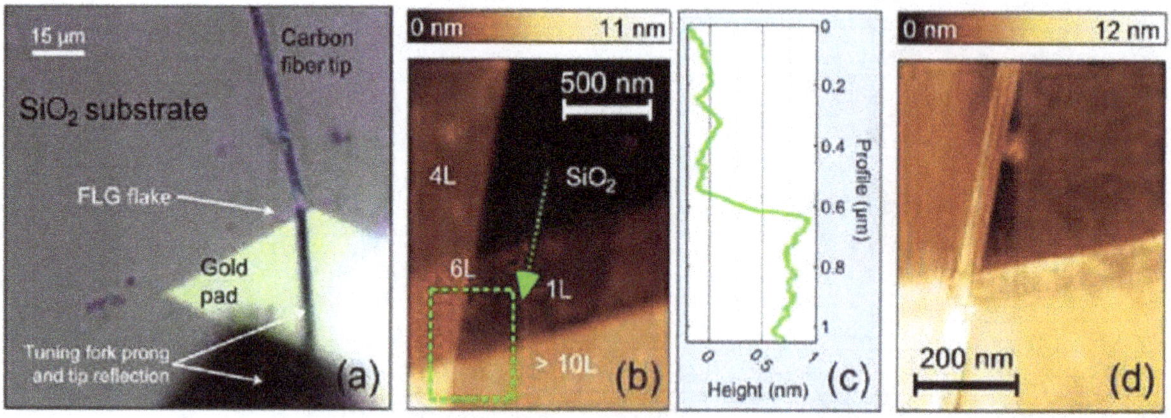

Figure 7. (a) Optical image of the coarse tip positioning on a few-layers graphene flake on the SiO_2 substrate, (b) AFM topography image of the interface between the few-layers graphene flake and the the SiO_2 substrate and areas with different number of layers (labeled as >10, 6, 4 and 1 L) are found, (c) Topographic line profile acquired along the dotted line in (b), showing the interface between the SiO_2 substrate and a monolayer (1L) graphene region, and (d) STM topography image of the regions marked by the dashed rectangle in (b) (Castellanos-Gomes, 2012; Arramel et al., 2012; Castellanos-Gomez, 2012; Smit et al., 2012).

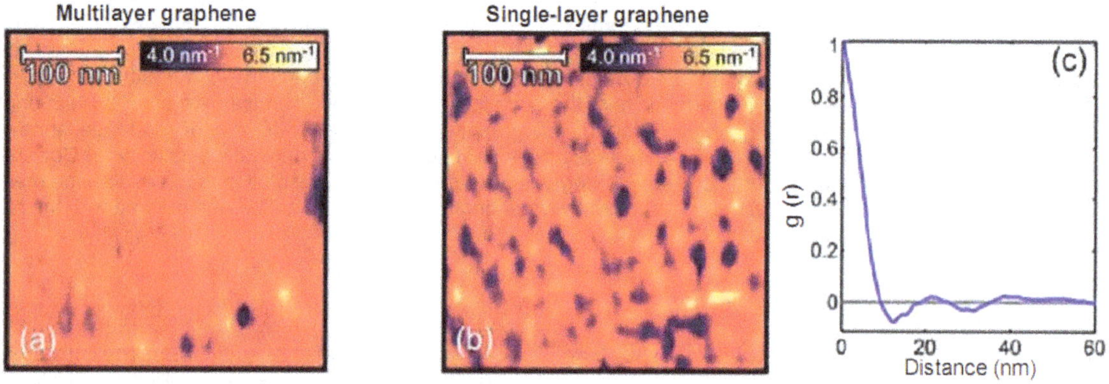

Figure 8. (a) and (b) show the local tunneling decay constant maps measured on a multilayer and a single-layer (1 L) region, respectively. (c) Radial autocorrelation function of the local tunneling decay image in (b) (Castellanos-Gomes, 2012; Arramel et al., 2012; Castellanos-Gomez, 2012; Smit et al., 2012).

$L_{def.} \approx$ 20 nm (Figure 8). It is reasonable to assume that the charge inhomogeneities (defects) are located at the interface between the SiO_2 layer (300 nm thick) and the graphene flake (Castellanos-Gomes, 2012; Arramel et al., 2012; Smit et al., 2012). A similar quantity[$L_{def.} \approx$ 11 - 17 nm, (Wojtaszek et al., 2011) for the hydrogen adsorption centers in the monolayer graphene flakes on the SiO_2 substrate has been above considered.

In Bocquet et al. (2012), hydrogenation of deuterium-intercalated quasi-free-standing monolayer graphene on SiC(0001) was obtained and studied with LEED and HREELS. While the carbon honeycomb structure remained intact, it has shown a significant band gap opening in the hydrogenated material. Vibrational spectroscopy evidences for hydrogen chemisorption on the quasi-free-standing graphene has been provided and its thermal stability has been studied (Figure 9). Deuterium intercalation, transforming the buffer layer in quasi-free-standing monolayer graphene (denoted as SiC-D/QFMLG), has been performed with a D atom exposure of ~5×10^{17} cm^{-2} at a surface temperature of 950K. Finally, hydrogenation up to saturation of quasi-free-standing monolayer graphene has been performed at room temperature with H atom exposure > 3×10^{15} cm^{-2}. The latter sample has been denoted as SiC-D/QFMLG-H to stress the different isotopes used.

According to a private communication from R. Bisson, the temperature indicated at each point in Figure 9 corresponds to successive temperature ramp (not linear) of 5 min. Within a formal kinetics approach for the first order reactions (Nechaev, 2010; Bazarov, 1976), one can

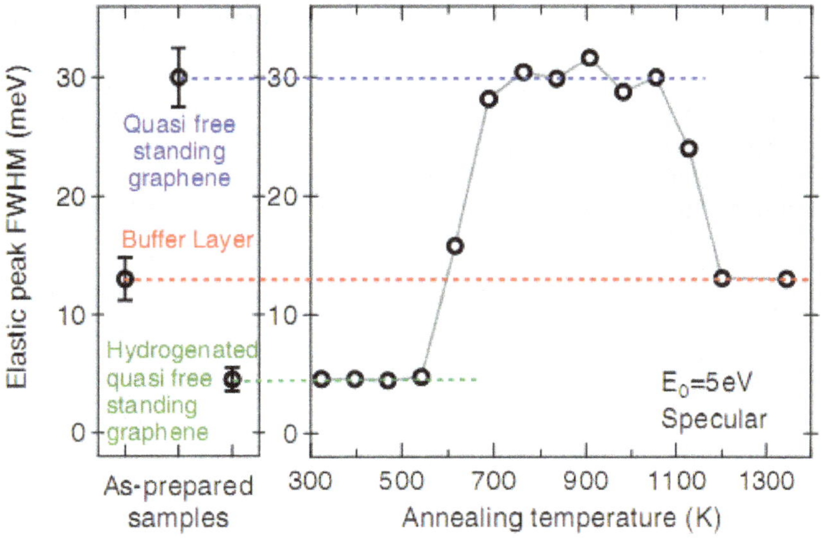

Figure 9. Evaluation of the HREELS elastic peak FWHM of SiC-D/QFMLG-H upon annealing. The uncertain annealing temperature is estimated to be ±5 %. Error bars represent the $\pm\sigma$ variation of FWHM measured across the entire surface of several samples (Bocquet et al., 2012).

treat the above noted points at T_i = 543, 611 and 686 K, by using Equation (8) transformed to a more suitable form (8′): $K_i \approx -(ln(C/C_{0i})/t)$, where t = 300 s, and the corresponding quantities C_{0i} and C are determined from Figure 9. It resulted in finding values of the reaction (hydrogen desorption from SiC-D/QFMLG-H samples) rate constant $K_{i(des.)}$ for 3 temperatures: T_i = 543, 611 and 686K. The temperature dependence is described by Equation (9). Hence, the desired quantities have been determined (Table 2) as the reaction (hydrogen desorption) activation energy $\Delta H_{(des.)(SiC-D/QFMLG-H)[55]}$= 0.7 ± 0.2 eV, and the per-exponential factor of the reaction rate constant $K_{0(des.)(SiC-D/QFMLG-H)[55]} \approx 9 \times 10^2\ s^{-1}$. The obtained value of $\Delta H_{(des.)(SiC-D/QFMLG-H)[55]}$ is close (within the errors) to the similar ones ($E_{TDS-peak\ \#1[5]}$ and $E_{TDS-peak\ \#2[5]}$) for TDS processes #1 and #2 (Table 1A). But the obtained value $K_{0des.(SiC-D/QFMLG-H)[55]}$ differs by several orders from the similar ones ($K_{0des.(TDS-peak\ \#1)[5]}$ and $K_{0des.(TDS-peak\ \#2)[5]}$) for TDS processes #1 and #2 (Table 1A). Nevertheless, these three desorption processes may be related to chemisorption models "H" and/or "G" (Figure 4).

These analytical results on characteristics of hydrogen desorption (dehydrogenation) from (of) SiC-D/QFMLG-H samples (Bocquet et al., 2012) may be also (as the previous results) interpreted within the models used for interpretation of the similar characteristics for the epitaxial graphenes (Elias et al., 2009) (Tables 1A and 2).

In the same way, one can treat the points from Figure 9 (at T_i = 1010, 1120 and 1200 K), which are related to the intercalated deuterium desorption from SiC-D/QFMLG samples. This results in finding the desired quantities (Table 2): the reaction (deuterium desorption) activation energy $\Delta H_{(des.)(SiC-D/QFMLG)[55]}$= 2.0 ± 0.6 eV, and the per-exponential factor of the reaction rate constant $K_{0(des.)(SiC-D/QFMLG)[55]} \approx 1 \times 10^6\ s^{-1}$.

Such a relatively low (in comparison with the vibration C-H or C-D frequencies) value of $K_{0(des.)(SiC-D/QFMLG)[55]}$, points out that the process cannot be described within the Polanyi-Wigner model (Nechaev, 2010; Nechaev and Veziroglu, 2013), related to the case of a non-diffusion rate limiting kinetics.

And as concluded in Bocquet et al. (2012), the exact intercalation mechanism of hydrogen diffusion through the anchored graphene lattice, at a defect or at a boundary of the anchored graphene layer, remains an open question.

Formally, this desorption process (obviously, of a diffusion-limiting character) may be described (as shown below) similarly to TDS process III (model "F*") (Table 1B), and the apparent diffusion activation energy may be close to the break-down energies of the C-H bonds.

Obviously such analytical results on characteristics of deuterium desorption from SiC-D/QFMLG samples (Bocquet et al., 2012) may not be interpreted within the models used for interpretation of the similar characteristics for the epitaxial graphenes (Elias et al., 2009) (Tables 1A and 2).

But these results (for SiC-D/QFMLG samples of Bocquet et al. (2012) may be quantitatively interpreted on the basis of using the characteristics of process III (Table 1B). Indeed, by using the quantities' values (from Table 1) of $\Delta H_{(des.)(SiC-D/QFMLG)[55]} \approx Q_{app.III} \approx$ 2.6 eV, $K_{0(des.)(SiC-D/QFMLG)[55]} \approx 6 \times 10^8\ s^{-1}$ and $D_{0app.III} \approx 3 \times 10^{-3}\ cm^2/s$, one can evaluate the quantity of $L \approx (D_{0app.III} / K_{0(des.)})^{1/2} = 22$ nm. The obtained value of L coincides (within the errors) with values of the quantities of $L_{def.} \approx 11 - 17$ nm [Equation (10)] and $L_{def.} \approx 20$ nm (Figure 8b). It shows that in the case under consideration, the intercalation mechanism of hydrogen (deuterium) diffusion through the anchored graphene lattice at the corresponding point type defects (Brito et al., 2011; Zhang et al., 2014; Banhart et al., 2011; Yazyev and Louie, 2010; Kim et al., 2011; Koepke et al., 2013; Zhang and Zhao, 2013; Yakobson and Ding, 2011; Cockayne et al., 2011; Zhang et al., 2012; Eckmann et al., 2012), of the anchored graphene layer may have place. And the desorption process of the intercalated deuterium may be rate-limited by diffusion of deuterium atoms to a nearest one of such point type defects of the anchored graphene layer.

It is reasonable to assume that the quasi-free-standing monolayer graphene on the SiC-D substrate is permeable to atomic hydrogen (at room temperature) in some defect nano-regions (probably, in vacancies and/or triple junctions (nodes) of the grain-boundary network (Brito et al., 2011; Zhang et al., 2014; Banhart et al., 2011; Yazyev and Louie, 2010; Kim et al., 2011; Koepke et al., 2013; Zhang and Zhao, 2013; Yakobson and Ding, 2011; Cockayne et al., 2011; Zhang et al., 2012; Eckmann et al., 2012).

It would be expedient to note that the HREELS data (Bocquet et al., 2012) on bending and stretching vibration C-H frequencies in SiC-D/QFMLG-H samples [153 meV (3.7×10^{13} s^{-1}) and 331 meV (8.0×10^{13} s^{-1}), respectively] are consistent with those (Xie et al., 2011) considered above, related to the HREELS data for the epitaxial graphene (Elias et al., 2009).

The obtained characteristics (Table 2) of desorption processes (Wojtaszek et al., 2011; Castellanos-Gomez, 2012; Wojtaszek et al., 2012; Bocquet et al., 2012) show that all these processes may be of a diffusion-rate-controlling character (Nechaev, 2010).

CONSIDERATION AND INTERPRETATION OF THE RAMAN SPECTROSCOPY DATA ON DEHYDROGENATION OF GRAPHENE LAYERS ON SIO_2 SUBSTRATE

In Luo et al. (2009), graphene layers on SiO_2/Si substrate have been chemically decorated by radio frequency hydrogen plasma (the power of 5 - 15 W, the pressure of 1 T or) treatment for 1 min. The investigation of hydrogen coverage by Raman spectroscopy and micro-x-ray photoelectron spectroscopy (PES) characterization demonstrates that the hydrogenation of a single layer graphene on SiO_2/Si substrate is much less feasible than that of bi-layer and multilayer graphene. Both the hydrogenation and dehydrogenation processes of the graphene layers are controlled by the corresponding energy barriers, which show significant dependence on the number of layers. These results (Luo et al., 2009) on bilayer graphene/SiO_2/Si are in contradiction to the results (Elias et al., 2009) on a negligible hydrogenation of bi-layer epitaxial graphene on SiO_2/Si wafer, when obviously other defects are produced.

Within a formal kinetics approach (Nechaev, 2010; Bazarov, 1976), the kinetic data from (Figure 10a) for single layer graphene samples (1LG-5W and 1LG-15W ones) can be treated. Equation (7) is used to transform into a more suitable form (7'): $K \approx -[(\Delta C/\Delta t)/C]$, where Δt = 1800 s, and ΔC and C are determined from Figure 10a. The results have been obtained for 1LG-15W sample 3 values of the #1 reaction rate constant $K_{1(1LG\text{-}15W)}$ for 3 temperatures (T = 373, 398 and 423K), and 3 values of the#2reaction rate constant $K_{2(1LG\text{-}15W)}$ for 3 temperatures (T = 523, 573 and 623K). Hence, by using Equation 9, the following quantities for 1LG-15W samples have been determined (Table 3): the #1 reaction activation energy $\Delta H_{des.1(1LG\text{-}15W)} = 0.6 \pm 0.2$ eV, the per-exponential factor of the #1 reaction rate constant $K_{0des.1(1LG\text{-}15W)} \approx 2 \times 10^4$ s^{-1}, the #2 reaction activation energy $\Delta H_{des.2[(1LG\text{-}15W)} = 0.19 \pm 0.07$ eV, and the per-exponential factor of the #2 reaction rate constant $K_{0des.2[(1LG\text{-}15W)} \approx 3 \times 10^{-2}$ s^{-1}.

This also resulted in finding for 1LG-5W sample 4 values of the #1 reaction rate constant $K_{I(1LG\text{-}5W)}$ for 4 temperatures (T = 348, 373, 398 and 423K), and 2 values of the #2 reaction rate constant $K_{2(1LG\text{-}5W)}$ for 2 temperatures (T = 523 and 573 K). Therefore, by using Equation 9, one can evaluate the desired quantities for 1LG-5W specimens (Table 3): the #1 reaction activation energy $\Delta H_{des.1(1LG\text{-}5W)} = 0.15 \pm 0.04$ eV, the per-exponential factor of the #1 reaction rate constant $K_{0des.1[(1LG\text{-}5W)} \approx 2 \times 10^{-2}$ s^{-1}, the #2 reaction activation energy $\Delta H_{des.2(1LG\text{-}5W)} = 0.31 \pm 0.07$ eV, and the per-exponential factor of the #2reaction rate constant $K_{0des.2(1LG\text{-}5W)} \approx 0.5$ s^{-1}.

A similar treatment of the kinetic data from (Figure 10c) for bi-layer graphene 2LG-15W samples resulted in obtaining 4 values of the #2reaction rate constant $K_{2(2LG\text{-}15W)}$ for 4 temperatures (T = 623, 673, 723 and 773K). Hence, by using Equation (9), the following desired values are found (Table 3): the #2 reaction activation energy $\Delta H_{des.2(2LG\text{-}15W)} = 0.9 \pm 0.3$ eV, the per-exponential factor of the #2 reaction rate constant $K_{0des.2(2LG\text{-}15W)} \approx 1 \times 10^3$ s^{-1}.

A similar treatment of the kinetic data from (Figure 6c) in Luo et al. (2009) for bi-layer graphene 2LG-5W samples results in obtaining 4 values for the #1 reaction rate constant $K_{1(2LG\text{-}5W)}$ for 4 temperatures (T = 348, 373, 398 and 423K), and 3 values for the #2 reaction rate constant $K_{2(2LG\text{-}5W)}$ for 3 temperatures (T = 573, 623 and 673K). Their temperature dependence is described by Equation (9). Hence, one can evaluate the following desired values (Table 3): the #1 reaction activation energy $\Delta H_{des.1[(2LG\text{-}5W)} = 0.50 \pm 0.15$ eV, the per-exponential factor of the #1 reaction rate constant $K_{0des.1(2LG\text{-}5W)} \approx 2 \cdot 10^3$ s^{-1}, the #2reaction activation energy $\Delta H_{des.2(2LG\text{-}5W)} = 0.40 \pm 0.15$ eV, and the per-exponential

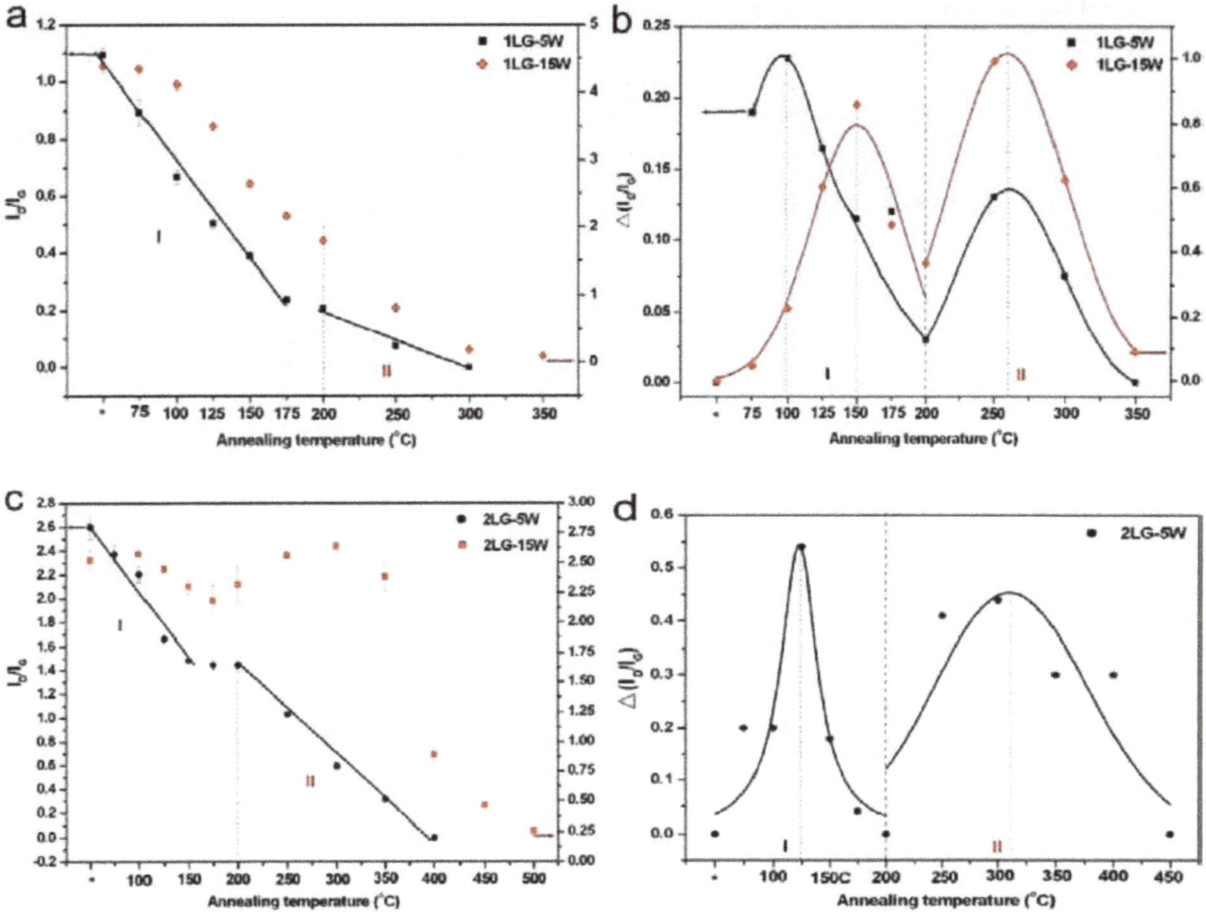

Figure 10. (a) The evoluation of the D and G band intensity ratio (I_D/I_G) with annealing temperatures of 1LG (single-layer graphene) hydrogenated by 5 and 15 W (the power), 1 Torr hydrogen plasma for 1 min (Luo et al. (2009)); (b) the evoluation of $\Delta(I_D/I_G)$ with annealing temperatures of 1 LG hydrogenated by 5 and 15 W, 1 Torr hydrogen plasma for 1 min; (c) the evoluation of the D and G band intensity ratio (I_D/I_G) with annealing temperatures of 2LG (bi-layer graphene) hydrogenated by 5 and 15 W, 1 Torr hydrogen plasma for 1 min; (d) the evoluation of $\Delta(I_D/I_G)$ with annealing temperatures of 2LG hydrogenated by 5 and 15 W, 1 Torr hydrogen plasma for 1 min. The asterisk (*) denotes the as-treated sample by H_2 plasma.

factor of the #2 reaction rate constant $K_{0des.2(2LG-5W)} \approx 1\ s^{-1}$.

The obtained analytical results (Table 3) on characteristics of desorption (dehydrogenation) processes #1and #2 (Luo et al., 2009) may be interpreted within the models used for interpretation of the similar characteristics for the epitaxial graphenes (Elias et al.,2009) (Table 1A). It shows that the desorption processes #1and #2 in Luo et al. (2009) may be of a diffusion-rate-controlling character.

CONSIDERATION AND INTERPRETATION OF THE TDS/STM DATA FOR HOPG TREATED BY ATOMIC DEUTERIUM

Hornekaer et al. (2006) present results of a STM study of HOPG samples treated by atomic deuterium, which reveals the existence of two distinct hydrogen dimer nano-states on graphite basal planes (Figures 11 and 12b). The density functional theory calculations allow them to identify the atomic structure of these nano-states and to determine their recombination and desorption pathways. As predicted, the direct recombination is only possible from one of the two dimer nano-states. In conclusion (Hornekaer et al., 2006), this results in an increased stability of one dimer nanospecies, and explains the puzzling double peak structure observed in temperature programmed desorption spectra (TPD or TDS) for hydrogen on graphite (Figure 12a).

By using the method of Nechaev (2010) of TDS peaks' treatment, for the case of TDS peak 1 (~65% of the total area, $T_{max\#1} \approx 473K$) in Figure 12), one can obtain values of the reaction #1 rate constant ($K_{(des.)1} = 1/\tau_{0.63(des.)1}$) for several temperatures (for instance, T = 458, 482 and 496K). Their temperature dependence can be described

Table 3. Analytical values of some related quantities.

Sample	Values/Quantities			
	$\Delta H_{(des.)1}$ (eV)	$K_{0(des.)1}$ (s^{-1}) {*L*}	$\Delta H_{(des.)2}$ (eV)	$K_{0(des.)2}$ (s^{-1}) {*L*}
1LG-15W (graphene) (Luo et al., 2009)	0.6 ± 0.2 (as processes ~I-II, ~model "G", Figure 4)	2×10^4 {$L \sim d_{sample}$}	0.19 ± 0.07 (as process~I, ~models "F","G", Figure 4)	3×10^{-2} {$L \sim d_{sample}$}
2LG-15W (bi-graphene) (Luo et al., 2009)			0.9 ± 0.3 (as processes~I-II, ~model"G",Figure 4)	1×10^3 {$L \sim d_{sample}$}
1LG-5W (graphene) (Luo et al., 2009)	0.15 ± 0.04 (as process~ I, ~ models "F","G",Figure 4)	2×10^{-2} {$L \sim d_{sample}$}	0.31 ± 0.07 (as process ~ I [14], ~models "F" ,"G", Figure 4)	5×10^{-1} {$L \sim d_{sample}$}
2LG-5W (bi-graphene) (Luo et al., 2009)	0.50 ± 0.15 (as processes ~I-II, ~model"G", Figure 4)	2×10^3 {$L \sim d_{sample}$}	0.40 ± 0.15 (as processes ~ I-II, ~model "G", Figure 4)	1.0 {$L \sim d_{sample}$}
HOPG (Hornekaer et al., 2006), TDS-peaks 1, 2	0.6 ± 0.2 (as processes ~ I - II, ~model"G", Figure 4)	1.5×10^4 {$L \sim d_{sample}$}	1.0 ± 0.3 (as processes ~ I-II, ~ model "G", Figure 4)	2×10^6 {$L \sim d_{sample}$}
Graphene/SiC (Watcharinyanon et al., 2011)			3.6 (as process ~IV [14],~models "C","D",Figure 4)	2×10^{14} $\sim \nu_{(C\text{-}H)}$ {$L \sim$ 17nm}
HOPG, TDS-peaks 1, 2 HOPG, TDS-peak 1 (Waqar et al., 2000)	2.4 (Waqar et al., 2000) (as process~III,~model "F*") 2.4 ± 0.5 (as process ~ III,~model "F*")	2×10^{10} {$L \sim$4 nm}	4.1 (Waqar et al., 2000) (as process~IV, ~models "C","D", Figure 4)	

by Equation (9). Hence, the desired values are defined as follows (Table 3): the #1 reaction (desorption) activation energy $\Delta H_{(des.)1}$ = 0.6 ± 0.2 eV, and the per-exponential factor of the #1 reaction rate constant $K_{0(des.)1} \approx 1.5 \times 10^4\ s^{-1}$.

In a similar way, for the case of TDS peak 2 (~35% of the total area, $T_{max\#2} \approx$ 588 K) in Figure 12a, one can obtain values of the #2 reactionrate constant ($K_{(des.)2} = 1/\tau_{0.63(des.)2}$) for several temperatures (for instance, T = 561 and 607K). Hence, the desired values are defined as follows (Table 3): the #2 reaction (desorption)activation energy $\Delta H_{(des.)2}$ = 1.0 ± 0.3 eV, and the per-exponential factor of the #2 reaction rate constant $K_{0(des.)2} \approx 2 \times 10^6\ s^{-1}$. The obtained analytical results (Table 3) on characteristics of desorption (dehydrogenation) processes #1and #2 in Hornekaer et al. (2006) (also as in Luo et al. (2009) may be interpreted within the models used above for interpretation of the similar characteristics for the epitaxial graphenes (Elias et al., 2009) (Table 1A). It shows that the desorption processes #1and #2 (in Hornekaer et al. (2006) and Luo et al. (2009) may be of a diffusion-rate-controlling character. Therefore, these processes cannot be described by using the Polanyi-Wigner equation (as it has been done in Hornekaer et al. (2006).

The observed "dimer nano-states" or "nano-protrusions" (Figures 11 and 12b) may be related to the defected nano-regions, probably, as grain (domain) boundaries (Brito et al., 2011; Zhang et al., 2014; Banhart et al., 2011; Yazyev and Louie, 2010; Kim et al., 2011; Koepke et al., 2013; Zhang and Zhao, 2013; Yakobson and Ding, 2011; Cockayne et al., 2011; Zhang et al., 2012; Eckmann et al., 2012), and/or triple and other junctions (nodes) of the grain-boundary network in the HOPG samples. Some defected nano-regions at the grain boundary network (hydrogen adsorption centres #1, mainly, the "dimer B" nano-structures) can be related to TPD (TDS) peak 1, the others (hydrogen adsorption

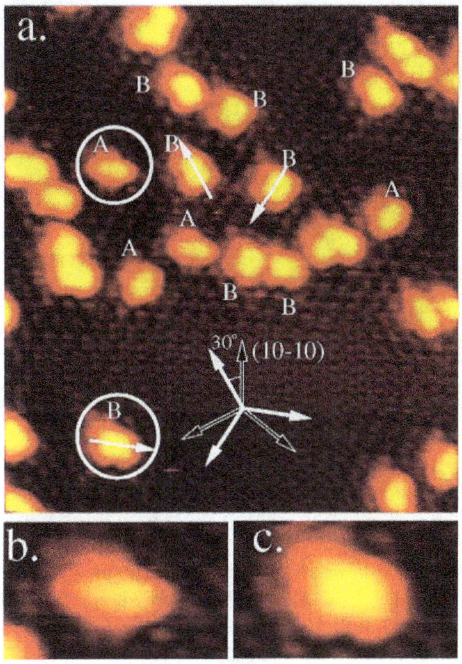

Figure 11. (a) STM image (103 × 114 Å^2) of dimer structures of hydrogen atoms on the graphite surface after a 1 min deposition at room temperature (Hornekaer et al., 2006). Imaging parameters: V_t = 884 mV, I_t = 160 pA. Examples of dimmer type A and B are marked. Black arrows indicate $\langle 21\bar{1}\bar{0}\rangle$ directions and white arrows indicate the orientation of the dimers 30° off. (c) Close up of dimer B structure in lower white circle in image (a).

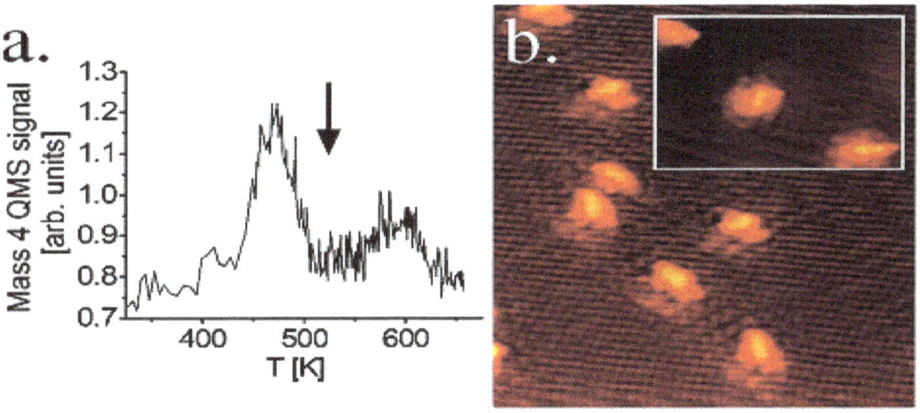

Figure 12. (a) A mass 4 amu, i.e., D_2, TPD spectrum from the HOPG surface after a 2 min D atom dose (ramp rate: 2 K / s below 450 K, 1 K / s above) (Hornekaer et al., 2006). The arrow indicates the maximum temperatue of the thermal anneal performed before recording the STM image in (b). (b) STM image (103 × 114 Å^2) of dimer structures of hydrogen atoms on the graphite surface after a 1 min deposition at room temperature and subsequent anneal to 525 K (ramp rate: 1 K / S, 30 s dwell at maximum temperature). Imaging parameters: V_t = 884 mV, I_t = 190 pA. The inset shows a higher resolution STM image of dimer structures of hydrogen atoms on the graphite surface after a 6 min deposition at room temperature and subsequent anneal to 550 K. Imaging parameters: V_t = -884 mV, I_t = -210 pA.

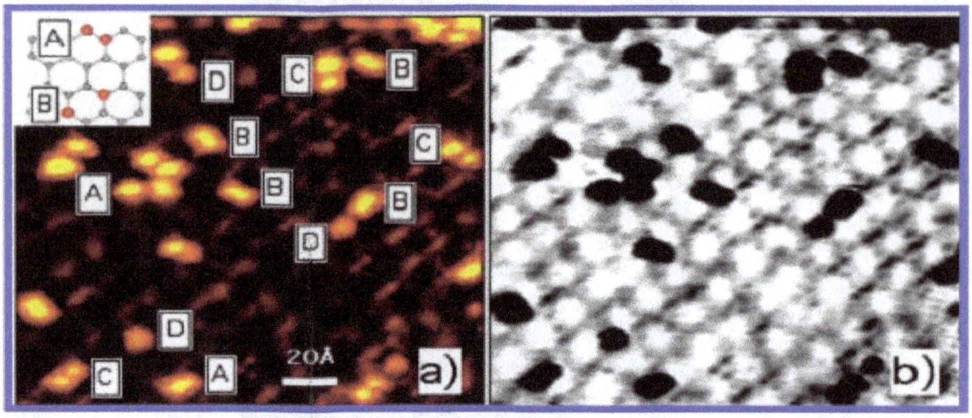

Figure 13. (a) Scanning tunneling microscopy (STM) image of hydrogenated graphene (Balog et al., 2009). The bright protrusions visible in the image are atomic hydrogen adsorbate structures identified as A = ortho-dimers, B = para-dimers, C = elongated dimers, D = monomers (imaging parameters: V_t = -0.245 V, I_t = -0.26 nA). Inset in (a); Schematic of the A ortho- and B para-dimer configuration on the graphene lattice. (b) Same image as in (a) with inverted color scheme, giving emphasis to preferential hydrogen adsorption along the 6 × 6 modulation on the SiC (0001)-(1 × 10 surface. Hydrogen dose at T_{beam} = 1600 K, t = 5 s, $F = 10^{12}$-10^{13} atoms/cm^2 s.

centres #2, mainly, the "dimer A" nano-structures) to TPD (TDS) peak 2.In Figures 11a and 12b, one can imagine some grain boundary network (with the grain size of about 2 - 5 nm) decorated (obviously, in some nano-regions at grain boundaries) by some bright nano-protrusions. Similar "nano-protrusions" are observed and in graphene/SiC systems (Balog et al., 2009; Watcharinyanon et al., 2011) (Figures 13 to 16).

In Balog et al. (2009), hydrogenation was studied by a beam of atomic deuterium 10^{12} - 10^{13} cm^{-2}s^{-1} (corresponding to $P_D \approx 10^{-4}$ Pa) at 1600K, and the time of exposure of 5 - 90 s, for single graphene on SiC-substrate. The formation of graphene blisters were observed, and intercalated with hydrogen in them (Figures 13 and 14), similar to those observed on graphite (Hornekaer et al., 2006) (Figures 11 and 12) and graphene/SiO_2 (Watcharinyanon et al., 2011) (Figures 15 and 16). The blisters (Balog et al., 2009) disappeared after keeping the samples in vacuum at 1073K (~ 15 min). By using Equation (8), one can evaluate the quantity of $\tau_{0.63(des.)1073K[58]} \approx$ 5 min, which coincides (within the errors) with the similar quantity of $\tau_{0.63(des.)1073K[17]} \approx$ 7 min evaluated for graphene/SiC samples (Watcharinyanon et al., 2011) (Table 3).

A nearly complete decoration of the grain boundary network (Brito et al., 2011; Zhang et al., 2014; Banhart et al., 2011; Yazyev and Louie, 2010; Kim et al., 2011; Koepke et al., 2013; Zhang and Zhao, 2013; Yakobson and Ding, 2011; Cockayne et al., 2011; Zhang et al., 2012; Eckmann et al., 2012), can be imagined in Figure 15b. Also, as seen in Figure 16, such decoration of the nano-regions obviously, located at the grain boundaries (Brito et al., 2011; Zhang et al., 2014; Banhart et al., 2011; Yazyev and Louie, 2010; Kim et al., 2011; Koepke et al., 2013; Zhang and Zhao, 2013; Yakobson and Ding, 2011; Cockayne et al., 2011; Zhang et al., 2012; Eckmann et al., 2012), has a blister-like cross-section height of about 1.7 nm and width of 10 nm order.

According to the thermodynamic analysis presented above, Equation (15), such blister-like decoration nano-regions (obviously, located at the grain boundaries (Brito et al., 2011; Zhang et al., 2014; Banhart et al., 2011; Yazyev and Louie, 2010; Kim et al., 2011; Koepke et al., 2013; Zhang and Zhao, 2013; Yakobson and Ding, 2011; Cockayne et al., 2011; Zhang et al., 2012; Eckmann et al., 2012), may contain the intercalated gaseous molecular hydrogen at a high pressure.

CONSIDERATION AND INTERPRETATION OF THE PES/ARPES DATA ON HYDROGENATION-DEHYDROGENATION OF GRAPHENE/SIC SAMPLES

In Watcharinyanon et al. (2011), atomic hydrogen exposures at a pressure of $P_H \approx 1 \times 10^{-4}$ Pa and temperature T = 973K on a monolayer graphene grown on the SiC(0001) surface are shown, to result in hydrogen intercalation. The hydrogen intercalation induces a transformation of the monolayer graphene and the carbon buffer layer to bi-layer graphene without a buffer layer. The STM, LEED, and core-level PES measurements reveal that hydrogen atoms can go underneath the graphene and the carbon buffer layer. This transforms the buffer layer into a second graphene layer. Hydrogen exposure (15 min) results initially in the formation of bi-layer graphene (blister-like) islands with a height of ~ 0.17 nm and a linear size of ~ 20 - 40 nm, covering about 40% of the sample (Figures 15b and e),

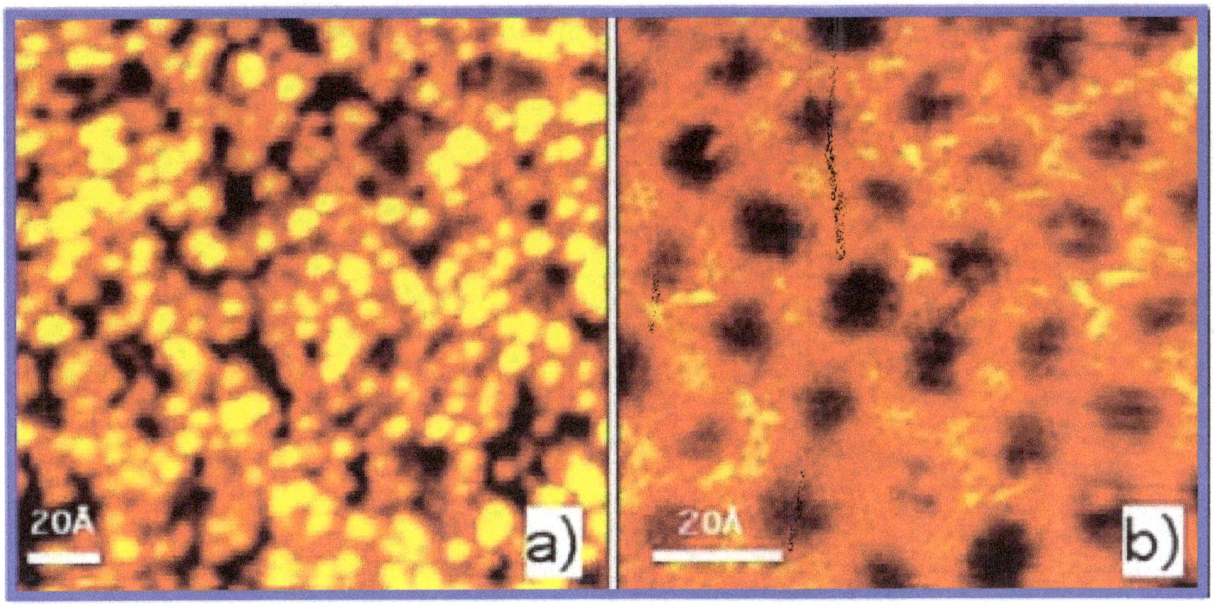

Figure 14. (a) STM image of the graphene surface after extended hydrogen exposure (Balog et al., 2009). The bright protrusions visible in the image are atomic hydrogen clusters (imaging parameters: V_t = -0.36 V, I_t = -0.32 nA). Hydrogen dose at T = 1600 K, t = 90 s, F = 10^{12}-10^{13} atoms/cm^2 s. (b) Large graphene area recovered from hydrogenation by annealing to 1073 K (imaging parameters: V_t = -0.38 V, I_t = -0.41 nA).

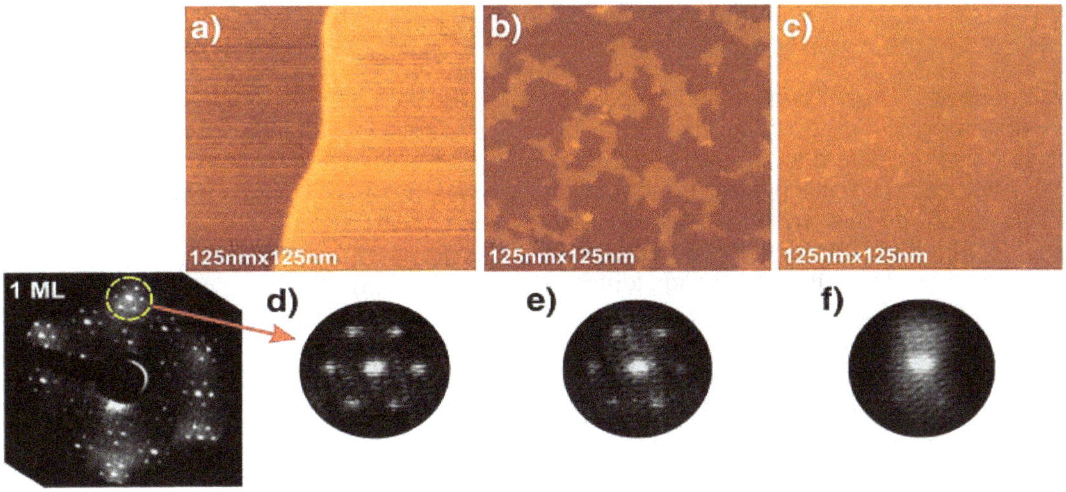

Figure 15. STM images (Watcharinyanon et al., 2011) collected at V = -1 V and I = 500 pA of a) monolayer graphene, b) after a small hydrogen exposure, and c) after a large hydrogen exposure. d) Selected part of the LEED patern collected at E = 107 eV from monolayer graphene, e) after a small hydrogen exposure, and f) after a large hydrogen exposure.

16a and b). With larger (additional 15 min) atomic hydrogen exposures, the islands grow in size and merge until the surface is fully covered with bi-layer grapheme (Figures 15c and 15f, 16c and d). A (√ 3 × √ 3) $R30°$ periodicity is observed on the bi-layer areas. Angle resolved PES and energy filtered X-ray photoelectron emission microscopy (XPEEM) investigations of the electron band structure confirm that after hydrogenation the single π-band characteristic of monolayer graphene is replaced by two π–bands that represent bi-layer graphene. Annealing an intercalated sample, representing bi-layer graphene, to a temperature of1123K or higher, re-establishes the monolayer graphene with a buffer layer on SiC (0001).

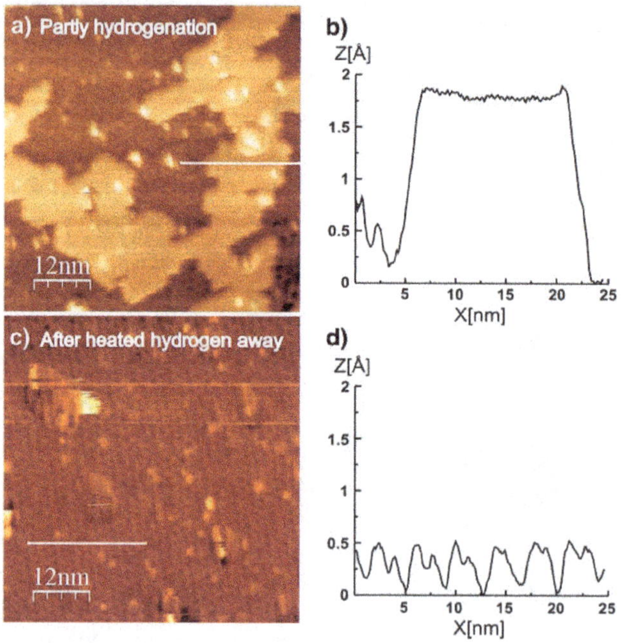

Figure 16. STM images (Watcharinyanon et al., 2011) of a) an island created by the hydrogen exposure (V = -1 V, I = 500 pA), b) line profile across the iland, c) a dehydrogenated sample showing mainly (6√3 × 6√3)R30° structure from the buffer layer (V = -2 V, I = 100 pA), and d) line profile across the (6√3 × 6√3)R30° structure.

The dehydrogenation has been performed by subsequently annealing (for a few minutes) the hydrogenated samples at different temperatures, from 1023 to 1273K. After each annealing step, the depletion of hydrogen has been probed by PES and ARPES (Figures 17 and 18). From this data, using Equations (8) and (9), one can determine the following tentative quantities: $\tau_{0.63(des.)}$ (at 1023 and 1123K), $\Delta H_{(des.)} \approx$ 3.6 eV and $K_{0(des.)} \approx 2 \times 10^{14}$ s^{-1} (Table 3).

The obtained value of the quantity of $\Delta H_{(des.)}$ coincides (within the errors) with values of the quantities of $Q_{app.IV} \approx$ 3.8 eV ≈ $\Delta H_{(C-H)"C","D"}$ (Table 1B), which are related to the diffusion-rate-limiting TDS process IV of a dissociative chemisorption of molecular hydrogen in defected regions in graphite materials (Table 1B), and to the chemisorption models "C" and "D"(Figure 4).

The obtained value of the quantity of $K_{0(des.)}$ may be correlated with possible values of the (C-H) bonds' vibration frequency ($\nu_{(C-H)"C","D"}$). Hence, by taking also into account that $\Delta H_{(des.)} \approx \Delta H_{(C-H)"C","D"}$, one may suppose the case of a non-diffusion-rate-controlling process corresponding to the Polanyi-Wigner model (Nechaev, 2010).

On the other hand, by taking also into account that $\Delta H_{(des.)} \approx \Delta H_{(C-H)"C","D"}$, one may suppose the case of a diffusion-rate-controlling process corresponding to the TDS process IV (Table 1B). Hence, by using the value (Nechaev, 2010) of $D_{0app.IV} \approx 6 \times 10^2$ cm^2/s, one can evaluate the quantity of $L \approx (D_{0app.IV} / K_{0(des.)})^{1/2}$ = 17 nm (Table 3). The obtained value of L (also, as and in the case of (SiC-D/QFMLG) (Bocquet et al., 2012), Table 2) coincides (within the errors) with values of the quantities of $L_{def.} \approx$ 11 - 17 nm [Equation (10)] and $L_{def.} \approx$ 20 nm (Figure 8b). The obtained value of L is also correlated with the STM data (Figures 15 and 16). It shows that the desorption process of the intercalated hydrogen may be rate-limited by diffusion of hydrogen atoms to a nearest one of the permeable defects of the anchored graphene layer.

When interpretation of these results, one can also take into account the model (proposed in (Watcharinyanon et al., 2011) of the interaction of hydrogen and silicon atoms at the graphene-SiC interface resulted in Si-C bonds at the intercalated islands.

CONSIDERATION AND INTERPRETATION OF THE TDS/STM DATA FOR HOPG TREATED BY ATOMIC HYDROGEN

In Waqar (2007), atomic hydrogen accumulation in HOPG samples and etching their surface under hydrogen TDS have been studied by using a STM and atomic force microscope (AFM). STM investigations revealed that the

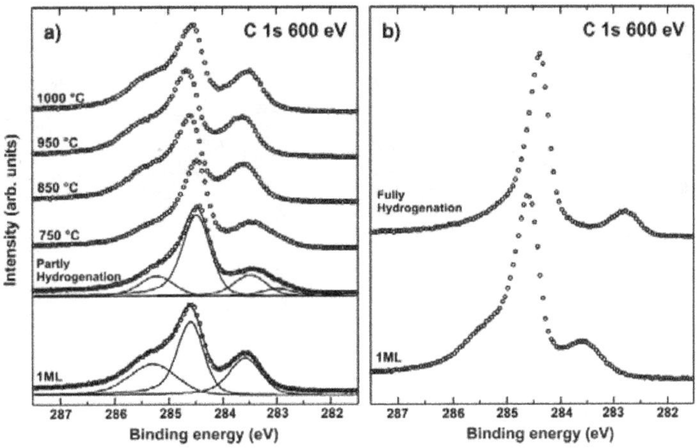

Figure 17. Normalized C 1s core level spectra of monolayer graphene (Watcharinyanon et al., 2011) before and after hydrogenation and subsequent annealing at 1023, 1123, 1223, and 1273 K. b) Fully hydrogenated graphene along with monolayer graphene before hydrogenation. The spectra were acquired at a photon energy of 600 eV.

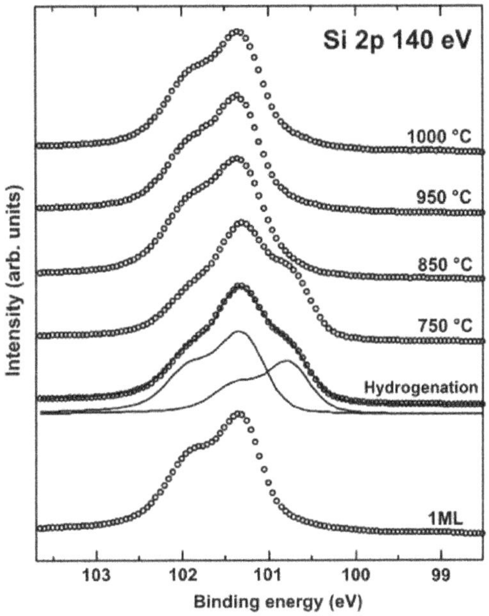

Figure 18. Normolized Si 2p core level spectra of monolayer graphene (Watcharinyanon et al., 2011) before and after hydrogenation and subsequent annealing at 1023, 1123, 1223, and 1273 K. The spectra were acquired at a photon energy of 140 eV.

surface morphology of untreated reference HOPG samples was found to be atomically flat (Figure 19a), with a typical periodic structure of graphite (Figure 19b). Atomic hydrogen exposure (treatment) of the reference HOPG samples (30 - 125 min at atomic hydrogen pressure $P_H \approx 10^{-4}$ Pa and a near-room temperature (~300K)) with different atomic hydrogen doses (D), has drastically changed the initially flat HOPG surface into a rough surface, covered with nanoblisters with an average radius of ~25 nm and an average height of ~4 nm

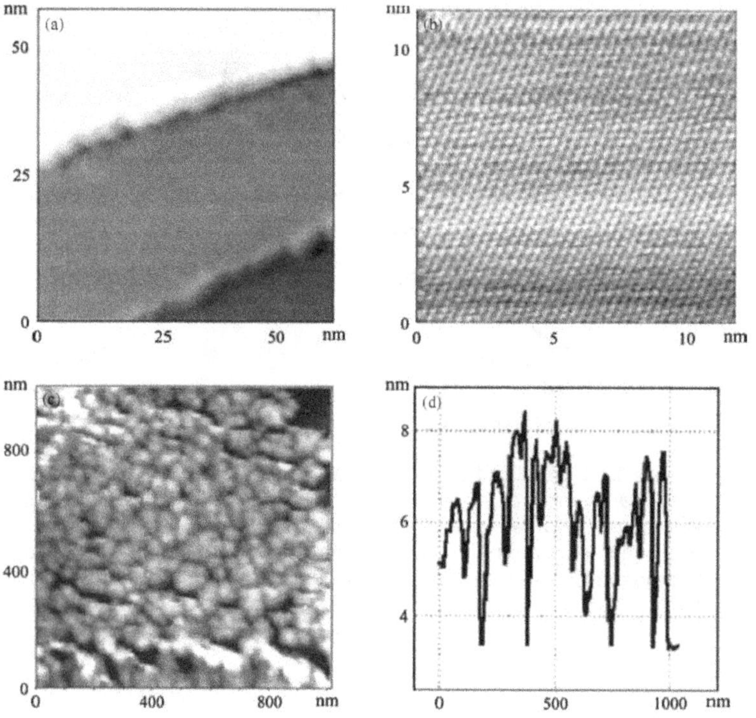

Figure 19. STM images of the untreated HOPG sample (Waqar, 2007) (under ambient conditions) taken from areas of (a) 60.8 x60.8 nm and (b) 10.9x10.9 nm (high resolution image of the square in image (a)). (c). AFM image (area of 1x1 nm) of the HOPG sample subjected to atomic hydrogen dose (D) of $1.8 \cdot 10^{16}$ H^0/cm^2. (d) Surface height profile obtained from the AFM image reported in (c). The STM tunnel V_{bias} and current are 50-100 mV and 1-1.5 mA, respectively.

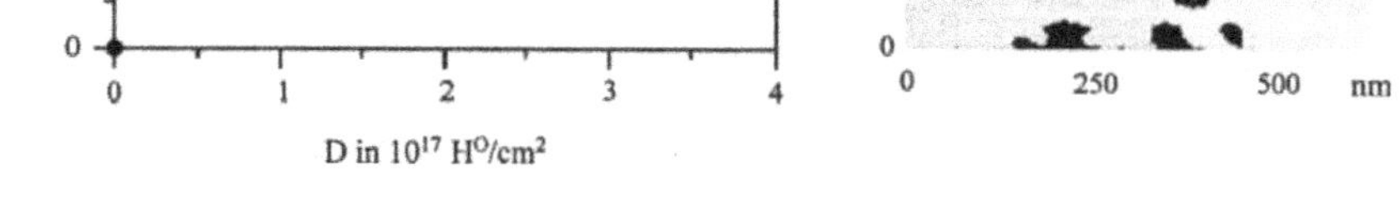

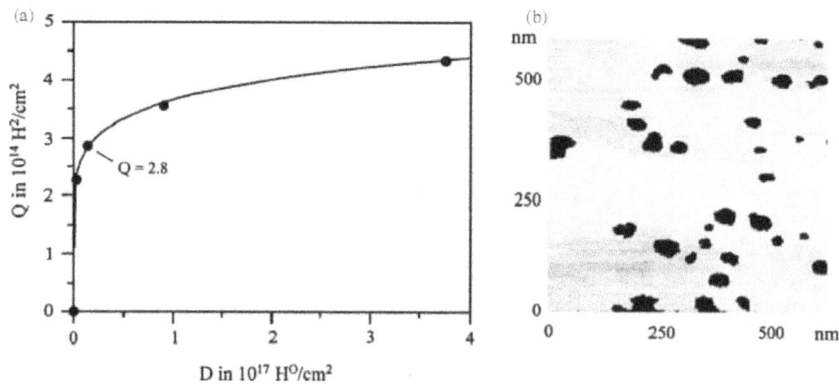

Figure 20. (a) Hydrogen storage efficiency of HOPG samples (Waqar, 2007), desorbed molecular hydrogen (Q) versus dose (D) of atomic hydrogen exposure. (b) STM image for 600x600 nm area of the HOPG sample subjected to atomic hydrogen dose of $1.8 \cdot 10^{16}$ H^0/cm^2, followed by hydrogen thermal desorption.

(Figures 19c and d).

TDS of hydrogen has been found in heating of the HOPG samples under mass spectrometer control. As shown in Figure 20a, with the increase of the total hydrogen doses (D) to which HOPG samples have been exposed, the desorbed hydrogen amounts (Q) increase and the percentage of D retained in samples approaches towards a saturation stage.

After TD, no nanoblisters were visible on the HOPG surface, the graphite surface was atomically flat, and

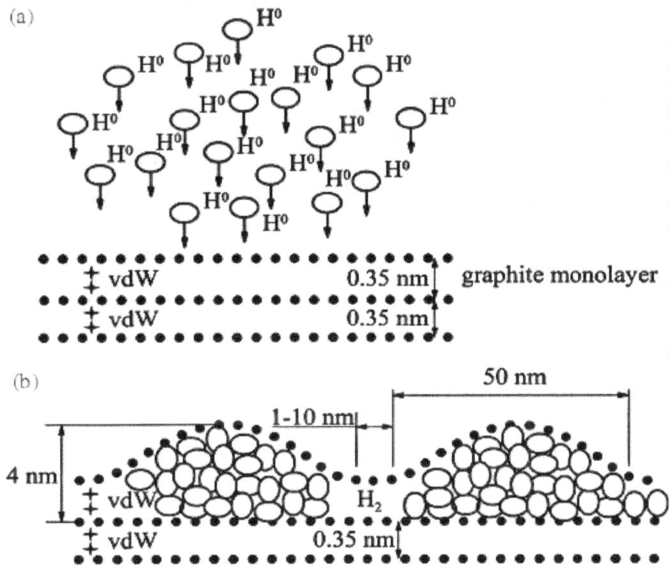

Figure 21. Model showing the hydrogen accumulation (intercalation) in HOPG, with forming blister-like nanostructures. (a) Pre-atomic hydrogen interaction step. (b) H_2, captured inside graphene blisters, after the interaction step. Sizes are not drawn exactly in scale (Waqar, 2007).

covered with some etch-pits of nearly circular shapes, one or two layers thick (Figure 20b). This implies that after release of the captured hydrogen gas, the blisters become empty of hydrogen, and the HOPG surface restores to a flat surface morphology under the action of corresponding forces.

According to the concept by Waqar (2007), nanoblisters found on the HOPG surface after atomic hydrogen exposure are simply monolayer graphite (graphene) blisters, containing hydrogen gas in molecular form (Figure 21). As suggested in Waqar (2007), atomic hydrogen intercalates between layers in the graphite net through holes in graphene hexagons, because of the small diameter of atomic hydrogen, compared to the hole's size, and is then converted to a H_2 gas form which is captured inside the graphene blisters, due to the relatively large kinetic diameter of hydrogen molecules.

However, such interpretation is in contradiction with that noted in Introduction results (Xiang et al., 2010; Jiang et al., 2009), that it is almost impossible for a hydrogen atom to pass through the six-member ring of graphene at room temperature.

It is reasonable to assume (as it has been done in some previous parts of this paper) that in HOPG (Waqar, 2007) samples atomic hydrogen passes into the graphite near-surface closed nano-regions (the graphene nanoblisters) through defects (perhaps, mainly through triple junctions of the grain and/or subgrain boundary network (Brito et al., 2011; Zhang et al., 2014; Banhart et al., 2011; Yazyev and Louie, 2010; Kim et al., 2011; Koepke et al., 2013; Zhang and Zhao, 2013; Yakobson and Ding, 2011; Cockayne et al., 2011; Zhang et al., 2012; Eckmann et al., 2012), in the surface graphene layer. It is also expedient to note that in Figure 20b, one can imagine some grain boundary network decorated by the etch-pits.

The average blister has a radius of ~25 nm and a height ~4 nm (Figure 19). Approximating the nanoblister to be a semi-ellipse form, results in the blister area of $S_b \approx 2.0 \times 10^{-11}$ cm^2 and its volume $V_b \approx 8.4 \times 10^{-19}$ cm^3. The amount of retained hydrogen in this sample becomes $Q \approx 2.8 \times 10^{14}$ H_2/cm^2 and the number of hydrogen molecules captured inside the blister becomes $n \approx (Q\ S_b) \approx 5.5 \times 10^3$. Thus, within the ideal gas approximation, and accuracy of one order of the magnitude, the internal pressure of molecular hydrogen in a single nanoblister at near-room temperature ($T \approx 300$ K) becomes $P_{H2} \approx \{k_B\ (Q\ S_b)\ T\ /\ V_b\} \approx 10^8$ Pa. The hydrogen molecular gas density in the blisters (at $T \approx 300$K and $P_{H2} \approx 1 \times 10^8$ Pa) can be estimated as $\rho \approx \{(QM_{H2}S_b)/V_b\} \approx 0.045$ g/cm^3, where M_{H2} is the hydrogen molecule mass. It agrees with data (Trunin et al., 2010) considered in Nechaev and Veziroglu (2013), on the hydrogen (protium) isotherm of 300K.

These results can be quantitatively described, with an accuracy of one order of magnitude, with the thermodynamic approach (Bazarov, 1976), and by using the condition of the thermo-elastic equilibrium for the reaction of ($2H_{(gas)} \rightarrow H_{2(gas\_in\_blisters)}$), as follows (Nechaev and Veziroglu, 2013):

$$(P_{H2}/P^0_{H2}) \approx (P_H/P^0_H)^2 \exp\{[\Delta H_{dis} - T\Delta S_{dis} - P^*_{H2}\,\Delta V)] / k_B T\} \quad (11)$$

Where P^*_{H2} is related to the blister "wall" back pressure (caused by P_{H2}) - the so called (Bazarov, 1976) surface pressure ($P^*_{H2} \approx P_{H2} \approx 1 \times 10^8$ Pa), P_H is the atomic hydrogen pressure corresponding to the atomic flux (Waqar, 2007) ($P_H \approx 1 \cdot 10^{-4}$ Pa), $P^0_{H2} = P^0_H = 1$ Pa is the standard pressure, $\Delta H_{dis} = 4.6$ eV is the experimental value (Karapet'yants and Karapet'yants, 1968) of the dissociation energy (enthalpy) of one molecule of gaseous hydrogen (at room temperatures), $\Delta S_{dis} = 11.8\ k_B$ is the dissociation entropy (Karapet'yants and Karapet'yants, 1968), $\Delta V \approx (S_b\ r_b\ /\ n)$ is the apparent volume change, r_b is the radius of curvature of nanoblisters at the nanoblister edge ($r_b \approx 30$ nm, Figures 19 and 21b), N_A is the Avogadro number, and $T$is the temperature ($T \approx 300$K). The quantity of ($P^*_{H2}\Delta V$) is related to the work of the nanoblister surface increasing with an intercalation of 1 molecule of H_2.

The value of the tensile stresses σ_b (caused by P^*_{H2}) in the graphene nanoblister "walls" with a thickness of d_b and a radius of curvature r_b can be evaluated from another condition (equation) of the thermo-elastic equilibrium of the system in question, which is related to Equation 11 as follows (Nechaev and Veziroglu, 2013):

$$\sigma_b \approx (P^*_{H2}\ r_b\ /\ 2\ d_b) \approx (\varepsilon_b\ E_b) \quad (12)$$

Where ε_b is a degree of elastic deformation of the graphene nanoblister walls, and E_b is the Young's modulus of the graphene nanoblister walls. Substituting in the first part of Equation (12), the quantities of $P^*_{H2} \approx 1 \times 10^8$ Pa, $r_b \approx 30$ nm and $d_b \approx 0.15$ nm results in the value of $\sigma_{b[15]} \approx 1 \times 10^{10}$ Pa.

The degree of elastic deformation of the graphene nanoblister walls, apparently reaches $\varepsilon_{b[15]} \approx 0.1$ (Figure 21b). Hence, with Hooke's law of approximation, using the second part of Equation (12), one can estimate, with the accuracy of one-two orders of the magnitude, the value of the Young's modulus of the graphene nanoblister walls: $E_b \approx (\sigma_b/\varepsilon_b) \approx 0.1$ TPa. It is close (within the errors) to the experimental value (Lee et al., 2008; Pinto and Leszczynski, 2014) of the Young's modulus of a perfect (that is, without defects) graphene ($E_{graphene} \approx 1.0$ TPa).

The experimental data (Waqar, 2007; Waqar et al., 2010) on the TDS (the flux J_{des}) of hydrogen from graphene nanoblisters in pyrolytic graphite can be approximated by three thermodesorption (TDS) peaks, that is, #1 with $T_{max\#1} \approx 1123$K, #2 with $T_{max\#2} \approx 1523$K, and #3 with $T_{max\#3} \approx 1273$K. But their treatment, with using the above mentioned methods (Nechaev, 2010), is difficult due to some uncertainty relating to the zero level of the J_{des} quantity.

Nevertheless, TDS peak #1 (Waqar et al., 2010) can be characterized by the activation desorption energy $\Delta H_{(des.)1[59]} = 2.4 \pm 0.5$ eV, and by the per-exponential factor of the reaction rate constant of $K_{0(des.)1[59]} \approx 2 \times 10^{10}$ s^{-1} (Table 3). It points that TDS peak 1 (Waqar et al., 2010) may be related to TDS peak (process) III, for which the apparent diffusion activation energy is $Q_{app.III} = (2.6 \pm 0.3)$ eV and $D_{0app.III} \approx 3 \times 10^{-3}$ cm^2/s (Table 1B). Hence, one can obtain (with accuracy of one-two orders of the magnitude) a reasonable value of the diffusion characteristic size of $L_{TDS\text{-}peak1[59]} \approx (D_{0app.III}/K_{0(des.)1[59]})^{1/2} \approx 4$ nm, which is obviously related to the separating distance between the graphene nanoblisters (Figure 21b) or (within the errors) to the separation distance between etch-pits (Figure 20b) in the HOPG specimens (Waqar, 2007; Waqar et al., 2010).

As noted in the previous parts of this paper, process III is related to model "F*" (Yang and Yang, 2002) (with $\Delta H_{(C\text{-}H)\text{“F*”}} = (2.5 \pm 0.3)$ eV (Nechaev, 2010), and it is a rate-limiting by diffusion of atomic hydrogen between graphene-like layers (in graphite materials and nanomaterials), where molecular hydrogen cannot penetrate (according to analysis (Nechaev, 2010) of a number of the related experimental data).

Thus, TDS peak (process) 1 (Waqar, 2007; Waqar et al., 2010) may be related to a rate-limiting diffusion of atomic hydrogen, between the surface graphene-like layer and neighboring (near-surface) one, from the graphene nanoblisters to the nearest penetrable defects of the separation distance $L_{TDS\text{-}peak1[59]} \sim 4$ nm.

As considered below, a similar (relevance to results (Waqar, 2007; Waqar et al., 2010) situation, with respect to intercalation of a high density molecular hydrogen into closed (in the definite sense) nanoblisters and/or nanoregions in graphene-layer-structures, may occur in hydrogenated GNFs.

A POSSIBILITY OF INTERCALATION OF SOLID H_2 INTO CLOSED NANOREGIONS IN HYDROGENATED GRAPHITE NANOFIBERS (GNFS) RELEVANT TO THE HYDROGEN ON-BOARD STORAGE PROBLEM

The possibility of intercalation of a high density molecular hydrogen (up to solid H_2) into closed (in the definite sense) nanoregions in hydrogenated GNFs is based both on the analytical results presented in the previous psrts of this study (Tables 1 to 3), and on the following facts (Nechaev and Veziroglu, 2013):

(1) According to the experimental and theoretical data (Trunin et al., 2010) (Figures 22 and 23), a solid molecular hydrogen (or deuterium) of density of $\rho_{H2} = 0.3 - 0.5$ $g/cm^3(H_2)$can exist at 300K and an external pressure of $P = 30 - 50$ Gpa.

(2) As seen from data in Figures 19 to 21and Equations 11 and 12, the external (surface) pressure of $P = P^*_{H2} = 30$ to 50 GPa at $T \approx 300$K may be provided at the expense of the association energy of atomic hydrogen ($T\Delta S_{dis} - \Delta H_{dis}$), into some closed (in the definite sense) nano-

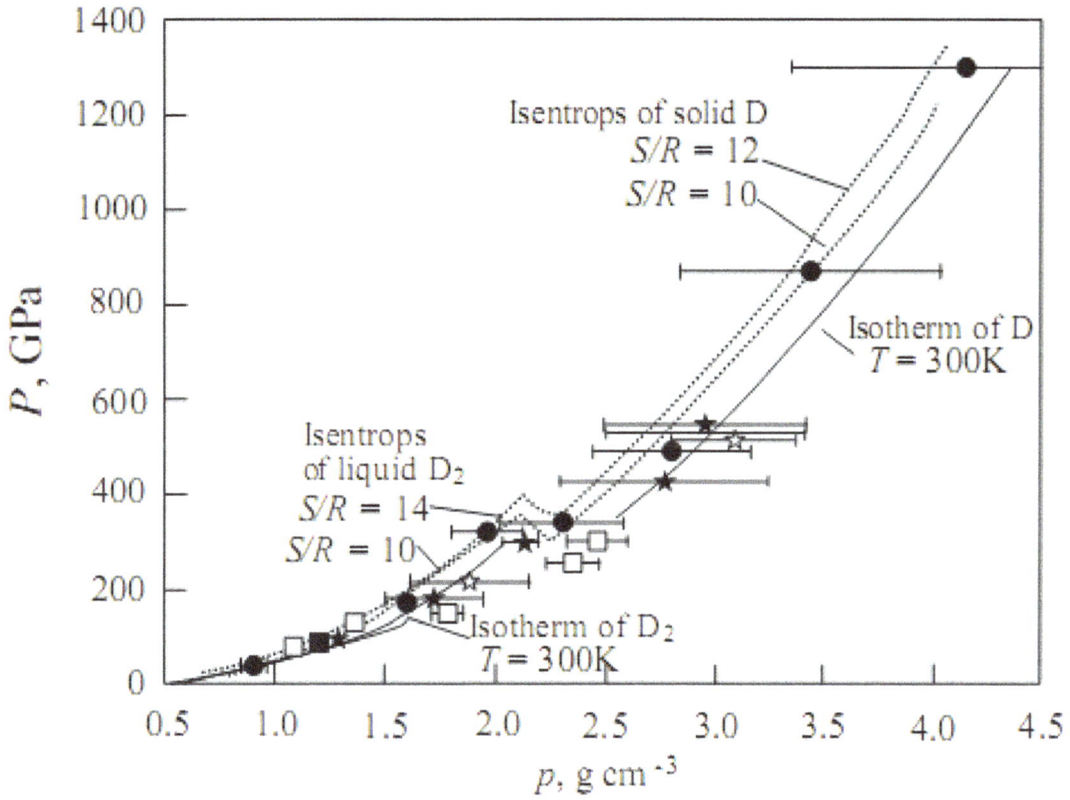

Figure 22. Isentropes (at entropies S/R = 10, 12 and 14, in units of the gas constant R) and isotherms (at T = 300 K) of molecular and atomic deuterium (Trunin et al., 2010). The symbols show the experimental data, and curves fit calculated dependences. The density (ρ) of protium was increased by a factor of two (for the scale reasons). Thickened portion of the curve is an experimental isotherm of solid form of molecular hydrogen (H_2). The additional red circle corresponds to a value of the twinned density $\rho \approx 1$ g/cm^3 of solid H_2 (at $T \approx 300$ K) and a near-megabar value of the external compression pressure $P \approx 50$ GPa (Nechaev and Veziroglu, 2013).

regions in hydrogenated (in gaseous atomic hydrogen with the corresponding pressure P_H) graphene-layer-nanostructures possessing of a high Young's modulus ($E_{graphene} \approx 1$ TPa).

(3) As shown in Nechaev and Veziroglu (2013), the treatment of the extraordinary experimental data (Gupta et al., 2004) (Figure 24) on hydrogenation of GNFs results in the empirical value of the hydrogen density ρ_{H2}= (0.5 ± 0.2) g(H_2)/cm^3(H_2) (or $\rho_{(H2\text{-}C\text{-}system)} \approx$ 0.2 g(H_2)/cm^3(H_2-C-system)) of the intercalated (at $T \approx$ 300K) high-purity reversible hydrogen (about 17 mass% H_2); it corresponds to the state of solid molecular hydrogen at the pressure of $P = P^*_{H2} \approx 50$ GPa, according to data from Figures 22 and 23.

(4) Substituting in Equation (12) the quantities of $P^*_{H2} \approx 5 \times 10^{10}$ Pa, $\varepsilon_b \approx 0.1$ (Figure 24), the largest possible value of $E_b \approx 10^{12}$ Pa (Lee et al., 2008; Pinto and Leszczynski (2014)), the largest possible value of the tensile stresses ($\sigma_b \approx 10^{11}$ Pa (Lee et al., 2008; Pinto and Leszczynski, 2014) in the edge graphene "walls" (of a thickness of d_b and a radius of curvature of r_b) of the slit-like closed nanopores of the lens shape (Figure 24), one can obtain the quantity of (r_b / d_b) ≈ 4. It is reasonable to assume $r_b \approx$ 20 nm; hence, a reasonable value follows of $d_b \approx$ 5 nm.

(5) As noted in (Nechaev and Veziroglu, 2013), a definite residual plastic deformation of the hydrogenated graphite (graphene) nano-regions is observed in Figure 24. Such plastic deformation of the nanoregins during hydrogenation of GNFs may be accompanied with some mass transfer resulting in such thickness (d_b) of the walls.

(6) The related data (Figure 25) allows us to reasonably assume a break-through in results (Nechaev and Veziroglu, 2013) on the possibility (and particularly, physics) of intercalation of a high density molecular hydrogen (up to solid H_2) into closed (in the definite sense) nanoregions in hydrogenated GNFs (Gupta et al., 2004; Park et al., 1999), relevant for solving of the current problem (Akiba, 2011; Zuettel, 2011; DOE targets, 2012) of the hydrogen on-board effective storage.

(7) Some fundamental aspects - open questions on engineering of "super" hydrogen storage carbonaceous nanomaterials, relevance for clean energy applications, are also considered in (Nechaev and Veziroglu, 2013) and in this study, as well.

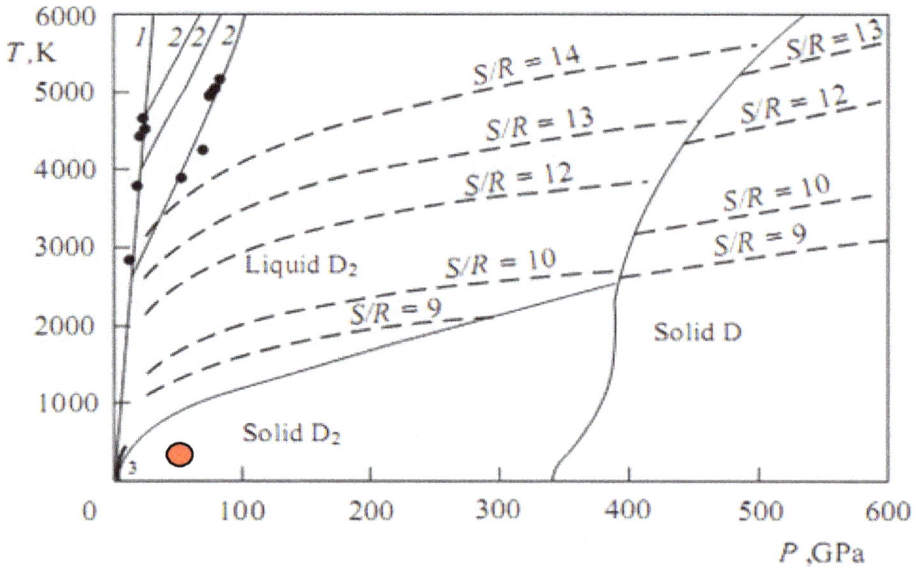

Figure 23. Phase diagram (Trunin et al., 2010), adiabats, and isentropes of deuterium calculated with the equation of state: *1* and *2* are a single and a doubled adiabat, • – the experimental data, *3* – melting curve, thickened portion of the curve – the experimental data. The additional red circle corresponds to a value of temperature $T \approx 300$ K and a near-megabar value of the external compression pressure $P \approx 50$ GPa (Nechaev and Veziroglu, 2013).

DISCUSSION

On the "thermodynamic forces" and/or energetics of forming (under atomic hydrogen treatment) of graphene nanoblisters in the surface HOPG layers and epitaxial graphenes

A number of researchers (Waqar, 2007; Watcharinyanon et al., 2011; Wojtaszek et al., 2011; Castellanos-Gomezet al., 2012; Bocquet et al., 2012; Hornekaer et al., 2006; Luo et al., 2009; Balog et al., 2009; Waqar et al., 2010) have not sufficiently considered the "thermodynamic forces" and/or energetics of forming (under atomic hydrogen treatment) graphene nanoblisters in the surface HOPG layers and epitaxial graphenes.

Therefore, in this study, the results of the thermodynamic analysis (Equations 11 and 12) are presented, which may be used for interpretation of related data (Figures 6 to 8, 11 to 16, 19 to 21).

On some nanodefects (grain boundaries, their triple junctions and others), penetrable for atomic hydrogen, in the surface HOPG graphene-layers and epitaxial graphenes

A number of researchers noted above have not taken into account (in a sufficient extent) the calculation results (Xiang et al., 2010) showing that the barrier for the penetration of a hydrogen atom through the six-member ring of a perfect graphene is larger than 2.0 eV. Thus, it is almost impossible for a hydrogen atom to pass through the six-member ring of a perfect (that is, without defects) graphene layer at room temperature.

Therefore, in this study, a real possibility of the atomic hydrogen penetration through some nanodefects in the graphene-layer-structures, that is, grain boundaries, their triple junctions (nodes) and/or vacancies (Brito et al., 2011; Zhang et al., 2014; Banhart et al., 2011; Yazyev and Louie, 2010; Kim et al., 2011; Koepke et al., 2013; Zhang and Zhao, 2013; Yakobson and Ding, 2011; Cockayne et al., 2011; Zhang et al., 2012; Eckmann et al., 2012), are considered. These analytical results may be used for interpretation of the related data (for instance, Figures 6 to 8, 11 to 16, 19 to 21).

On finding and interpretation of the thermodynamic characteristics of "reversible" hydrogenation-dehydrogenation of epitaxial graphenes and membrane ones

A number of researchers, for instance ones noted above have not treated and compared their data on "reversible" hydrogenation-dehydrogenation of membrane graphenes and epitaxial ones, with the aim of finding and interpretation of the thermodynamic characteristics. Therefore, in this analytical study, the thermodynamic approaches (particularly, Equations 1 to 12), such

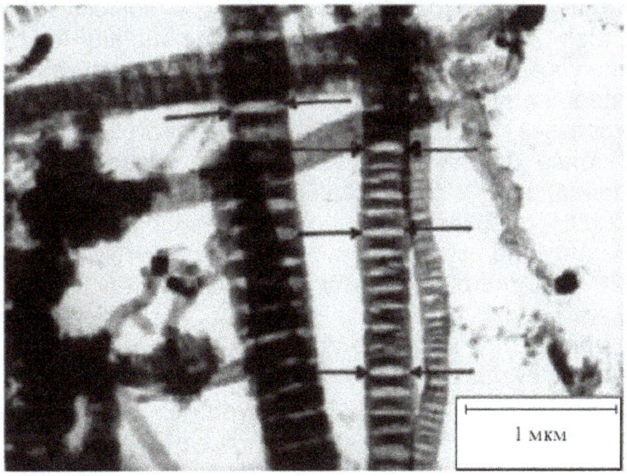

Figure 24. Micrographs (Gupta et al., 2004) of hydrogenated graphite nanofibers (GNFs) after release from them (at ~300 K for ~10 min (Park et al., 1999) of intercalated high-density hydrogen (~17 mass.% - the gravimetrical reversible hydrogen capacity). The arrows in the picture indicate some of the slit-like closed nanopores of the lens shape, where the intercalated high-density solid hydrogen nanophase (Nechaev and Veziroglu, 2013) was localized.

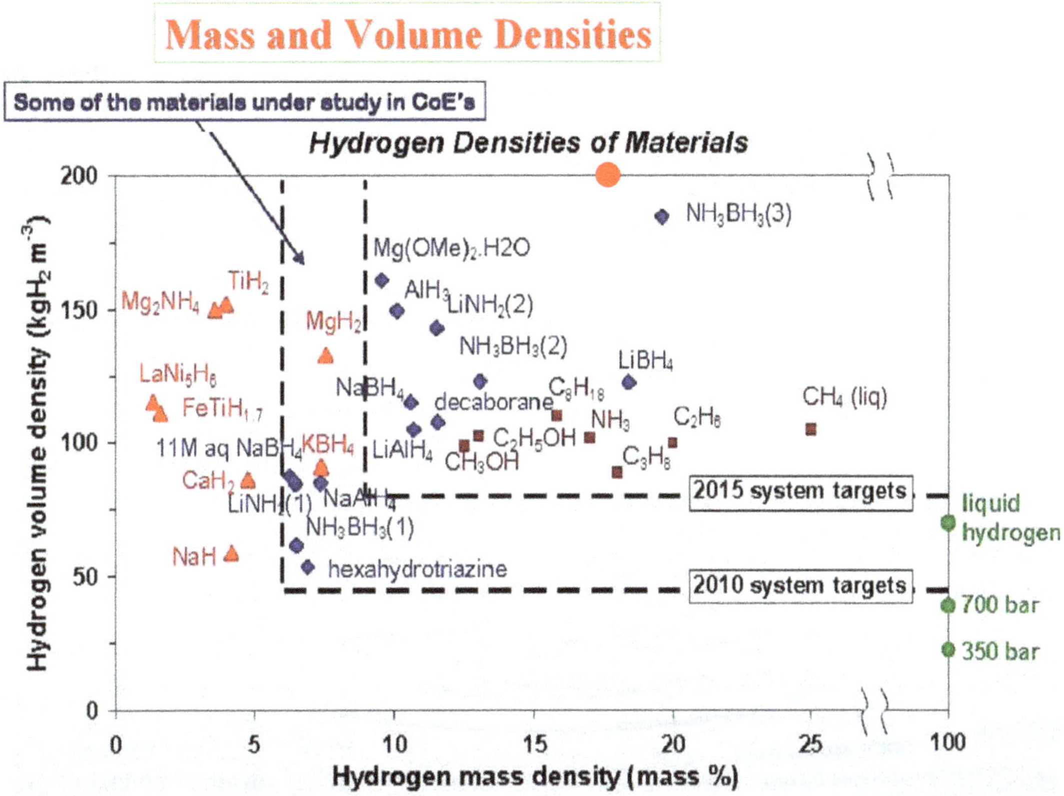

Figure 25. It is shown (in the face of known achievements) U.S. DOE system targets for 2010 and 2015, relevant to gravimetric and volumetric hydrogen on-board storage densities. The additional red circle is related to the solid hydrogen nanophase (Nechaev and Veziroglu, 2013) intercalated into the hydrogenated GNFs (Figure 24).

treatment results of related theoretical and experimental data (Tables 1 to 3) and their interpretation are presented. As shown, these analytical results may be used for a more detailed understanding and revealing of the atomic mechanisms of the processes.

There is a considerable difference (in the declared errors and without any explanation) in the theoretical values of the energetic graphene (CH) quantities ($\Delta H_{(C\text{-}H)}$, $\Delta H_{(bind.)}$, $\Delta H_{(C\text{-}C)}$) obtained in different theoretical studies, for instance, in (Sofo et al., 2007; Dzhurakhalov and Peeters, 2011) (Table 1A).

Unfortunately, the theoretical values of the graphene quantity of $\Delta H_{(C\text{-}C)}$ is usually not evaluated by the researchers, and not compared by them with the much higher values of the graphene (both theoretical, and experimental) quantity of $\Delta H_{(C\text{-}C)}$ (Table 1A). It could be useful, for instance, when considering the fundamental strength properties of graphane and graphene structures. As far as we know, most researchers have not taken into account the alternative possibility supposed in (Elias et al., 2009) that (i) the experimental graphene membrane (a free-standing one) may have "a more complex hydrogen bonding, than the suggested by the theory", and that (ii) graphane (CH) (Sofo et al., 2007) may be the until now theoretical material.

In this connection, it seems expedient to take into account also some other approaches and results (Sorokin and Chernozatonskii, 2013; Davydov and Lebedev, 2012; Khusnutdinov, 2012; Chernozatonskii et al., 2012; Data et al., 2012).

On the thermodynamic characteristics and atomic mechanisms of "reversible" hydrogenation-dehydrogenation of free-standing graphene membranes

The thermodynamic analysis of experimental data (Elias et al., 2009) on "reversible" hydrogenation-dehydrogenation of free-standing graphene membranes have resulted in the following conclusive suppositions and/or statements:

(1) These chemisorption processes are related to a non-diffusion-rate-limiting case. They can be described and interpreted within the physical model of the Polanyi-Wigner equation for the first order rate reactions (Nechaev, 2010; Nechaev and Veziroglu, 2013), but not for the second order rate ones (Zhao et al., 2006).

(2) The desorption activation energy is of $\Delta H_{des.(membr.[5])} = \Delta H_{C\text{-}H(membr.[5])} = 2.6 \pm 0.1$ eV (Table 1A). The value of the quantity of $\Delta H_{C\text{-}H(membr.[5])}$ coincides (within the errors), in accordance with the Polanyi-Wigner model, with the values of the similar quantities for theoretical graphenes (Sofo et al., 2007; Openov and Podlivaev, 2010) (Table1A) possessing of a diamond-like distortion of the graphene network. The value of the quantity of $\Delta H_{C\text{-}H(membr.[5])}$ coincides (within the errors) with the value of the similar quantity for model "F*" (Table 1B) manifested in graphitic structures and nanostructures not possessing of a diamond-like distortion of the graphene network (an open theoretical question).

(3) The desorption frequency factor is of $K_{0des.(membr.[5])} = \nu_{C\text{-}H(membr.[5])} \approx 5 \times 10^{13}$ s^{-1} (Table 1A); it is related to the corresponding vibration frequency for the C-H bonds (in accordance with the Polanyi-Wigner model for the first order rate reactions.

(4) The adsorption activation energy (in the approximation of $K_{0ads.} \approx K_{0des.}$) is of $\Delta H_{ads.(membr.[5])} = 1.0 \pm 0.2$ eV (Table 1A). The heat of adsorption of atomic hydrogen by the free standing graphene membranes (Elias et al., 2009) can be evaluated as: $(\Delta H_{ads.(membr.[5])} - \Delta H_{des.(membr.[5])}) = -1.5 \pm 0.2$ eV (an exothermic reaction).

(5) Certainly, these tentative analytical results could be directly confirmed and/or modified by receiving and treating (within Equations (8) and (9) approach) of the experimental data on $\tau_{0.63}$ at several annealing temperatures.

On the thermodynamic characteristics and atomic mechanisms of "reversible" hydrogenation-dehydrogenation of epitaxial graphenes

The thermodynamic analyses of experimental data (Waqar, 2007; Watcharinyanon et al., 2011; Wojtaszek et al., 2011; Castellanos-Gomez et al., 2012; Bocquet et al., 2012; Luo et al., 2009) on "reversible" hydrogenation-dehydrogenation of epitaxial graphenes have resulted in the following conclusive suppositions and/or statements:

(1) These chemisorption processes for all 16 considered epitaxial graphenes (Tables 1A, 2 and 3), unlike ones for the free-standing graphene membranes (Table 1A), are related to a diffusion-rate-limiting case. They can be described and interpreted within the known diffusion approximation of the first order rate reactions (Nechaev, 2010; Nechaev and Veziroglu, 2013), but not within the physical models of the Polanyi-Wigner equations for the first (Hornekaer et al., 2006) or for the second (Zhao et al., 2006) order rate reactions.

(2) The averaged desorption activation energy for 14 of 16 considered epitaxial graphenes (Tables 1A, 2 and 3) is of $\Delta H_{des.(epitax.)} = 0.5 \pm 0.4$ eV, and the averaged quantity of $\ell n K_{0des.(epitax.)} = 5 \pm 8$, that is, $K_{0des.(epitax.)} \approx 1.5 \times 10^2$ s^{-1} (or $5 \times 10^{-2} - 5 \times 10^5$ s^{-1}); the adsorption activation energy (in a rough approximation of $K_{0ads.} \approx K_{0des.}$) is of $\Delta H_{ads.(epitax.)} = 0.3 \pm 0.2$ eV.

(3) The above obtained values of characteristics of dehydrogenation of the epitaxial graphenes can be presented, as follows: $\Delta H_{des.} \sim Q_{app.I}$, $K_{0des.} \sim (D_{0app.I} / L^2)$, where $Q_{app.I}$ and $D_{0app.I}$ are the characteristics of process I (Table 1B), $L \sim d_{sample}$, that is, being of the order of diameter (d_{sample}) of the epitaxial graphene samples. The

diffusion-rate-limiting process I is related to the chemisorption models "F" and "G" (Figure 4). These results unambiguously point that in the epitaxial graphenes the dehydrogenation processes are rate-limiting by diffusion of hydrogen, mainly, from chemisorption "centers" (of "F" and/or "G" types (Figure (4) localized on the internal graphene surfaces to the frontier edges of the samples. These results also point that the solution and the diffusion of molecular hydrogen may occur between the graphene layer and the substrate, unlike for a case of the graphene neighbor layers in graphitic structures and nanostructures, where the solution and the diffusion of only atomic hydrogen (but not molecular one) can occur (process III (Nechaev, 2010), Table 1B).

(4) The above formulated interpretation (model) is direct opposite to the supposition (model) of a number of researchers, those believe in occurrence of hydrogen desorption (dehydrogenation) processes, mainly, from the external epitaxial graphene surfaces.And it is direct opposite to the supposition - model of many scientists that the diffusion of hydrogen along the graphene-substrate interface is negligible.

(5) In this connection, it is expedient to take into account also some other related experimental results, for instance (Stolyarova et al., 2009; Riedel et al., 2009; Riedel et al., 2010; Goleret al., 2013; Jones et al., 2012; Lee et al., 2012), on the peculiarities of the hydrogenation-dehydrogenation processes in epitaxial graphenes, particularly, in the graphene-substrare interfaces.

Conclusion

(1) The chemisorption processes in the free-standing graphene membranes are related to a non-diffusion-rate-limiting case. They can be described and interpreted within the physical model of the Polanyi-Wigner equation for the first order rate reactions, but not for the second order rate reactions.

The desorption activation energy is of $\Delta H_{des.(membr.)} = \Delta H_{C\text{-}H(membr.)} = 2.6 \pm 0.1$ eV. It coincides (within the errors), in accordance with the Polanyi-Wigner model, with the values of the similar quantities for theoretical graphanes (Table 1A) possessing of a diamond-like distortion of the graphene network. It also coincides (within the errors) with the value of the similar quantity [process III, model "F*" (Table 1B)] manifested in graphitic structures and nanostructures, not possessing of a diamond-like distortion of the graphene network (an open theoretical question).

The desorption frequency factor is of $K_{0des.(membr.)} = \nu_{C\text{-}H(membr.)} \approx 5 \times 10^{13}$ s^{-1} (Table 1A). It is related to the corresponding vibration frequency for the C-H bonds (in accordance with the Polanyi-Wigner model).

The adsorption activation energy (in the approximation of $K_{0ads.} \approx K_{0des.}$) is of $\Delta H_{ads.(membr.)} = 1.0 \pm 0.2$ eV (Table 1A). The heat of adsorption of atomic hydrogen by the free standing graphene membranes (Elias et al., 2009) may be as $(\Delta H_{ads.(membr.)} - \Delta H_{des.(membr.)}) = -1.5 \pm 0.2$ eV (an exothermic reaction).

(2) The hydrogen chemisorption processes in epitaxial graphenes (Tables 1A, 2 and 3), unlike ones for the free-standing graphene membranes (Table 1A), are related to a diffusion-rate-limiting case. They can be described and interpreted within the known diffusion approximation of the first order rate reactions, but not within the physical models of the Polanyi-Wigner equations for the first or for the second order rate reactions.

The desorption activation energy is of $\Delta H_{des.(epitax.)} = 0.5 \pm 0.4$ eV. The quantity of $\ell n K_{0des.(epitax.)}$ is of 5 ± 8, and the per-exponential factor of the desorption rate constant is of $K_{0des.(epitax.)} \approx 1.5 \times 10^2$ s^{-1} (or $5 \times 10^{-2} - 5 \times 10^5$ s^{-1}). The adsorption activation energy (in a rough approximation of $K_{0ads.} \approx K_{0des.}$) is of $\Delta H_{ads.(epitax.)} = 0.3 \pm 0.2$ eV.

The above obtained values of characteristics of dehydrogenation of the epitaxial graphenes can be presented as $\Delta H_{des.} \sim Q_{app.I}$ and $K_{0des.} \sim (D_{0app.I} / L^2)$, where $Q_{app.I}$ and $D_{0app.I}$ are the characteristics of process I (Table 1B), $L \sim d_{sample}$, that is, being of the order of diameter (d_{sample}) of the epitaxial graphene samples. The diffusion-rate-limiting process I is related to the chemisorption models "F" and "G" (Figure 4). These results unambiguously point that in the epitaxial graphenes the dehydrogenation processes are rate-limiting by diffusion of hydrogen, mainly, from chemisorption "centers" [of "F" and/or "G" types (Figure 4)] localized on the internal graphene surfaces to the frontier edges of the samples. These results also point that the solution and the diffusion of molecular hydrogen occurs in the interfaces between the graphene layers and the substrates. It differs from the case of the graphene neighbor layers in graphitic structures and nanostructures, where only atomic hydrogen solution and diffusion can occur (process III, model "F*", Table 1B). Such an interpretation (model) is direct opposite, relevance to the supposition (model) of a number of researchers, those believe in occurrence of hydrogen desorption processes, mainly, from the external epitaxial graphene surfaces.And it is direct opposite to the supposition-model of many scientists that the diffusion of hydrogen along the graphene-substrate interface is negligible.

(3) The possibility, and particularly, the physics of intercalation of a high density molecular hydrogen (up to solid H_2) in closed nanoregions, in hydrogenated GNFs have been discussed, in connection to the analytical results (Tables 1 to 3) and the empirical facts considered in this paper.

It is relevant for developing of a key breakthrough nanotechnology of the hydrogen on-board efficient and compact storage (Figure 25) - the very current problem.

Such a nanotechnology may be developed within a reasonable (for the current hydrogen energy demands and predictions) time frame of several years. International cooperation is necessary.

Conflict of Interest

The author(s) have not declared any conflict of interest.

ACKNOWLEDGMENTS

The authors are grateful to A. Yürüm, A. Tekin, N. K. Yavuz and Yu. Yürüm, participants of the joint RFBR-TUBAK project, for helpful and fruitful discussions. This work has been supported by the RFBR (Project #14-08-91376 CT) and the TUBITAK (Project # 213M523).

REFERENCES

Akiba E (2011). Hydrogen related R&D and hydrogen storage materials in Japan. In: Materials of Int. Hydrogen Research Showcase 2011, University of Birmingham, UK, April 13-15, 2011; the UK-SHEC website: http://www.uk-shec.org.uk/uk-shec/showcase/ShowcasePresentations.html

Balog R, Jørgensen B, Wells J, Lægsgaard E, Hofmann P, Besenbacher F, Hornekær L (2009). Atomic hydrogen adsorbate structures on graphene. J. Am. Chem. Soc. 131(25): 8744-8745.

Banhart F, Kotakovski J, Krasheninnikov AV (2011). Structural defects in graphene (Review). ACS Nano. 5(1):26-41.

Bauschlicher CW(Jr), So CR (2002). High coverages of hydrogen on (10.0), (9.0) and (5.5) carbon nanotubes. Nano Lett. 2(4):337-341.

Bazarov IP (1976). Thermodynamics. "Vysshaya Shkola", Moscow.

Bocquet FC, Bisson R, Themlin JM, Layet JM, Angot T (2012). Reversible hydrogenation of deuterium-intercalated quasi-free-standing graphene on SiC(0001). Phys. Rev. B - Condensed Matter Mater. Phys. 85(20):article # 201401.

Boukhvalov DW, Katsnelson MI, Lichtenstein AI (2008). Hydrogen on graphene: total energy, structural distortions and magnetism from first-principles calculations. Phys. Rev. B. 77:035427-1-7.

Brito WH, Kagimura R, Miwa RH (2011).Hydrogen grain boundaries in grapene. Appl. Phys. Lett. 98(21): article # 213107.

Castellanos-Gomes A, Arramel, Wojtaszek M, Smit RHM, Tombros N, Agrait N, Van Wees BJ, Rubio-Bollinger G (2012). Electronic inhomogeneities in graphene: the role of the substrate interaction and chemical doping. Boletin Grupo Español Carbón. 25:18-22.

Castellanos-Gomez A, Smit RHM, Agraït N, Rubio-Bollinger G (2012). Spatially resolved electronic inhomogeneities of graphene due to subsurface charges. Carbon. 50(3): 932-938.

Castellanos-Gomez A,Wojtaszek M, Arramel, Tombros N,Van Wees BJ (2012). Reversible hydrogenation and bandgap opening of graphene and graphite surfaces probed by scanning tunneling spectroscopy. Small 8(10):1607-1613.

Chernozatonskii LA, Mavrin BN, Sorokin PB (2012). Determination of ultrathin diamond films by Raman spectroscopy. Physica Status Solidi B. 249(8):1550-1554.

Cockayne E, Rutter GM,Guisinger NP, Crain JN, First PN, Stroscio JA (2011). Grain boundary loops in graphene. Phys. Rev. B- Condensed Matter Mater. Phys. 83(19):article # 195425.

Data J, Ray NR, Sen P, Biswas HS, Wogler EA (2012). Structure of hydrogenated diamond-like carbon by Micro-Raman spectroscopy. Mater. Lett. 71:131-133.

Davydov SYu, Lebedev AA (2012). Epitaxial single-layer graphene on the SiC substrate. Material Science Forum. 717-720:645-648.

DOE targets for on-board hydrogen storage systems for light-duty vehicles (2012). (http://wwwl.eere.energy.gov/hydrogenandfuelcells/storage/pdfs/targets_onboard_hydro_storage.pdf.).

Dzhurakhalov AA, Peeters FM (2011). Structure and energetics of hydrogen chemisorbed on a single graphene layer to produce graphane. Carbon 49:3258-3266.

Eckmann A, Felten A, Mishchenko A, Brintell L, Krupke R, Novoselov KS, Casiraghi C (2012). Probing the nature of defects in graphene by Raman spectroskopy. Nano Lett. 12(8):3925-3930.

Elias DC, Nair RR, Mohiuddin TMG, Morozov SV, Blake P, Halsall MP, Ferrari AC, Boukhvalov DW, Katsnelson MI, Geim AK, Novoselov KS (2009). Control of graphene's properties by reversible hydrogenation: evidence for graphane. Science 323(5914):610-626.

Geim AK, Novoselov KS (2007).The rise of graphene. Nature Mater. 6(3):183-191.

Goler S, Coletti C, Piazza V, Pingue P, Colangelo F, Pallegrini V, Emtsev KV, Forti S, Starke U, Beltram F, Heun S (2013). Revealing the atomic structure of the buffer layer between SiC(0001) and epitaxial graphene. Carbon 51(1):249-254.

Gupta BK, Tiwari RS, Srivastava ON (2004). Studies on synthesis and hydrogenation behavior of graphitic nanofibers prepared through palladium catalyst assisted thermal cracking of acetylene. J. Alloys Compd. 381:301-308.

Han SS, Jung H, Jung DH, Choi S-H, Park N (2012). Stability of hydrogenation states of graphene and conditions for hydrogenspillover. Phys. Rev. B – Condens. Matter. Mater. Phys. 85(15):article # 155408.

Hornekaer L, Šljivančanin Ž, Xu W, Otero R, Rauls E, Stensgaard I, Lægsgaard E, Hammer B, Besenbacher F (2006). Metastable structures and recombination pathways for atomic hydrogen on the graphite (0001) surface. Phys. Rev. Lett. 96:article # 156104.

Jiang D, Cooper VR, Dai S (2009). Porous graphene as the ultimate membrane for gas separation. Nano Lett. 9: 4019-4024.

Jones JD, Morris CF, Verbeck GF, Perez JM (2012). Oxidative pit formation in pristin, hydrogenated and dehydrogenated graphene. Appl. Surface Sci. 10: 1-11.

Karapet'yants MK, Karapet'yants ML (1968). Fundamental Thermodynamic Constants of Inorganic and Organic Substances. "Khimiya", Moscow.

Khusnutdinov NR (2012). The thermal Casimir–Polder interaction of an atom with a spherical plasma shell. J. Phys. A: Math. Theor. 45:265-301 [arXiv:1203.2732].

Kim K, Lee Z, Regan W, Kisielowski C, Crommie MF, Zettl A (2011). Grain boundary mapping in polycrystalline graphene. ACS Nano. 5(3):2142-2146.

Koepke JC, Wood JD, Estrada D, Ong ZY, He KT, Pop E, Lyding JW (2013). Atomic-scale evidence for potential barriers and strong carrier scattering at graphene grain boundaries: A scanning tunneling microscopy study. ACS Nano. 7(1):75-86.

Lebegue S, Klintenberg M, Eriksson O, Katsnelson MI (2009). Accurate electronic band gap of pure and functionalized graphane from GW calculations. Phys. Rev. B – Condensed Matter Mater. Phys. 79(24): article # 245117.

Lee C, Wei X, Kysar JW,Hone J(2008). Measurement of the elastic properties and intrinsic strength of monolayer graphene. Science 321(5887):385-388.

Lee MJ, Choi JS, Kim JS, Byun I-S, Lee DH, Ryu S, Lee C, Park BH (2012). Characteristics and effects of diffused water between graphene and a SiO_2 substrste. Nano Res. 5(10): 710-717.

Lehtinen PO, Foster AS, Ma Y, Krasheninnikov AV, Nieminen RM (2004). Irradiation-induced magnetism in graphite: A density functional study. Phys. Rev. Lett. 93: 187202-1-4.

Luo Z, Yu T, Kim KJ, Ni Z, You Y, Lim S, Shen Z, Wang S, Lin J (2009). Thickness-dependent reversible hydrogenation of graphene layers. ACS Nano. 3(7):1781-1788.

Nechaev YuS (2010). Carbon nanomaterials, relevance to the hydrogen storage problem. J. Nano Res. 12:1-44.

Nechaev YuS, Veziroglu, TN (2013). Thermodynamic aspects of the stability of the graphene/graphane/hydrogen systems, relevance to the hydrogen on-board storage problem. Adv. Mater. Phys. Chem. 3:255-280.

Openov LA, Podlivaev AI (2010). Thermal desorption of hydrogen from graphane. Tech. Phys. Lett. 36(1): 31-33.

Openov LA, Podlivaev AI (2012). Thermal stability of single-side hydrogenated graphene. Tech. Phys. 57(11):1603-1605.

Palerno V (2013). Not a molecule, not a polymer, not a substrate the

many faces of graphene as a chemical platform. Chem. Commun. 49(28):2848-2857.
Park C, Anderson PE, Chambers A, Tan CD, Hidalgo R, Rodriguez NM (1999). Further studies of the interaction of hydrogen with graphite nanofibers. J. Phys. Chem. B. 103:10572-10581.
Pimenova SM, Melkhanova SV, Kolesov VP, Lobach AS (2002). The enthalpy of formation and C-H bond enthalpy hydrofullerene $C_{60}H_{36}$. J. Phys. Chem. B. 106(9):2127-2130.
Pinto HP, Leszczynski J (2014). Fundamental properties of graphene. In: Handbook of Carbon Nano Materials. Volume 5 (Graphene – Fundamental Properties), Eds. F. D'Souza, K. M. Kadish, Word Scientific Publishing Co, New Jersey et al., pp. 1-38.
Podlivaev AI, Openov LA (2011). On thermal stability of graphone. Semiconductors 45(7): 958-961.
Pujari BS, Gusarov S, Brett M, Kovalenko A (2011). Single-side-hydrogenated graphene: Density functional theory predictions. Phys. Rev. B. 84:041402-1-6.
Riedel C, Coletti C, Iwasaki T, Starke U (2010). Hydrogen intercalation below epitaxial graphene on SiC(0001). Mater. Sci. Forum. 645-648:623-628.
Riedel C, Coletti C, Iwasaki T, Zakharov AA, Starke U (2009). Quazi-free-standing epitaxial graphene on SiC obtained by hydrogen intercalation. Phys. Rev. Lett. 103:246804-1-4.
Sessi P, Guest JR, Bode M, Guisinger NP (2009). Pattering graphene at the nanometer scale via hydrogen desorption. Nano Lett. 9(12):4343-4347.
Showcase 2011, University of Birmingham, UK, April 13-15, 2011; the UK-SHEC website: http://www.uk-shec.org.uk/uk-shec/showcase/ShowcasePresentations.html
Sofo JO, Chaudhari AS, Barber GD (2007). Graphane: A two-dimensional hydrocarbon. Phys. Rev. B. 75:153401-1-4.
Sorokin PB, Chernozatonskii LA (2013). Graphene based semiconductor nanostructures. Physics-Uspekhi. 56(2):113-132.
Stolyarova E, Stolyarov D, Bolotin K, Ryu S, Liu L, Rim KT, Klima M, Hybrtsen M, Pogorelsky I, Pavlishin I, Kusche K, Hone J, Kim P, Stormer HL, Yakimenko V, Flynn G (2009). Observation of graphene bubbles and effective mass transport under graphene films. Nano Lett. 9(1):332-337.
Trunin RF, Urlin VD, Medvedev AB (2010). Dynamic compression of hydrogen isotopes at megabar pressures. Phys. Usp. 53:605-622.
Waqar W, Klusek Z, Denisov E, Kompaniets T, Makarenko I, Titkov A, Saleem A (2000). Effect of atomic hydrogen sorption and desorption on topography and electronic properties of pyrolytic graphite. Electrochemical Soc. Proc. 16:254-265.
Waqar Z (2007). Hydrogen accumulation in graphite and etching of graphite on hydrogen desorption. J. Mater. Sci. 42(4):1169-1176.
Watcharinyanon S, Virojanadara C, Osiecki JR, Zakharov AA, Yakimova R, Uhrberg RIG, Johanson LI (2011). Hydrogen intercalation of graphene grown 6H-SiC(0001). Surface Sci. 605(17-18):1662-1668.
Wojtaszek M, Tombros N, Garreta A, Van Loosdrecht PHM,Van Wees BJ (2011). A road to hydrogenating graphene by a reactive ion etching plasma. J. Appl. Phys. 110(6):article # 063715.
Xiang H, Kan E, Wei S-H, Whangbo MH, Yang J (2009). "Narrow" graphene nanoribbons made easier by partial hydrogenation. Nano Lett. 9(12): 4025-4030.
Xiang HJ, Kan EJ, Wei S-H, Gong XG, Whangbo M-H (2010). Thermodynamically stable single-side hydrogenated graphene. Phys. Rev. B. 82:165425-1-4.
Xie L, Wang X, Lu J, Ni Z, Luo Z, Mao H, Wang R, Wang Y, Huang H, Qi D, Liu R, Yu T, Shen Z, Wu T, Peng H, Oezyilmaz B, Loh K, Wee ATS, Ariando S,Chen W (2011). Room temperature ferromagnetism in partially hydrogenated epitaxial graphene. Appl. Phys. Lett. 98(19):article # 193113.
Yakobson BI, Ding F (2011). Observational geology of graphene, at the nanoscale (Review). ACS Nano 5(3):1569-1574.
Yang FH, Yang RT (2002). Ab initio molecular orbital study of adsorption of atomic hydrogen on graphite: Insight into hydrogen storage in carbon nanotubes. Carbon 40:437-444.
Yazyev OV, Helm L (2007). Defect-induced magnetism in graphene. Phys. Rev. B. 75:125408-1-5.
Yazyev OV, Louie SG (2010).Topological defects in graphene: Dislocations and grain boundaries. Phys. Rev. B - Condensed Matter Mater. Phys. 81(19): article # 195420.
Zhang J, Zhao J (2013). Structures and electronic properties of symmetric and nonsymmetric graphene grain boundaries. Carbon. 55:151-159.
Zhang J, Zhao J, Lu J (2012). Intrinsic strength and failure behaviours of graphene grain boundaries. ACS Nano. 6(3):2704-2711.
Zhang T, Li X, Gao H (2014). Defects controlled wrinkling and topological design in graphene. J. Mech. Phys. Solids 67:2-13.
Zhao X, Outlaw RA, Wang JJ, Zhu MY, Smith GD, Holloway BC (2006). Thermal desorption of hydrogen from carbon nanosheets. J. Chem. Phys. 124:194704-1-6.
Zhou J, Wang Q, Sun Q, Chen XS, Kawazoe Y, Jena P (2009). Ferromagnetism in semihydrogenated graphene sheet. Nano Letters. 9(11):3867-3870.
Zuettel A (2011). Hydrogen the future energy carrier. In: Materials of Int. Hydrogen Research

13

Effect of an axial magnetic field on the heat and mass transfer in rotating annulus

Sofiane ABERKANE[1], Malika IHDENE[2], Mourad MODERES[3] and Abderahmane GHEZAL[4]

[1]Département Energétique, Faculté des sciences de l'ingénieur, Université M'Hamed Bougara de Boumerdés-35000, Algérie.
[2]Université de Yahia Farès, Médéa- 26000, Algérie.
[3]Faculté des hydrocarbures et de la chimie, Université M'Hamed Bougara de Boumerdés-35000, Algérie.
[4]Laboratoire de mécanique des fluides théorique et appliquée, Faculté de physique, Université des sciences et de la technologie de Houari Boumediene Bab Ezzouar, Alger-16111, Algérie.

This study is interested in the effect of an axial magnetic field imposed on incompressible flow of electrically conductive fluid between two horizontal coaxial cylinders. The imposed magnetic field is assumed uniform and constant. The effect of heat generation due to viscous dissipation is also taken into account. The inner and outer cylinders are maintained at different uniform temperatures and concentrations. The movement of the fluid is due to rotation of the cylinder with a constant speed. An exact solution of the governing equations for momentum and energy are obtained in the form of Bessel functions. A finite difference implicit scheme was used in the numerical solution to solve the governing equations of convection flow and mass transfer. The velocity, concentration and temperature distributions were obtained with and without the magnetic field. The results show that for different values of the Hartmann number, the velocity and concentration between the two cylinders decreases as the Hartmann number increases. On the other hand, the Hartmann number does not affect the temperature. Also, it is found that by increasing the Hartmann number, the Nusselt and Sherwood numbers decreases.

Key words: Rotating cylinders, viscous dissipation, heat transfer, mass transfer, magnetic field, Bessel function, finite difference.

INTRODUCTION

The study of flow of electrically conductive fluids, called magnetohydrodynamic (MHD) has attracted much attention due to its various applications. In astrophysics and geophysics, it is applied to the study of stellar structures, terrestrial cores and solar plasma. In industrial processes, it finds its application in MHD pumps, nuclear reactors, the extraction of geothermal energy, metallurgical and crystal growth in the field of semiconductors, the control of the behavior of fluid flow and the stability of convective flows. The analysis of flow and heat and mass transfer, known as the double-diffusive convection, in cylindrical annuli has been investigated in several pieces of literature. However, to the author's knowledge a few studies have been conducted on double-diffusive convection in a rotating annulus in the presence of a magnetic field. Recently, the effect of magnetic field on the laminar convection in either vertical or horizontal rotating concentric annuli has been investigated. Ben and Henry. (1996) investigated numerically the effect of a constant magnetic field on a three-dimensional buoyancy-induced flow in a cylindrical cavity, they put in light the structural changes of the flow induced by the magnetic field for each field orientation. Singh et al. (1997) presented exact solutions for fully developed natural convection in open-ended vertical concentric annuli under a radial magnetic field. El Amin (2003) studied the effects of both first- and second-order resistance due to the solid matrix on forced convective flow from a horizontal circular cylinder in the presence of a magnetic field and viscous dissipation, with a variable surface temperature boundary condition. Hayat and Kara (2006) investigated the Couette time-dependent flow of an incompressible third-grade fluid subjected to a magnetic field of variable strength analytically. Group theoretic methods were employed to analyze the nonlinear problem and a solution for the velocity field was obtained analytically. Sankar et al. (2006) studied numerically a natural convection of a low Prandtl number electrically conducting fluid under the influence of either axial or radial magnetic field in a vertical cylindrical annulus. They showed that the magnetic field can be suppress the flow and heat transfer. Bessaïh et al. (2009) studied the MHD stability of an axisymmetric rotating flow in a cylindrical enclosure containing liquid metal ($Pr = 0.015$), with an aspect ratio equal to 2, and subjected to a vertical temperature gradient and an axial magnetic field. Azim et al. (2010) studied numerically the effect of magnetic field and Joule heating on the coupling of convection flow along and conduction inside a vertical flat plate in the presence viscous dissipation and heat generation. Ellahi et al. (2010) determined analytic solutions for a nonlinear problem governing the MHD flowof a third grade fluid in the annulus of rotating concentric cylinders. Makinde and Onyejekwe (2011) investigated a steady flow and heat transfer of an electrically conducting fluid with variable viscosity and electrical conductivity between two parallel plates in the presence of a transverse magnetic field. Kakarantzas et al. (2011) studied numerically the combined effect of a horizontal magnetic field and volumetric heating on the natural convection flow and heat transfer of a low Prandtl number fluid in a vertical annulus. Seth et al. (2011) studied the effects of rotation and magnetic field on unsteady Couette flow of a viscous incompressible electrically conducting fluid between two horizontal parallel porous plates in a rotating medium. Mozayyeni and Rahimi (2012) investigated numerically the problem of mixed convection of a fluid in the fully developed region between two horizontally concentric cylinders with infinite lengths, in the presence of a constant magnetic field with a radial MHD force direction, considering the effects of viscous heat dissipation in the fluid in both steady and unsteady states. Seth and Singh (2013) studied theoretically the effect of Hall current and a uniform transverse magnetic field on unsteady MHD Couette flow of class-II in a rotating system. Takhar et al. (2003) studied the unsteady mixed convection flow over a rotating vertical cone in the presence of a magnetic field. Recently, some attention has been paid by Ashorynejad (2013) to the effect of magnetic field convection on natural convection heat transfer in a horizontal cylindrical annulus enclosure filled with nanofluid using the Lattice Boltzmann method. Also Sheikholeslami et al. (2013) solved the problem of heat and fluid flow of a nanofluid in a half-annulus enclosure with one wall under constant heat flux in presence of magnetic flied using control volume based finite element method. In another publication Sheikholeslami et al. (2014) applied Lattice Boltzmann Method to simulate the effect of magnetic field on free convection of nanofluid, in an eccentric semi-annulus. In a recent paper Aminfar et al. (2014) experimentally studied the effects of using magnetic nanofluid and also applying an external magnetic field on the critical heat flux of subcooled flow boiling in vertical annulus.

Some surveyed studies in the literature were concerned primarily with the double diffusive convection, Teamah (2007) carried out a numerical study of double-diffusive laminar mixed convection within a two-dimensional, horizontal annulus rotating cylinders. The results for both average Nusselt and Sherwood numbers were correlated in terms of Lewis number, thermal Rayleigh number and buoyancy ratio. Moreover, Molki et al. (1990) applied the naphthalene sublimation technique to an annulus with a rotating inner cylinder in order to study heat transfer in the entrance region to obtain heat transfer data for laminar flows and compare them with results of mass transfer. Kefeng and Wen-Qiang (2006) simulated numerically the characteristics of transient double-diffusive convection in a vertical cylinder using a finite element method. Recently, Venkatachalappa et al. (2011) carried out numerical computations to investigate the effect of axial or radial magnetic field on the double-diffusive natural convection in a vertical cylindrical annular cavity.

Although the exact solutions for the Hartmann flow and the MHD Couette flow have been achieved for more than seventy years, the solutions for a heat transfer in flow between concentric rotating cylinders, also known as Taylor Couette flows, under external magnetic field have been restricted to high Hartmann numbers.

The aim of the present study is to examine analytically and numerically the effects of an external axial magnetic field applied to the forced convection flow of an electrically conducting fluid between two horizontal

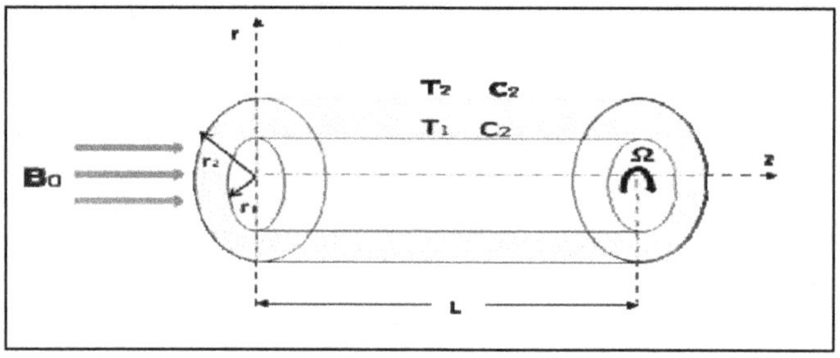

Figure 1. Geometry of the problem.

concentric cylinders, considering the effects of viscous heat dissipation in the fluid. Also we investigated numerically the effects of the magnetic field on the mass transfer in the annular cavity. It should be noted that the natural convection is supposed negligible in this work, which is not always the case of the vertical cylinder. The forced flow is induced by the rotating inner cylinder, in slow constant angular velocity and the other is fixed.

FORMULATION OF THE PROBLEM

Consider a laminar flow of a viscous incompressible electrically conductive fluid between two coaxial cylinders. The inner cylinder of radius r_1 is rotated at a constant speed Ω_1 and the outer cylinder of radius r_2 is kept fixed. The inner and outer walls are maintained at a constant and different temperatures and concentrations, but their values for the inner are higher than the outer, while the top and bottom walls are insulated and impermeable. The two cylinders are electrically isolated. The flow is subjected to a constant uniform and axially magnetic field B_0. Geometry of the problem is presented in Figure 1. We assume that the magnetic Reynolds number is neglected. When the magnetic field is uniform and externally applied, its time variations can be neglected and the set of flow equations further simplified to involve only the Navier-Stokes equations and the conservation of the electric current. Also we assume that the electric field is zero. In this study the viscous dissipation term in the energy equation is considered.

ANALYTICAL STUDY

The flow is assumed to be steady, laminar and unidirectional, therefore the radial and axial components of the velocity and the derivatives of the velocity with respect to θ and z are zero. Under these assumptions and in cylindrical coordinates, the governing equations for the flow following the azimuthal direction can be written as follows:

$$\upsilon\left(\frac{\partial^2 v}{\partial r^2}+\frac{1}{r}\frac{\partial v}{\partial r}-\frac{v}{r^2}\right)-\frac{\sigma v B_0^{\,2}}{\rho}=0 \tag{1}$$

$$\frac{k}{r}\frac{\partial}{\partial r}\left(r\frac{\partial T}{\partial r}\right)=-\mu\left(\frac{\partial v}{\partial r}-\frac{v}{r}\right)^2 \tag{2}$$

$$r=r_1 : v(r)=\Omega_1 r_1, T=T_1 \tag{3}$$

$$r=r_2 : v(r)=\Omega_2 r_2, T=T_2 \tag{4}$$

The governing equation and boundary conditions, Equations (1) to (4), which are in non-adimensional form, become:

$$\frac{\partial^2 v^*}{\partial r^{*2}}+\frac{1}{r^*}\frac{\partial v^*}{\partial r^*}-\left(\frac{Ha^2}{(1-\eta)^2}+\frac{1}{r^{*2}}\right)v^*=0 \tag{5}$$

$$\frac{1}{r^*}\frac{\partial}{\partial r^*}\left(r^*\frac{\partial\theta}{\partial r^*}\right)=-Ec\Pr\left(\frac{\partial v^*}{\partial r^*}-\frac{v^*}{r^*}\right)^2 \tag{6}$$

$$r^*=\eta : v^*(r^*)=1, \theta=1 \tag{7}$$

$$r^*=1 : v^*(r^*)=b, \theta=0 \tag{8}$$

Where

$$r^*=\frac{r}{r_2}, v^*=\frac{v}{\Omega_1 r_1}, \eta=\frac{r_1}{r_2}, b=\frac{\Omega_2 r_2}{\Omega_1 r_1}, Ha=B_0 d\sqrt{\frac{\sigma}{\rho\upsilon}}, \theta=\frac{T-T_2}{T_1-T_2}, Pr=\frac{\upsilon}{a}, Ec=\frac{(\Omega_1 r_1)^2}{Cp\Delta T}$$

Where, the stars are dropped for convenience. The velocity profile in the annular space is obtained by solving the Equation (5) as follows:

$$v(r)=C_1 I_1\left(\frac{Ha}{1-\eta}r\right)+C_2 K_1\left(\frac{Ha}{1-\eta}r\right)=0 \tag{9}$$

Where:

$$M=\frac{Ha}{1-\eta}$$

C_1 and C_2 are the constants of integration, which are determined from the boundary conditions on the velocity.

$$C_1=\frac{K_1(M)-bK_1(\eta M)}{I_1(\eta M)K_1(M)-K_1(\eta M)I_1(M)}$$

$$C_2=\frac{bI_1(\eta M)-I_1(M)}{I_1(\eta M)K_1(M)-K_1(\eta M)I_1(M)}$$

I_1 is the modified Bessel function of the first kind of order 1, and K_1 is the modified Bessel function of the second kind of order 1. To obtain the temperature field from Equation (6), we performed calculations by using the expansions with three terms of the modified Bessel functions $I_1(Mr)$ and $K_1(Mr)$ used by Omid et al. (2012), for small values of Ha. It can be used as following:

$$I_1(Mr) \cong \frac{1}{2}Mr + \frac{(Mr)^3}{16} + \frac{(Mr)^5}{384} \quad (10)$$

$$K_1(Mr) \cong \frac{1}{Mr} + \left[\frac{1}{2}\ln\left(\frac{Mr}{2}\right) - \frac{1}{4}(-2\gamma+1)\right](Mr) + \left[\frac{1}{16}\ln\left(\frac{Mr}{2}\right) - \frac{1}{32}\left(\frac{5}{2}-2\gamma\right)\right](Mr)^3 \quad (11)$$

Where γ is Euler's constant defined by:

$$\gamma = \lim_{x\to\infty}\left[1+\frac{1}{2}+\frac{1}{3}+\frac{1}{4}+.....+\frac{1}{m}-\ln(m)\right] = 0,5772156649...$$

By substituting the values of $I_1(Mr)$ and $K_1(Mr)$ from the above expansions in the velocity equation, Equation (9), and using the new velocity distribution in Equation (6) to find the temperature field. The temperature gradient is given then by the following equation:

$$\frac{\partial\theta}{\partial r} = \frac{C_3}{r} - \frac{Br}{r}\left[\begin{array}{l} 2C_2^2\ln\left(\frac{Mr}{2}\right) + \frac{2C_2^2}{M^2r^2} + C_5(Mr)^6\ln\left(\frac{Mr}{2}\right) + \\ C_6(Mr)^6 - \frac{1}{384}C_2^2(Mr)^6\left(\ln\left(\frac{Mr}{2}\right)\right)^2 \\ +C_7(Mr)^2 + C_8(Mr)^4 - \frac{1}{32}C_2^2(Mr)^4\ln\left(\frac{Mr}{2}\right) \\ +C_9(Mr)^8 - \frac{1}{3072}C_1C_2(Mr)^8\ln\left(\frac{Mr}{2}\right) \\ -\frac{1}{92160}C_1^2(Mr)^{10} \end{array}\right] \quad (12)$$

Where the constants C_5 to C_9 are given in terms of C_1 and C_2 as follows:

$$C_5 = \frac{11}{2304}C_1^2 - \frac{1}{192}C_1C_2 - \frac{1}{192}C_2^2\gamma$$

$$C_6 = \frac{11}{2304}C_2^2\gamma + \frac{7}{2304}C_1C_2 - \frac{1}{384}C_2^2\gamma - \frac{1}{384}C_1^2 - \frac{125}{55296}C_2^2 - \frac{1}{192}C_1C_2\gamma$$

$$C_7 = \frac{1}{4}C_2^2\gamma - \frac{7}{16}C_2^2 + \frac{1}{4}C_1C_2$$

$$C_8 = \frac{1}{32}C_2^2 - \frac{1}{32}C_2^2\gamma - \frac{1}{48}C_1C_2$$

$$C_9 = \frac{7}{24576}C_1C_2 - \frac{1}{3072}C_1C_2\gamma - \frac{1}{3072}C_1^2$$

The solution of the energy equation is:

$$\theta = C_4 + C_3\ln(r) + Br\left[\begin{array}{l} C_{10}(Mr)^6 + C_{11}(Mr)^8 + C_{12}(Mr)^4 + \frac{1}{2}C_7(Mr)^2 - \frac{1}{921600}C_1^2(Mr)^{10} \\ +C_2^2\left(\ln\left(\frac{Mr}{2}\right)\right)^2 2C_2^2\ln\left(\frac{Mr}{2}\right) + \frac{C_2^2}{(Mr)^2} + \frac{1}{6}C_5(Mr)^6\ln\left(\frac{Mr}{2}\right) \\ +C_6(Mr)^6 - \frac{1}{2304}C_2^2(Mr)^6\left(\ln\left(\frac{Mr}{2}\right)\right)^2 + \frac{1}{6912}C_2^2(Mr)^6\ln\left(\frac{Mr}{2}\right) \\ -\frac{1}{128}C_2^2(Mr)^4\ln\left(\frac{Mr}{2}\right) - \frac{1}{24576}C_1C_2(Mr)^8\ln\left(\frac{Mr}{2}\right) \end{array}\right] \quad (13)$$

Where the Constants C_{10}, C_{11} and C_{12} are given as follows:

$$C_{10} = \frac{1}{36}C_5 + \frac{1}{6}C_6 - \frac{1}{41472}C_2^2$$

$$C_{11} = \frac{1}{8}C_9 + \frac{1}{196608}C_1C_2$$

$$C_{12} = \frac{1}{4}C_8 + \frac{1}{512}C_2^2$$

NUMERICAL STUDY

In this numerical study, we consider two-dimensional and axisymmetric unsteady flow. We opted for the velocity - pressure formulation due to its rapidity of prediction, its lower cost, and its ability to simulate real conditions. The finite difference scheme adopted for the resolution is very similar to that used by Peyrret (1976), and Ghezal and Porterie. (2011), this is a semi implicit scheme of Crank-Nicholson type, iterative process based on the perturbation of the continuity equation by introducing an artificial compressibility. The spatial discretization using the marker and cell (MAC) is shown in Figure 2. The iterative procedure is assumed converged when the following test is verified

$$\max\left(|L_u|,|L_v|,|L_w|,|L_\theta|,|L_C||D|\right) < \varepsilon$$

where L_u, L_v, L_w, L_θ, L_C and D represents operators differences relating to system equations corresponding to the problem variables u, v, w, θ, C and Π respectively, ε is of the order of 10^{-5} depending on the considered case.

We then proceeded to a study of the mesh sensitivity of the field of study. This study led us to retain a mesh of 336 nodes along the direction r and 48 nodes in the z direction.

Mathematical equations

Based on these dimensionless variables, the conservation equations of mass, momentum and energy are written in non rotating frame cylindrical coordinates as follows (where the stars are dropped for convenience):

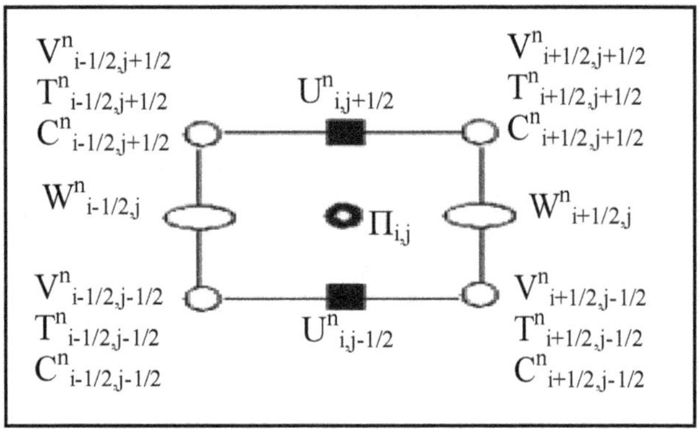

Figure 2. M.A.C cell.

$$\frac{\partial u}{\partial r}+\frac{u}{r}+\frac{\partial w}{\partial z}=0 \quad (14)$$

$$\frac{\partial u}{\partial t}+u\frac{\partial u}{\partial r}-\frac{v^2}{r}+w\frac{\partial u}{\partial z}=-\frac{\partial \Pi}{\partial r}+\frac{1-\eta}{Ta}\left(\frac{\partial^2 u}{\partial r^2}+\frac{1}{r}\frac{\partial u}{\partial r}+\frac{\partial^2 u}{\partial z^2}-\frac{u}{r^2}\right)-\frac{Ha^2 u}{(1-\eta)Ta} \quad (15)$$

$$\frac{\partial v}{\partial t}+u\frac{\partial v}{\partial r}+\frac{vu}{r}+w\frac{\partial v}{\partial z}=\frac{1-\eta}{Ta}\left(\frac{\partial^2 v}{\partial r^2}+\frac{1}{r}\frac{\partial v}{\partial r}+\frac{\partial^2 v}{\partial z^2}-\frac{v}{r^2}\right)-\frac{Ha^2 v}{(1-\eta)Ta} \quad (16)$$

$$\frac{\partial w}{\partial t}+u\frac{\partial w}{\partial r}+w\frac{\partial w}{\partial z}=-\frac{\partial \Pi}{\partial z}+\frac{1-\eta}{Ta}\left(\frac{\partial^2 w}{\partial r^2}+\frac{1}{r}\frac{\partial w}{\partial r}+\frac{\partial^2 w}{\partial z^2}\right) \quad (17)$$

$$\frac{\partial \theta}{\partial t}+u\frac{\partial \theta}{\partial r}+w\frac{\partial \theta}{\partial z}=\frac{1-\eta}{Pr\,Ta}\left(\frac{\partial^2 \theta}{\partial r^2}+\frac{1}{r}\frac{\partial \theta}{\partial r}+\frac{\partial^2 \theta}{\partial z^2}\right)+\frac{(1-\eta)Ec\Phi}{Ta} \quad (18)$$

$$\frac{\partial C}{\partial t}+u\frac{\partial C}{\partial r}+w\frac{\partial C}{\partial z}=\frac{1-\eta}{ScTa}\left(\frac{\partial^2 C}{\partial r^2}+\frac{1}{r}\frac{\partial C}{\partial r}+\frac{\partial^2 C}{\partial z^2}\right) \quad (19)$$

Where:

$Ha=Bd\sqrt{\frac{\sigma}{\rho\upsilon}}$ is the Hartmann number,

$Ta=\frac{\Omega_1 r_1 d}{\nu}$ is the Taylor number,

$Pr=\frac{\upsilon}{a}$ is the Prandtl number

$Sc=\frac{\upsilon}{D}$ is the Schmidt number,

$d=r_1-r_2$ is the width of the annular space,

$\Phi=2\left[\left(\frac{\partial u}{\partial r}\right)^2+\left(\frac{u}{r}\right)^2+\left(\frac{\partial w}{\partial z}\right)^2\right]+\left(\frac{\partial u}{\partial z}+\frac{\partial w}{\partial r}\right)^2+\left(\frac{\partial v}{\partial r}-\frac{v}{r}\right)^2+\left(\frac{\partial v}{\partial z}\right)^2$ is the viscous dissipation function.

The rate of heat transfer in non – dimensional for the inner and outer cylinder is given by:

$$Nu_i(z)=-\xi\left.\frac{\partial \theta}{\partial r}\right|_{r=\eta}$$

$$Nu_e(z)=-\xi\left.\frac{\partial \theta}{\partial r}\right|_{r=1}$$

With: $\xi=1-\eta$

The average Nusselt number on the inner and outer cylinders is given by:

$$\overline{Nu_i}=\frac{1}{L}\int_0^z Nu_i(z)dz$$

$$\overline{Nu_e}=\frac{1}{L}\int_0^z Nu_e(z)dz$$

Similarly, we can calculate both local Sherwood number as follows:

$$Sh_i(z)=-\xi\left.\frac{\partial C}{\partial r}\right|_{r=\eta}$$

$$Sh_e(z)=-\xi\left.\frac{\partial C}{\partial r}\right|_{r=1}$$

Initial and boundary conditions

At the time t=0:

$$u(r,z,0)=v(r,z,0)=w(r,z,0)=\Pi(r,z,0)=\theta(r,z,0)=C(r,z,0)=0 \quad (20)$$

The boundary conditions are as follows:

$$r=\eta \;\; z\geq 0:\; u(r_1,z)=v(r_1,z)=w(r_1,z)=0$$
$$\theta(r_1,z)=C(r_1,z)=1 \quad (21)$$

$$r=1 \quad z\geq 0:\; u(r_2,z)=v(r_2,z)=w(r_2,z)=0$$
$$\theta(r_2,z)=C(r_2,z)=0 \quad (22)$$

$$\eta<r<1 \quad z=0: \quad u=v=w=0,\; \frac{\partial C}{\partial z}=\frac{\partial \theta}{\partial z}=0 \quad (23)$$

$$z=L: \quad u=v=w=0,\; \frac{\partial \theta}{\partial z}=\frac{\partial C}{\partial z}=0 \quad (24)$$

RESULTS AND DISCUSSION

In order to understand the physical situation of the problem and the effects of the Hartmann and Eckert numbers, we have found the numerical and analytical

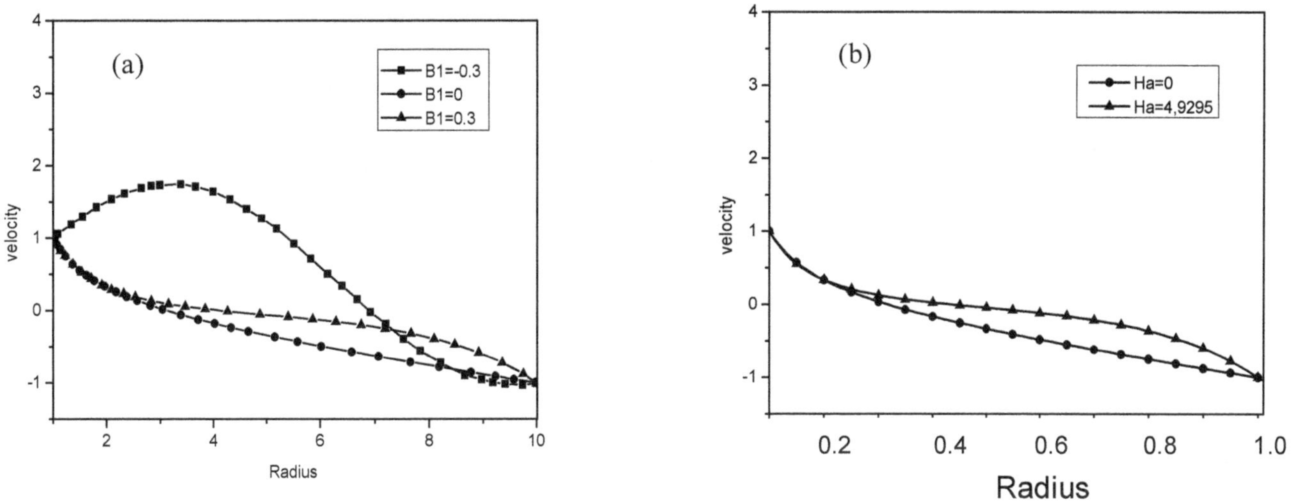

Figure 3. Velocity profile of the fluid in the annulus for b = -1, (a) Result of Dizaji Feiz et al. (2008), (b) Results of the present analytical study.

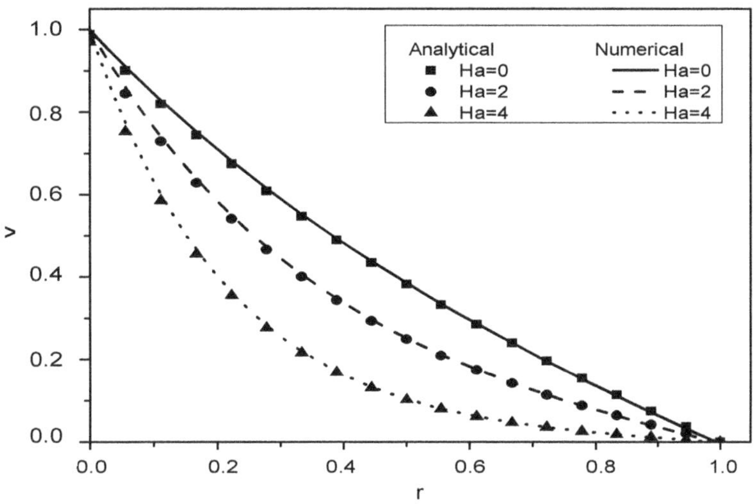

Figure 4. Comparison of analytical and numerical results of velocity profile, for η = 0.5,b=0, Ta=20,t^*=120.

values of the velocity, temperature, concentration, the Nusselt number and Sherwood number. The analytical method developed in the present work has been compared in Figure 3 with the results obtained by Feiz-Dizaji (2008), for the velocity profiles of the fluid in the annulus of concentric cylinders with velocity ration b =-1. The results are found to be in good agreement except for the negative values of magnetic field B_1 which cannot exist in our non- dimensional study, Hartmann number (Ha) is a strictly positive non-dimensional number.

The results obtained through the developed code in FORTRAN based on an implicit finite-difference method described earlier, are compared with those calculated using the analytical approach for small value of Hartmann number. The velocity, temperature and average Nusselt numbers are evaluated analytically and numerically for different values of Hartmann number in Figures 4, 5 and 6.

Obviously, the velocity and temperature profiles, for various Ha obtained via these two different methods, agree with each other reasonably well. We can notice in Figure 4 that the velocity profile without magnetic field Ha=0 is quasi-linear, and an increase in Hartman number, which causes a reduction of the velocity in the annular space because the centrifugal force is counter- productive

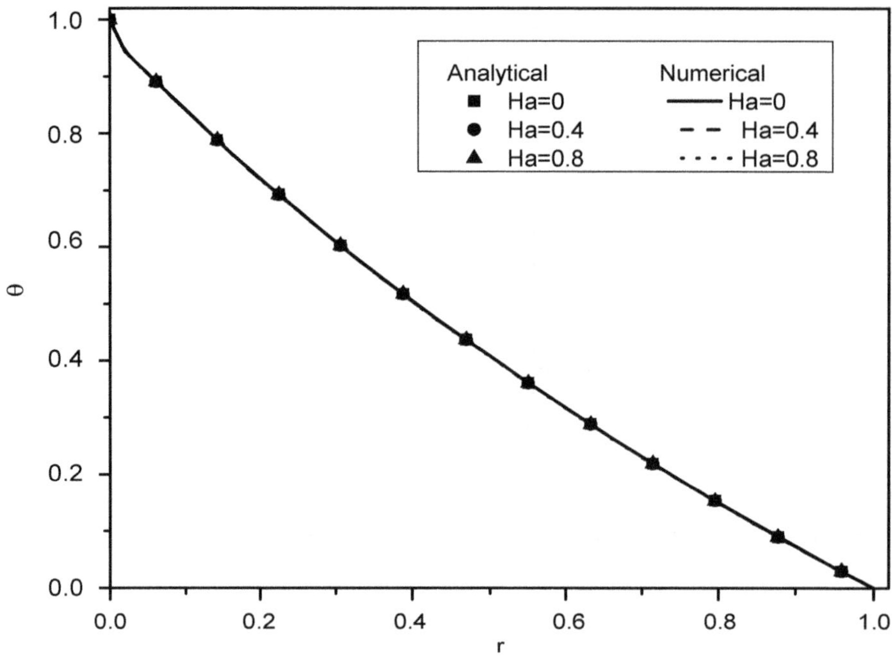

Figure 5. Comparison of analytical and numerical results of temperature profile, for η = 0.5, *b=0,* Ta=20, Pr = 0.02, Ec=0.5, t^**=120.*

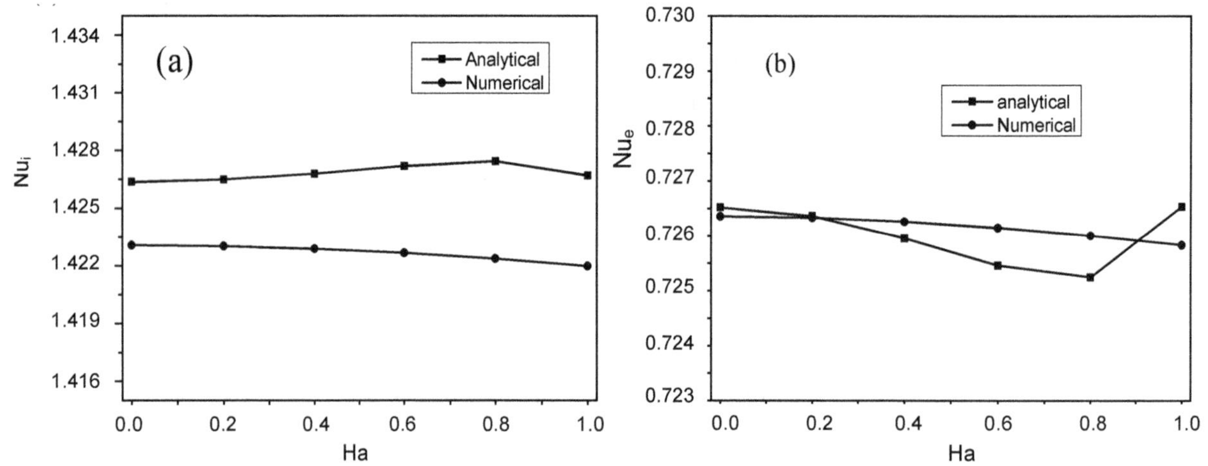

Figure 6. Comparison of analytical and numerical results of average Nusselt number on (a) inner and (b) outer surfaces of the cylinder against the Hartmann number, for *η = 0.5, b=0, Pr = 0.02, Ec=0.5.*

and the Lorentz electromagnetic force acts as a flow damper. It is observed from Figure 5 that the effect of weak magnetic field on the radial profile of temperature is insignificant for small value of Prandtl numbers (Pr=0.02) which is appropriate for liquid metal. It is valid in the case of low and high values of Hartmann number. Figure 6 displays the effect of Hartmann number on the average Nusselt number on inner and outer surfaces. As can be seen, from this figure that the analytic approach corresponding to the expansion with three terms of the modified Bessel functions is closer to the numerical approach. The difference between the analytical and numerical values is approximately 10^{-3}, even the results obtained from the present theoretical analysis are restricted to a one-dimensional flow and the numerical results are calculated using two-dimensional axisymmetric flow. Figure 7 shows the effect of Hartmann number on the local Nusselt number on the inner and

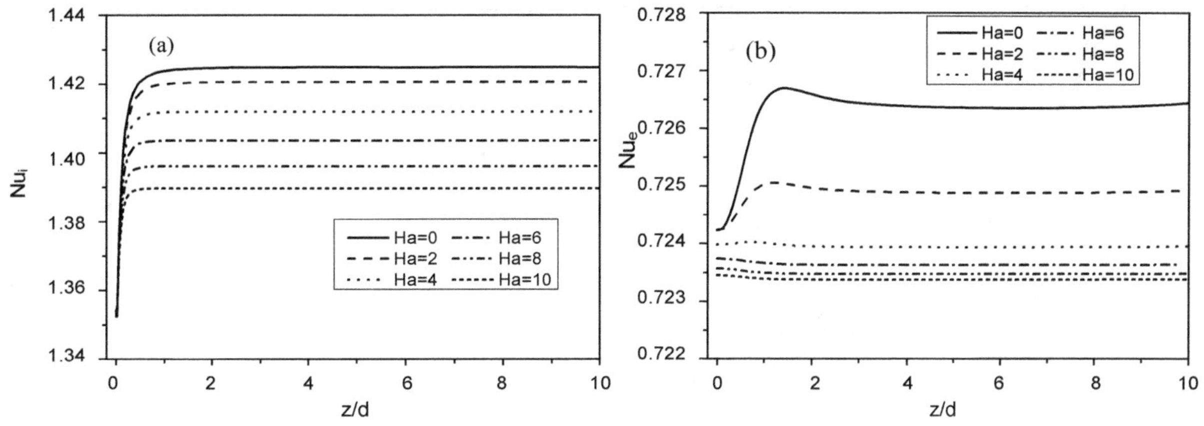

Figure 7. Effect of Hartman number on local Nusslet number distribution on (a) inner and (b) outer cylinders, for η = *0.5, Pr = 0.02, Ec=0.5,* t^{*}*=120.*

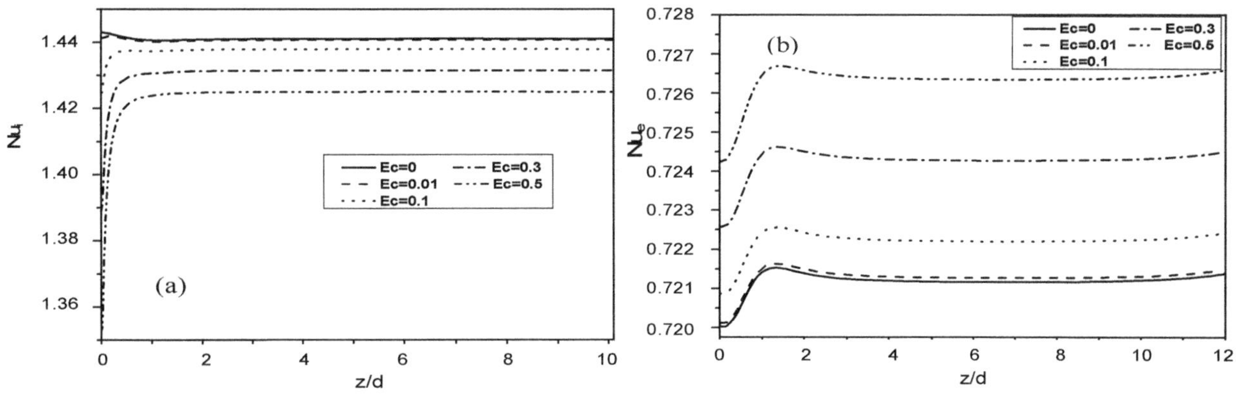

Figure 8. Effect of Eckert number on local Nusselet number distribution on (a) inner and (b) outer cylinders, for η = *0.5, Pr = 0.02, Ha=0,* t^{*}*=120.*

outer surfaces, for an Eckert number Ec = 0.5. It is found that for high values of Hartmann number, the local Nusselt number on the inner and outer surfaces decreases.

In fact when the Eckert number is considerable, the heat generation in the fluid increases because the viscous dissipation. Thus the temperature of the fluid in the annular space increases causing a decrease in the temperature gradient in the vicinity of the inner cylinder and an increase of the gradient in the vicinity of the outer cylinder. A significant increase in the Hartmann number causes a reduction of the centrifugal force, which results in a gradual decrease in the Nusselt number. It is worth to mention that this phenomenon is in full accordance with what was previously observed by Mozayyeni (2013) for a horizontal cylindrical annulus, also by El-Amin (2003) for horizontal cylinder in a porous medium and by Takhar et al. (2003). for rotating vertical cone.

The analysis of the variation of local Nusselt number on the inner and outer cylinder shows that this number tends to a limit value. It can be noticed that the Nusselt number on the outer cylinder is lower than on the inner cylinder, because the velocity and temperature gradient are higher for the cold inner cylinder than for the outer cylinder. It should be also noted that the effect of magnetic field on the temperature distribution is insignificant, whereas the changes induced by the magnetic field on the temperature gradient and therefore on the Nusselt number is considerable.

Effect of Eckert number on the distribution of local Nusselt number on the inner and outer cylinders isdisplayed in Figure 8, for Ha = 0. As can be seen, with increase of Eckert number, the influence of heat transfer due to the viscous dissipation in the annular space is improved, which leads to the increase in the average temperature of the fluid at this region, especially near the

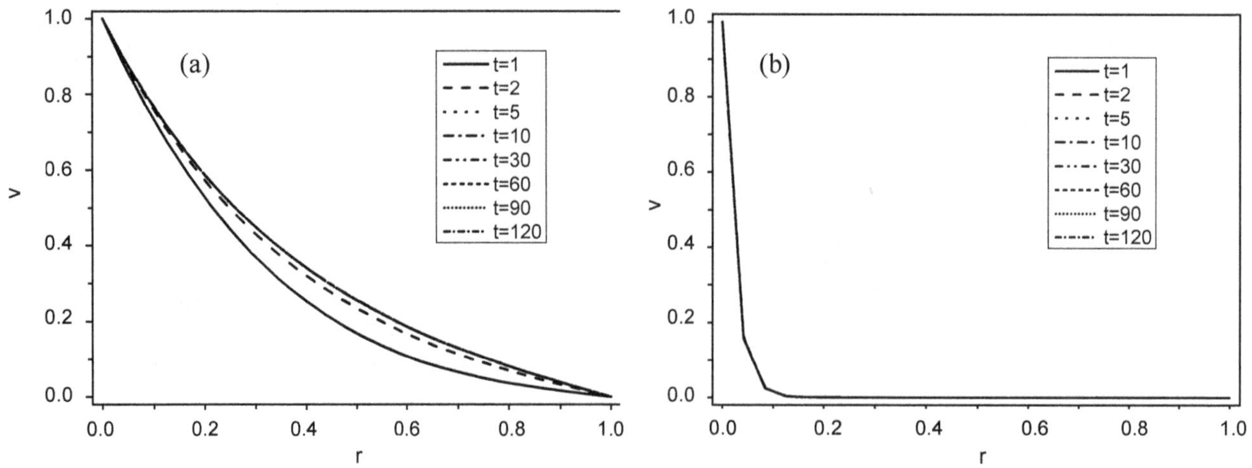

Figure 9. Velocity distribution at different times at z/d=7 and for Ta=20, (a) Ha = 2 and (b) Ha = 50.

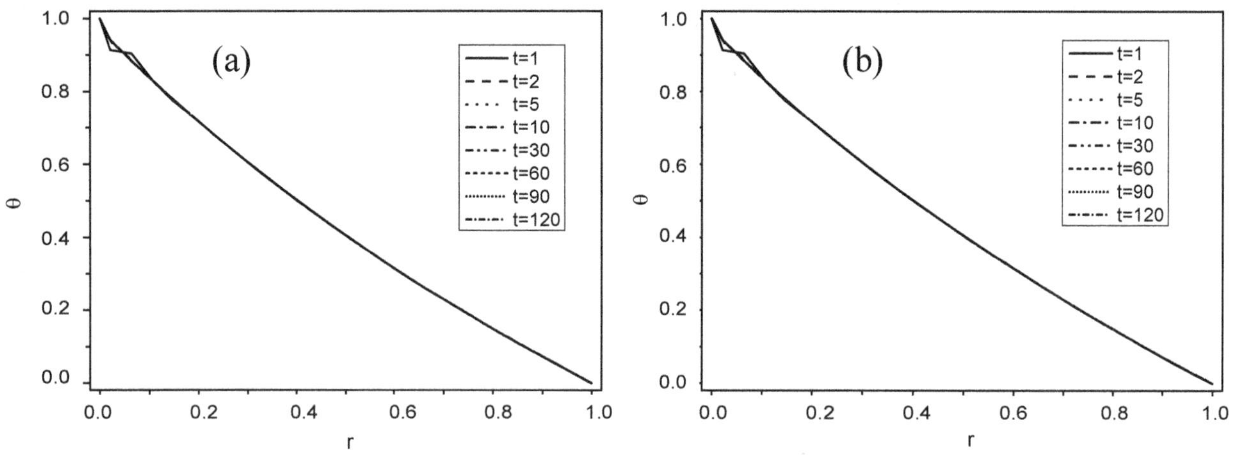

Figure 10. Temperature distribution at different times for z/d=7, Ta=20, Ec=0, Pr=0.02, for (a)Ha = 2 and (b)Ha=50.

inner cylinder, to be more than in the past. The dimensionless temperatures of inner and outer cylinders are maintained at 1.0 and 0.0, respectively.

It is evident that by increasing the average temperature of fluid in annular space, the rate of heat transfer between the fluid and inner cylinder decreases due to the reduction of the temperature difference between them. In contrast to the other case, the local Nusselt number on the outer cylinder increases as the Eckert number increases because of the enhancement of temperature differences between the fluid and the outer cylinder.

In this part, some results are presented in different non-dimensional time values for the distribution of velocity and temperature in the annulus Figures 9, 10, and 11. From Figure 10, we can notice that for a small value of Prandtl number (Pr=0,02), The effect of the time variation is found to be not significant on the temperature, it reaches faster a steady-state to the point that we can't notice the difference between the steady and unsteady states flows. As we know, for larger fluid Prandtl number, the momentum flow transfer is faster than heat transfer. This can be seen clearly in Figure 11 (for a fluid with Pr =7) and the distribution of the azimuthal component of velocity reaching a steady-state quicker than the temperature at the mid-length. There is not much difference in velocity at t= 10, compared to t=120, but comparing temperature distribution at t=10 with values greater than 10, it indicates that much more time is still needed to reach steady-state.

The effects of Hartmann number on the concentration

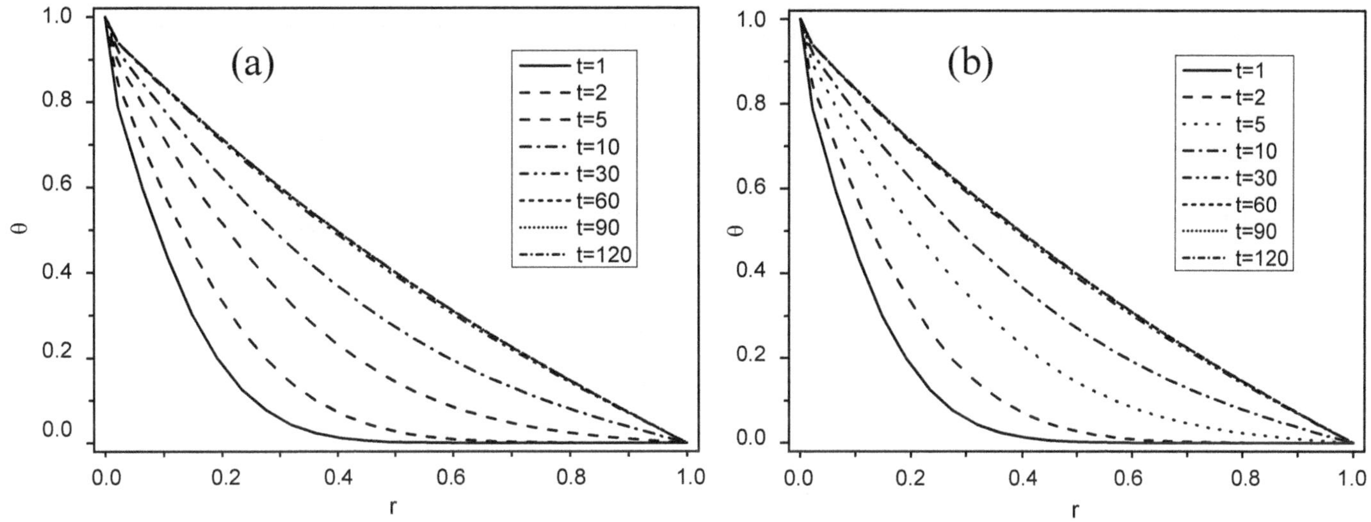

Figure 11. Temperature distribution at different times at z/d=7, for Ta=20, Ec=0, Pr=7, for (a) Ha = 2 and (b) Ha = 5.

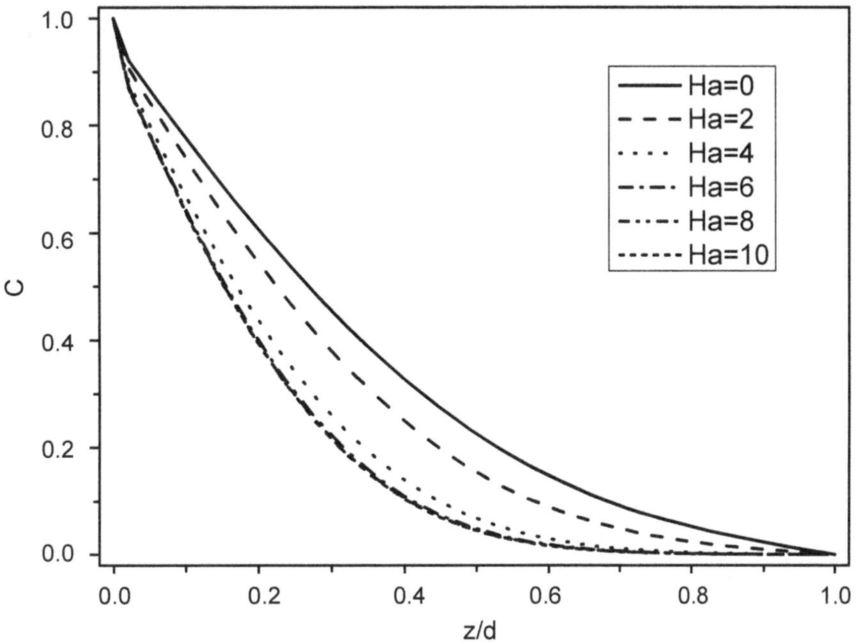

Figure 12. Concentration profile as a function of Hartmann number, for η = 0.5, Ta=20, Pr= 0.02, Ec=0, *t=120, z/d=7.*

At the annulus center are shown in Figure 12. It is observed that the concentration decrease from the inner cylinder where is considered to be a source for the concentration to tends towards almost zero value in the outer cylinder. As can be also seen from this figure, for increasing Hartmann number the concentration decreases in the annular cavity.

The mass transfer rate across the annular cavity is investigated using the computed local Sherwood numbers in the inner and outer cylinders, for different Hartmann number and Sc=10 in Figure 13. It can be noticed that the rate of mass transfer is higher on the inner cylinder than on the outer cylinder. This is reasonable to expect since the velocity and concentration

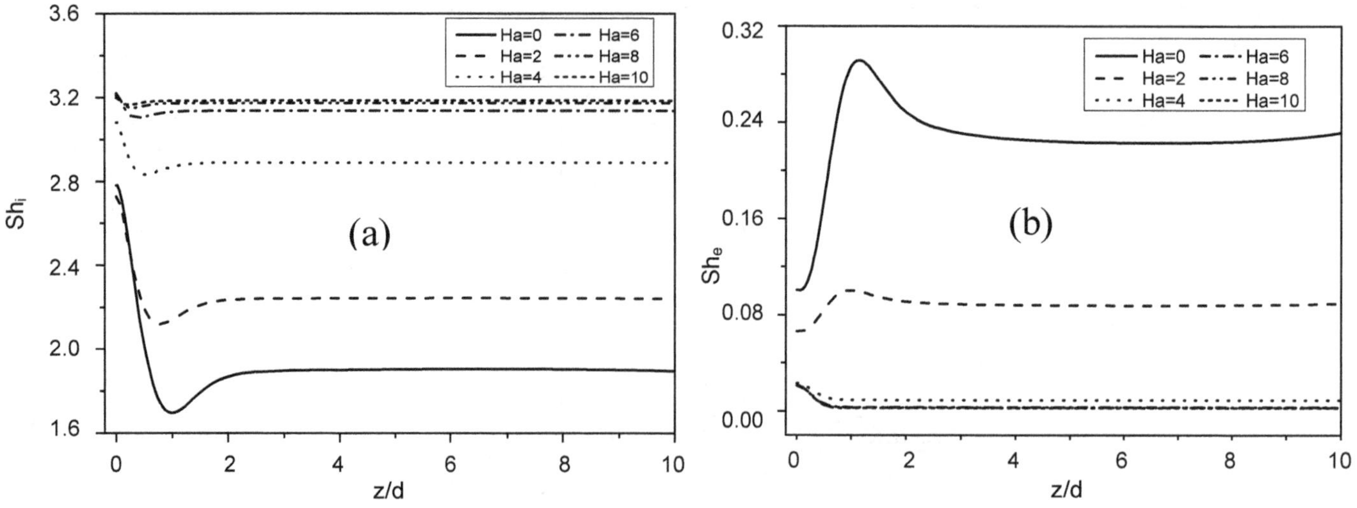

Figure 13. Effect of Hartman number on local Sherwood number distribution on (a) inner and (b) outer cylinders, for *η = 0.5, Sc=10, Ec=0, t=120.*

gradient are higher for the inner cylinder than for the outer cylinder. The rate of mass transfer profile on the outer surface is decreased with decreasing the values of the magnetic field parameter. Another interesting point is the unexpected behavior of local Schmidt number with an increase in Hartman number, the application of an axial magnetic field tends to decrease the mass transfer on the inner cylinder.

CONCLUSION

In this study, the MHD forced convection flow and mass transfer of an electrically conducting fluid between two horizontal concentric cylinders in the presence of an axial magnetic field considering the effects of viscous heat dissipation in the fluid has been investigated numerically and analytically. The velocity distribution in the annulus is obtained analytically in terms of the modified Bessel functions whose argument contains Hartmann number and radial coordinate. To obtain the temperature, the expansions of the modified Bessel functions, with three terms are used in the energy equation.

It is found that the velocity and concentration decreases in the annulus with increase of Hartmann number. However, an increase in Hartmann number does not affect the temperature. The effects of magnetic field strength and Eckert number on local Nusselt number have been examined. The results show that an increase in Hartmann number reduces the Nusselt number on both surfaces of the cylinders. Also it was noticed that as the Eckert number increases local Nusselt number increases on the outer cylinder, but opposite trend is observed on the inner cylinder. The application of a magnetic field generates some interesting changes in mass transfer, an increasing in Hartmann number causes a reduction on the locale Sherwood number.

Conflict of Interest

The authors have not declared any conflict of interest.

REFERENCES

Aminfar H, Mohammadpourfard M, Maroofiazar R (2014). Experimental study on the effect of magnetic field on critical heat Flux of ferrofluid flow boiling in a vertical annulus. Exper. thermal. fluid sci. ://dx.doi.org/10.1016/j.expthermflusci.2014.06.023.

Ashorynejad Hamid RA, Mohamad A, Sheikholeslami M (2013), Magnetic field effects on natural convection flow of a nanofluid in a horizontal cylindrical annulus using Lattice Boltzmann method. Int. J. Thermal Sci. pp. 240-250. http://dx.doi.org/10.1016/j.ijthermalsci.2012.08.006

Azim MA, Mamun AA, Rahman MM (2010). Viscous Joule heating MHD–conjugate heat transfer for a vertical flat plate in the presence of heat generation, Int. Commun. Heat Mass Transfer 37:666–67. http://dx.doi.org/10.1016/j.icheatmasstransfer.2010.02.002

Ben HH, Henry D (1996). Numerical simulation of convective three-dimensional flows in a horizontal cylinder under the action of a constant magnetic field. J. Cryst. Growth 166:436-673. http://dx.doi.org/10.1016/0022-0248 (96)00044-9

Bessaïh R, Boukhari A, Marty PH (2009). Magnetohydrodynamics stability of a rotating flow with heat transfer. Int. Commun. Heat. Mass Transfer 36:893-901. http://dx.doi.org/10.1016/j.icheatmasstransfer.2009.06.009

Hayat T, Kara AH (2006). Couette flow of a third-grade fluid with variable magnetic field. Math. Compute. Modeling 43:132–137. http://dx.doi.org/ 10.1016/j.mcm.2004.12.009

El-Amin MF (2003). Combined effect of viscous dissipation and joule heating on MHD forced convection over a non-isothermal horizontal cylinder embedded in a fluid saturated porous medium. J. Magnet. Magnetic Mater. 263:337–343. http://dx.doi.org/ 10.1016/S0304-8853(03)00109-4.

Ellahi R, Hayat T, Mahomed FM, Zeeshan A (2010). Analytic solutions for MHD flow in an annulus. Commun. Nonlinear. Sci. Numer. Simulat. 15:1224–1227. http://dx.doi.org/10.1016/j.cnsns.2009.05.050

Feiz-Dizaji A, Salimpour MR, Jam F (2008). Flow field of a third-grade non-Newtonian fluid in the annulus of rotating concentric cylinders in the presence of magnetic field. J. Math. Anal. Appl. 337:32–645. http://dx.doi.org/ 10.1016/j.jmaa.2007.03.110.

Ghezal A, Porterie B (2011). Loraud J.C., Etude dynamique et thermique d'un écoulement pulsé en présence d'un solide chauffé en rotation. Mécanique Industries 12:45-65. http://dx.doi.org/10.1051/meca/2011004

Molki M, Astill KN, Leal E (1990). Convective heat-mass transfer in the Convective heat-mass transfer in the entrance region of a concentric annulus having a rotating inner cylinder entrance. Int. J. Heat. Fluid Flow, 11:2. http://dx.doi.org/ 10.1016/0142-727X(90)90005-V.

Kakarantzas SC, Sarris IE, Vlachos NS (2011). Natural convection of liquid metal in a vertical annulus with lateral and volumetric heating in the presence of a horizontal magnetic field. Int. J. Heat. Mass Transfer 54:3347-3356. http://dx.doi.org/10.1016/j.ijheatmasstransfer.2011.03.051

Makinde OD, Onyejekwe OO (2011). A numerical study of MHD generalized Couette flow and heat transfer with variable viscosity and electrical conductivity. J. Mag. Magn. Mater. 323:2757–2763. http://dx.doi.org/ 10.1016/j.jmmm.2011.05.040

Mozayyeni HR, Rahimi AB (2012). Mixed convection in cylindrical annulus with rotating outer cylinder and constant magnetic field with an effect in the radial direction. Scientia Iranica B 19(1)91–105. http://dx.doi.org/ 10.1016/j.scient.2011.12.006

Omid M, Shohel M, Ioan P (2012). Analysis of first and second laws of thermodynamics between two isothermal cylinders with relative rotation in the presence of MHD flow. Int. J. Heat Mass Transfer 55:4808–4816. http://dx.doi.org/10.1016/j.ijheatmasstransfer.2012.04048

Peyrret R (1976). Unsteady evolution of horizontal jet in a stratified fluid. J. Fluid mech. 78:(1):49-63. http://dx.doi.org/10.1017/S0022112076002322

Sankar M, Venkatachalappa M, Shivakumara IS (2006). Effect of magnetic field on natural convection in a vertical cylindrical annulus. International J. Eng. Sci. 44:1556–1570. http://dx.doi.org/10.1016/j.ijengsci.2006.06.004

Seth GS, Ansari MdS, Nandkeolyar R (2011). Effects of rotation and magnetic field on unsteady Couette flow in a porous channel. J. Appl. Fluid Mech. 4(2):95-103. Available online at www.jafmonline.net

Seth GS, Singh JK (2013). Effects of Hall current of unsteady MHD Couette flow of class-II in a rotating system. J. Appl. Fluid Mech. 6(4):473-484. Available online at www.jafmonline.net

Sheikholeslami M, Gorji-Bandpy M, Ganji DD, Soleimani S (2013). Effect of a magnetic field on natural convection in an inclined half-annulus enclosure filled with Cu–water nanofluid using CVFEM. Advan. Powder Technol. 24(6):980–991. http://dx.doi.org/ 10.1016/j.apt.2013.01.012

Sheikholeslami M, Gorji-Bandpy M, Ganji DD (2014). MHD free convection in an eccentric semi-annulus filled with nanofluid. J.Taiwan Instit. Chem. Eng. 1204–1216. http://dx.doi.org/ 10.1016/j.jtice.2014.03.010.

Singh SK, Jha BK, Singh AK (1997). Natural convection in vertical concentric annuli under a radial magnetic field. Heat. Mass Transfer, 32:399–401,Springer-Verlag. http://dx.doi.org/10.1007/s002310050137

Venkatachalappa M, Do Younghae Sankar M (2011). Effect of magnetic field on the heat and mass transfer in a vertical annulus. Int. J. Eng. Sci.49:262–278. http://dx.doi.org/ 10.1016/j.ijengsci.2010.12.002

Kefeng Shi, Wen-Qiang Lu (2006). Time evolution of double-diffusive convection in a vertical cylinder with radial temperature and axial solutal gradients. Int. J. Heat. Mass Transfer 49:995-1003. http://dx.doi.org/10.1016/j.ijheatmasstransfer.2005.0.009

Takhar HS, Chamkha AJ, Nath G (2003). Unsteady mixed convection flow from a rotating vertical cone with a magnetic field, Heat Mass Transfer 39:297–304. http://dx.doi.org/ 10.1007/s00231-002-0400-1

Teamah MA (2007). Numerical simulation of double diffusive laminar mixed convection in a horizontal annulus with hot, solutal and rotating inner cylinder. Int. J. Thermal Sci. 46:637–648. http://dx.doi.org/ 10.1016/j.ijthermalsci.2006.09.002.

14

Magnetohydrodynamics (MHD) boundary layer stagnation point flow with radiation and chemical reaction towards a heated shrinking porous surface

E. J. Christian, Y. I. Seini and E. M. Arthur

Department of Mathematics, Faculty of Mathematical Sciences, University for Development Studies, P. O. Box 1350, Tamale –Ghana.

This paper is an investigation of the effects of chemical reaction on two dimensional steady stagnation point flow of an electrically conducting, incompressible, and viscous fluid with radiation towards a heated shrinking porous surface. A chemically reactive species is emitted from the vertical surface into the flow field. The governing partial differential equations are solved using the Newton-Raphson shooting method along with the fourth-order Runge-Kutta integration. Velocity, temperature and concentration profiles are presented graphically. Numerical results for the skin friction coefficient, the rate of heat transfer represented by the local Nusselt number and the rate of mass transfer represented by the local Sherwood number are presented in tables and discussed quantitatively. The effects of magnetic field parameter, the velocity ratio parameter, the radiation parameter, the suction parameter, Schmidt number, Eckert number, Prandtl number and reaction rate parameter on the flow field are discussed.

Key words: Porous medium, stagnation point, magnetohydrodynamics (MHD), shrinking surface, radiation.

INTRODUCTION

Heat transfer is an important area of fluid dynamic research. The presence of magnetic field in a body of fluid has now been known to have significant practical applications in science, engineering and industry. It is commonly encountered in nuclear power plants, cooling of transmission lines and in electric transformers. Some investigations have been conducted to study the effects of radiation on electrically conducting fluids due to its wide applications in space technology.

The problem of radiation on magnetohydrodynamic (MHD) free convection flows under different surface or boundary conditions using different mathematical techniques have been reported in the literature. For instance, Seini and Makinde (2013) investigated the MHD boundary layer flow due to exponential stretching surface with radiation and chemical reaction and observed that the rate of heat transfer at the surface was adversely affected by increases in the transverse magnetic field parameter and the radiation parameter. Mahapatra and Nandy (2011) presented a momentum and heat transfer solution to MHD axisymmetric stagnation-point flow over a shrinking sheet. Mahapatra et al. (2011) analysed the steady two-dimensional MHD stagnation-point flow of an electrically conducting fluid over a shrinking sheet with a uniform transverse magnetic field whilst Jafar et al. (2011) investigated the MHD stagnation point flow over a nonlinearly stretching/shrinking sheet. In a related study, Javed et al. (2012) analysed the heat transfer in a viscous fluid over a non-linear shrinking sheet in the presence of a magnetic field and obtained dual solutions for the shrinking sheet problem whilst Zeeshan et al. (2012) investigated the porosity and magnetohydrodynamic flow of non-Newtonian nanofluid in coaxial cylinders using the Homotopy Analyses Method (HAM). Similar results have been reported by many authors including Ibrahim and Makinde (2010a, b; 2011a, b), who considered the MHD flow under varied boundary conditions.

Makinde and Charles (2010) conducted computational dynamics on hydromagnetic stagnation point flow towards a stretching sheet and observed that the cooling rate of the stretching sheet in an electrically conducting fluid subjected to magnetic field could be controlled and a final product with desired characteristics can be achieved. Ibrahim and Makinde (2010a) investigations on MHD boundary layer flow of chemically reacting fluid with heat and mass transfer past a stretching sheet also concluded that both the magnetic field strength and the uniform heat source had significant impact in cooling surfaces. Rana and Bhargava (2012) analysed the steady laminar boundary layer flow resulting from non-linear stretching of a flat surface in a nanofluid whilst Hameed (2012) numerically analysed the steady non-Newtonian flows with heat transfer analysis, MHD and nonlinear slip effects. Ellahi (2013) provided analytical solutions to the effects of MHD and temperature dependent viscosity on the flow of non-Newtonian nanofluid in a pipe whilst Sheikholeslami et al. (2014) analysed the effects of heat transfer in the flow of nanofluids over a permeable stretching wall in a porous medium and concluded that increasing the nanoparticle volume fraction had the effect of decreasing the momentum boundary layer thickness and entropy generation rate but increases the thermal boundary layer thickness.

The presence of chemical reaction and non-uniform heat source over an unsteady stretching surface was investigated by Seini (2013) who observed that the heat and mass transfer rates and the skin friction coefficient increases as the unsteadiness parameter increased and decreases as the space-dependent and temperature-dependent parameters for heat source/sink increased. Alireza et al. (2013) then presented an analytical solution for MHD stagnation point flow and heat transfer over a permeable stretching sheet with chemical reaction. Arthur and Seini (2014) recently analyzed the MHD thermal stagnation point flow towards a stretching porous surface whilst Seini and Makinde (2014) analyzed the boundary layer flow problem near stagnation-points on a vertical surface with slip in the presence of transverse magnetic field. Sheikholeslami et al. (2014) then investigated the effects of MHD on Cu-water nanofluid flow and heat transfer by means of Control-Volume Finite-Element Method (CVFEM). Similarly, Ellahi et al. (2013) studied the non-Newtonian nanofluid flow through a porous medium between two coaxial cylinders with heat transfer and variable viscosity using the Homotopy Analysis Method (HAM). Furthermore, Ellahi et al. (2014a, b) investigated the effects of heat and mass transfer on peristaltic flow in a non-uniform rectangular duct and also analyzed the steady flows in viscous fluid with heat and mass transfer with slip effects using the Spectral Homotopy Analyses Method (SHAM) and obtained interesting results for the generalised Couette flow problem.

To the best knowledge of the authors, only a limited number of researchers have attempted to solve the problem of radiation effects on MHD boundary layer stagnation point flow towards a heated shrinking sheet, notably Muhammad and Shahzad (2011). Stagnation point flow with radiation towards a shrinking sheet is quite useful and important from the practical point of view. In this paper, an attempt is made to investigate the effect of radiation on chemically reacting MHD boundary layer flow towards a heated shrinking porous surface due to its numerous industrial applications involving cooling.

FORMULATION OF THE PROBLEM

Consider a steady two-dimensional flow of an incompressible and electrically conducting fluid towards a stagnation point on a porous stretching sheet in the presence of radiation and magnetic field of strength B_0, applied in the positive y direction as shown in Figure 1. The tangential velocity u_w, and the free stream velocity U_∞ were assumed to vary proportional to the distance x from the stagnation point so that $u_w(x) = ax$ and $U_\infty(x) = bx$. The induced magnetic field due to the motion of the electrically conducting fluid and the pressure gradient are neglected. The tangential temperature is maintained at the prescribed constant value T_w.

The boundary layer equations for a steady incompressible viscous hydrodynamic fluid are:

$$\frac{\partial u}{\partial x} + \frac{\partial v}{\partial y} = 0, \tag{1}$$

$$u\frac{\partial u}{\partial x} + v\frac{\partial u}{\partial y} = -\frac{1}{\rho}\frac{\partial p}{\partial x} + \frac{\mu}{\rho}\frac{\partial^2 u}{\partial y^2} + \frac{\sigma B_0^2}{\rho}(U - u), \tag{2}$$

$$u\frac{\partial T}{\partial x} + v\frac{\partial T}{\partial y} = \frac{k_0}{\rho c_p}\frac{\partial^2 T}{\partial y^2} + \frac{\sigma B_0^2}{\rho c_p}(U - u)^2 - \frac{\partial q_r}{\partial y}, \tag{3}$$

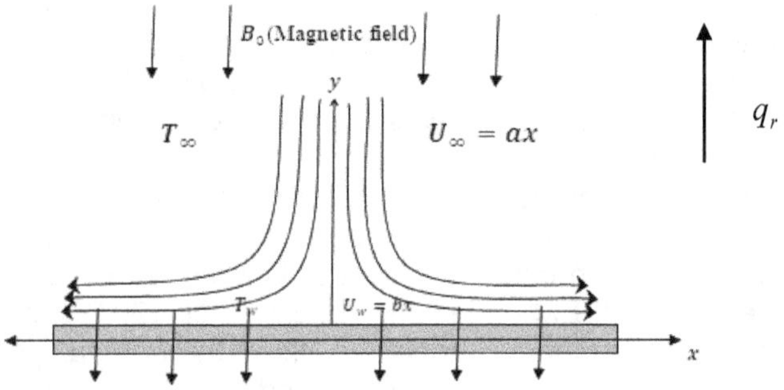

Figure 1. Schematic diagram of the problem.

$$u\frac{\partial C}{\partial x}+v\frac{\partial C}{\partial y}=D_m\frac{\partial^2 C}{\partial y^2}-\gamma(C-C_\infty). \tag{4}$$

Boundary conditions

$$y=0,\ u(x,0)=bx,\ v(x,0)=-V,\ T(x,0)=T_w,\ C(x,0)=C_w, \tag{5}$$

$$y\to\infty,\ u(x,\infty)\to U=ax,\ T(x,\infty)=T_\infty,\ C(x,\infty)=C_\infty,$$

where b > 0 is the shrinking rate, T_w is the heated wall surface temperature and T_∞ is the temperature of the fluid outside the boundary layer $(T_w > T_\infty)$, C_w is the wall surface concentration and C_∞ is the concentration of the fluid outside the boundary layer $(C_w > C_\infty)$.

Using the Rosseland approximation for radiation, Ibrahim and Makinde (2011b) simplified the heat flux as

$$q_r=-\frac{4\sigma}{3k}\frac{\partial T^4}{\partial y} \tag{6}$$

where k and σ are the Stefan-Boltzmann constant and the mean absorption coefficient respectively. We assume that the temperature differences within the flow such as the term T^4 may be expressed as a linear function of temperature. Hence, expanding T^4 in a Taylor series about T_∞ and neglecting higher order terms, we get;

$$T^4 \cong 4T_\infty^3 T - 3T_\infty^4. \tag{7}$$

Using Equations (6) and (7), Equation (3) reduces to

$$u\frac{\partial T}{\partial x}+v\frac{\partial T}{\partial y}=\left(\frac{k_0}{\rho c_p}+\frac{16}{3}\frac{\sigma T_\infty^3}{\rho c_p k}\right)\frac{\partial^2 T}{\partial y^2}+\frac{\sigma B_0^2}{\rho c_p}(U-u)^2, \tag{8}$$

Where $k=\dfrac{3Nr}{3Nr+4}$

In order to obtain the velocity, temperature and concentration fields for the problem, Equations (1), (2), (4) and (8) are solved subject to the appropriate boundary conditions given in Equation (5) by employing the following similarity variables.

$$\eta=y\sqrt{\frac{a}{\nu}},\quad u(x,y)=axf'(\eta),\quad v=-\sqrt{a\nu}f(\eta),\quad \theta(\eta)=\frac{T-T_\infty}{T_w-T_\infty},$$
$$p(x,\infty)=P_0-\frac{\rho a^2}{2}(x^2+y^2),\quad \phi(\eta)=\frac{C-C_\infty}{C_w-C_\infty}, \tag{9}$$

where η is a similarity parameter, P_0 is the stagnation pressure, a is the strength of the stagnation point with dimension $\frac{1}{t}$. Equation (9) satisfies Equation (1) identically by noting the usual definition for a stream function as:

$$u=\frac{\partial\psi}{\partial y} \text{ and } v=-\frac{\partial\psi}{\partial x},$$

hence, the velocity field represents a possible fluid motion. Equations (2), (4) and (9) are then transformed into

$$f'''+M^2(1-f')+1=f'^2-ff'', \tag{10}$$

$$\theta''+\Pr kf\theta'+\Pr EcM^2f'^2=0, \tag{11}$$

$$\phi''+Sc\phi'-Sc\beta\phi=0, \tag{12}$$

where $M=\sqrt{\dfrac{\sigma_e B_0^2}{\rho a}}$ and $\Pr=\dfrac{\mu c_p}{k_0}$ are the magnetic parameter and Prandtl number respectively. Ec, Sc and β are the Eckert number, the Schmidt number and the reaction rate parameter. $k=\dfrac{3Nr}{3Nr+4}$ and Nr is the radiation parameter. The boundary conditions given in Equation (5) are also transformed into

$$f'(0)=\frac{b}{a}=B,\ f(0)=fw,\ \theta(0)=1,\ \phi(0)=1,$$
$$f'(\infty)=1,\ \theta(\infty)=0,\ \phi(\infty)=0. \tag{13}$$

NUMERICAL PROCEDURE

The governing Equations (10), (11) and (12) are highly non-linear. Most physical systems are inherently nonlinear in nature and are of great interest to engineers, physicist and mathematicians, problems involving nonlinear ordinary differential equations are difficult to solve and give rise to interesting phenomena such as chaos. We employ the Runge–Kutta integration along with the Newton Raphson algorithm to obtain approximate solutions. In this method, we let:

Table 1. Comparison of shear stress and heat transfer for $M = 1$ $Nr = 3$ $\Pr = 0.7$ with various values of B.

B	Shear stress $(f''(0))$		Rate of Heat Transfer $(-\theta'(0))$	
	Muhammad and Shahzad (2011)	**Present Study**	**Muhammad and Shahzad (2011)**	**Present Study**
0.25	1.877455	1.877560	0.412813	0.413620
0.50	2.120114	2.120190	0.378822	0.379139
0.75	2.307090	2.307127	0.342202	0.342030
1.0	2.429972	2.429962	0.302334	0.301637

Table 2. Results of skin friction coefficient, Nusselt number and Sherwood number for varying parameter values when $\Pr = 0.71$, $Ec = 0.1$, $Sc = 0.24$.

M	β	F_w	B	Nr	$-f''(0)$	$-\theta'(0)$	$-\phi'(0)$
1	0.1	0.1	0.1	3	1.505368	0.327923	0.323899
2	0.1	0.1	0.1	3	2.178774	-0.17930	0.323899
3	**0.1**	0.1	0.1	3	2.977204	-1.09646	0.323899
1	**0.5**	0.1	0.1	3	1.505368	0.327923	0.487086
1	**1.0**	**0.1**	0.1	3	1.505368	0.327923	0.624423
1	0.1	**0.5**	0.1	3	1.717241	0.438848	0.323893
1	0.1	**1.0**	**0.1**	3	2.008972	0.594699	0.323899
1	0.1	0.1	**0.5**	3	0.896951	0.354065	0.487086
1	0.1	0.1	**0.8**	**3**	0.375867	0.369433	0.487086
1	0.1	0.5	0.1	**4**	1.505368	0.366143	0.487086
1	0.1	0.5	0.1	**5**	1.505368	0.351018	0.487086

$$
\begin{aligned}
x_1 &= f, \\
x_1' &= x_2 = f', \\
x_2' &= x_3 = f'' \\
x_3' &= f''' = x_2^2 - x_1 x_3 - 1 - M^2(1 - x_2), \\
x_4' &= x_5 = \theta', \\
x_5' &= -\Pr kfx_5 - \Pr EcM^2 x_2^2, \\
x_6' &= x_7 = \phi, \\
x_7' &= x_8 = \phi', \\
x_8' &= \phi'' = -Scx_8 + Sc\beta x_7.
\end{aligned}
\tag{14}
$$

RESULTS AND DISCUSSION

Table 1 compares results of this study and that of Muhammad and Shahzad (2011). It is observed that the numerical results are consistent with their work and hence validate our numerical procedure.

The results of varying parameter values on the local skin friction coefficient, the local Nusselt number and the local Sherwood number, which are respectively proportional to $-f''(0)$, $-\theta'(0)$ and $-\phi'(0)$ are shown in Table 2. It is observed that increasing the magnetic field strength parameter increases the skin friction coefficient at the surface due to the presence of the Lorenz force. It however reduces the rate of heat transfer and does not affect the rate of mass transfer for obvious reasons. With the case of increasing the reaction rate parameter, the rate of mass transfer at the surface increases, however, the skin friction coefficient and the rate of heat transfer are not affected by the reaction rate parameter. It is interesting to note that increasing the suction parameter do not only increase the skin friction coefficient but also the rate of heat and mass transfer. Increasing the velocity ratio parameter decreases the skin friction coefficient and increases the rate of heat transfer on the surface. Furthermore, increasing the radiation parameter causes a reduction in the rate of heat transfer at the surface whereas both the coefficient of skin friction and the rate of mass transfer are not affected.

Effects of parameter variation on the velocity profiles

The effect of varying various parameters on the velocity

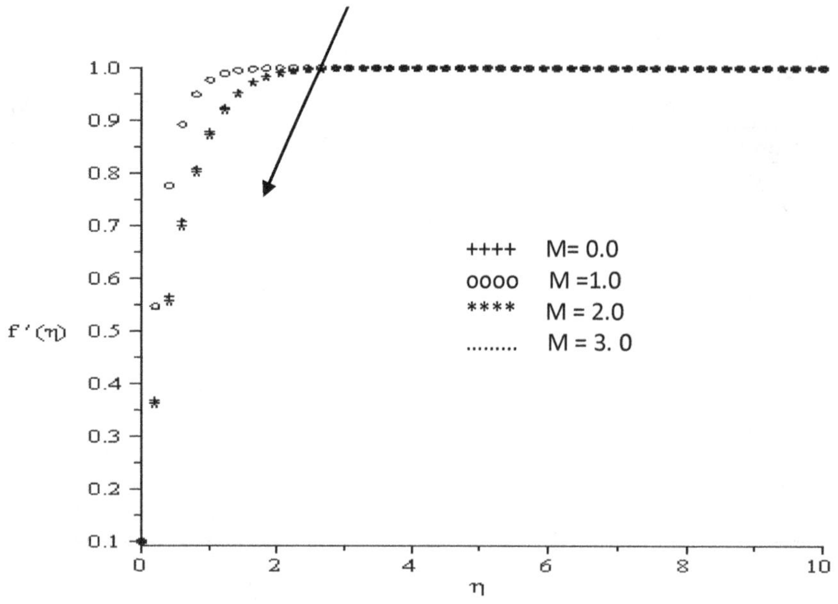

Figure 2. Velocity profiles for varying values of magnetic parameter (M).

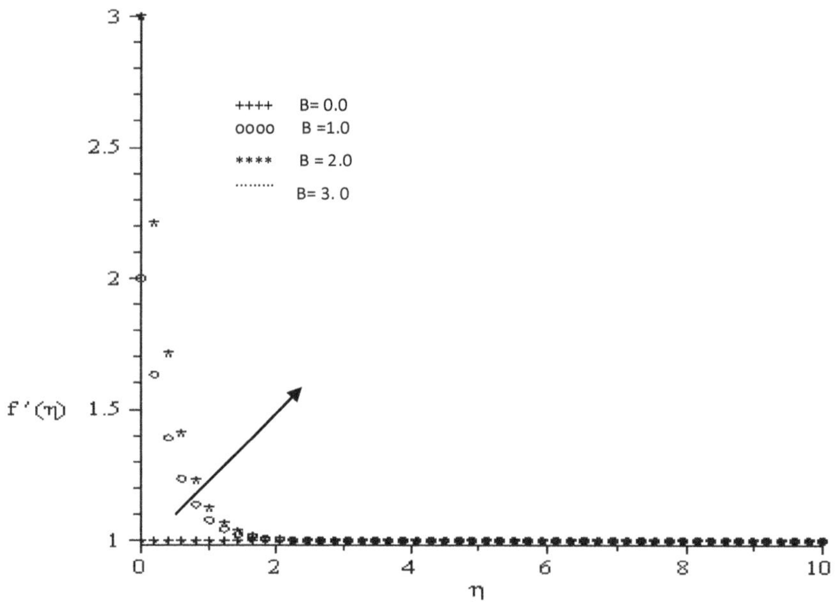

Figure 3. Velocity profiles for varying values of velocity ratio parameter (B).

boundary layer are depicted in Figures 2 to 4. The velocity profiles are observed to decrease when the magnetic field parameter is increased (Figure 2). This phenomenon is due to the fact that increasing the magnetic field strength increases the Lorenz force which creates an opposing force to the fluid transport. In Figures 3 and 4, it is observed that increasing the velocity ratio parameter (*B*) and the suction parameter (*fw*) increased the velocity boundary layer.

Effects of parameter variation of temperature profiles

Figures 5 to 10 depicts the effects of the magnetic field parameter, velocity ratio parameter, radiation parameter, Eckert number and Prandtl number respectively on the temperature profiles. It is observed that both the Eckert number and the magnetic parameter contribute to greater thermal boundary layer thickness. The reverse is true for increasing radiation parameter, velocity ratio parameter,

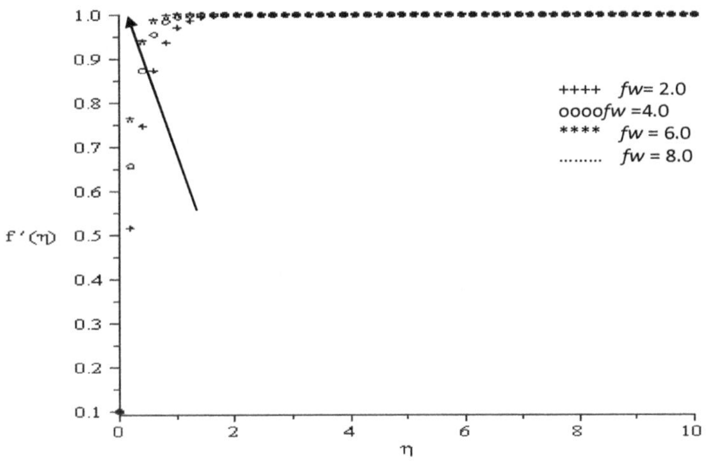

Figure 4. Velocity profiles for varying values of Suction parameter (f_w).

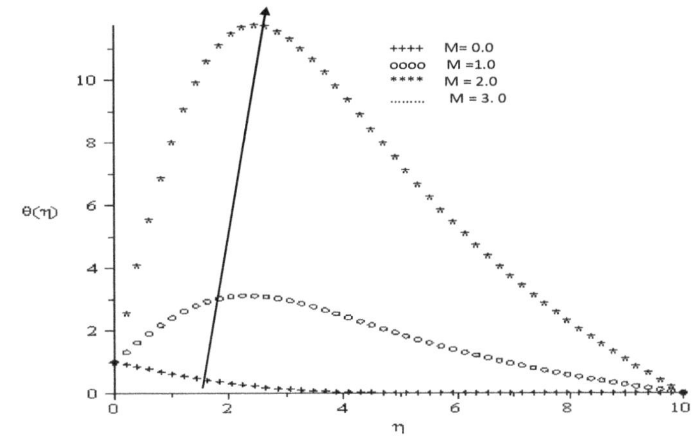

Figure 5. Temperature profiles for varying values of magnetic parameter (M).

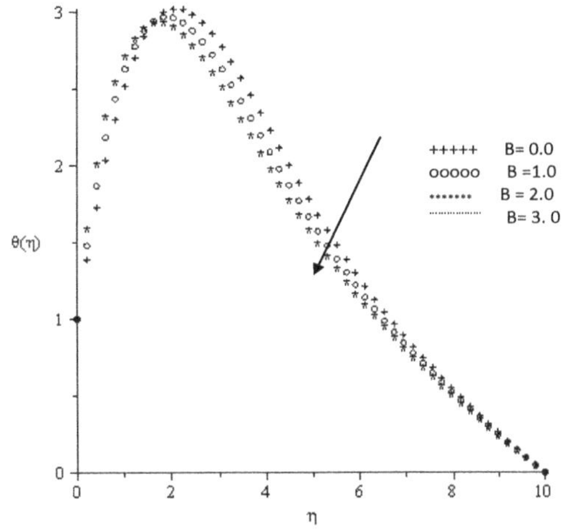

Figure 6. Temperature profiles for varying values of velocity ratio parameter (B).

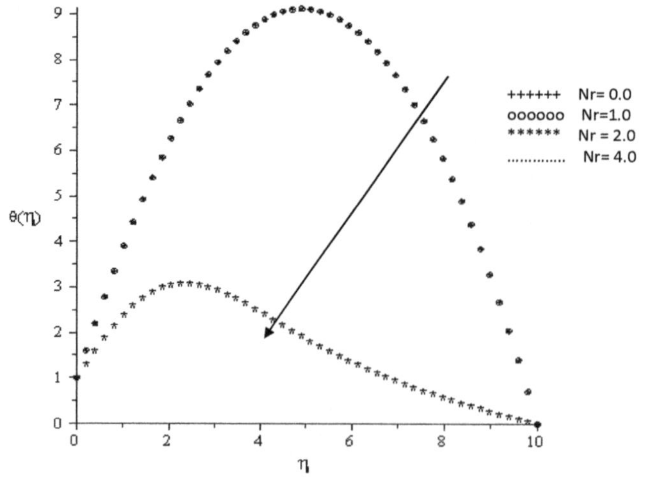

Figure 7. Temperature profiles for varying values of radiation parameter (Nr).

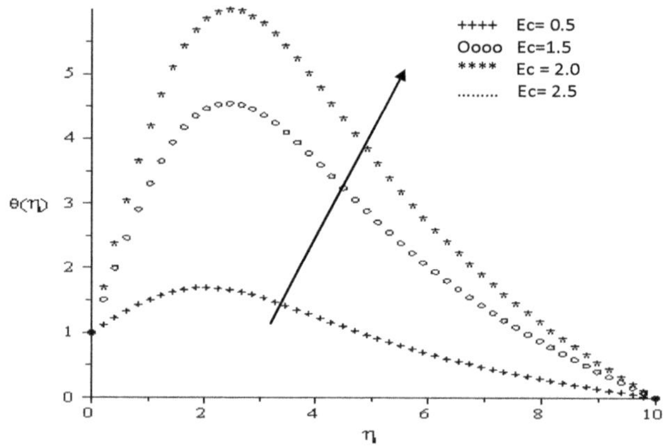

Figure 8. Temperature profiles for varying values of Eckert number (Ec).

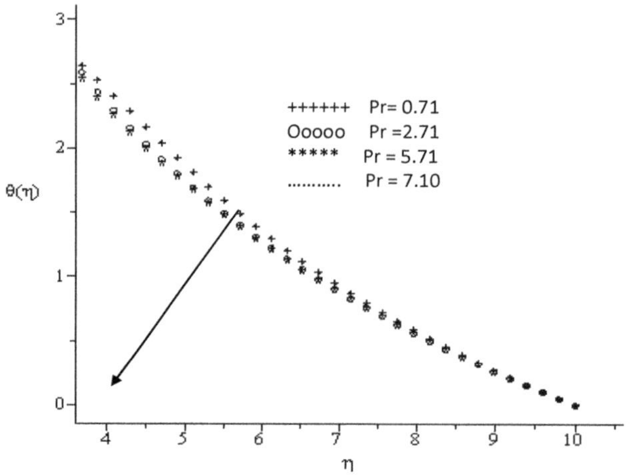

Figure 9. Temperature profiles for varying values of Prandtl number (Pr).

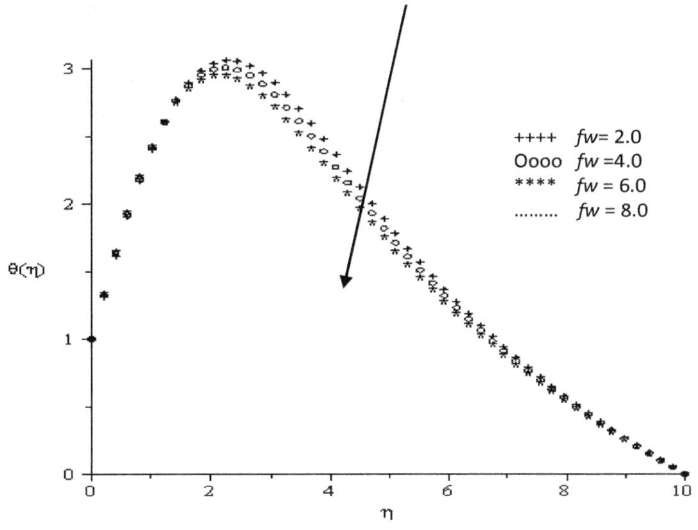

Figure 10. Temperature profiles for varying values of suction parameter (f_w).

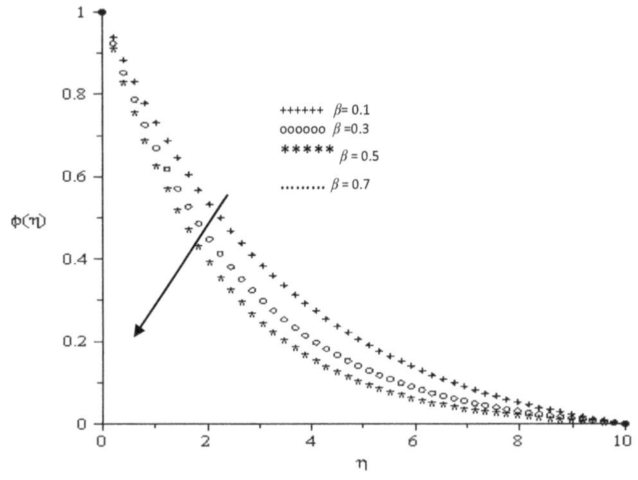

Figure 11. Concentration profiles for varying values of the reaction rate parameter (β).

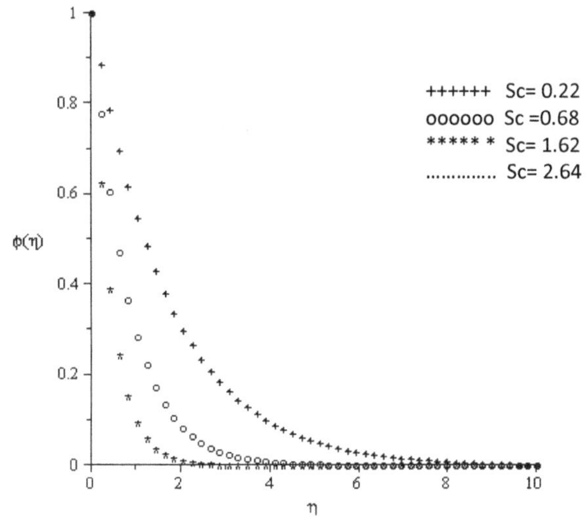

Figure 12. Concentration profiles for varying Schmidt number.

suction parameter and the Prandtl number. For the Prandtl number, it could be due to the fact that decreasing the thermal diffusivity results in the heat being diffused away from the surface which leads to increases in the temperature gradient at the surface.

Effects of parameter variation on the concentration profiles

It is observed in Figures 11 and 12 that increasing the reaction rate parameter and the Schmidt numbers reduces the concentration boundary layers.

CONCLUSIONS

The MHD boundary layer stagnation point flow with radiation and chemical reaction towards a heated shrinking porous surface has been investigated. Numerical results have been compared to earlier results published in the literature and the agreement was good. Our results reveal that:

(i) The velocity profiles of the flow increases with increasing suction parameter(f_w) and the velocity ratio (B). It however decreases with increasing values of the magnetic parameter (M).

(ii) The thermal boundary layer is observed to decrease with increasing values of $\Pr$, f_w, Nr and B. It however observed to increase with increasing values of M and Ec.

(iii) The concentration boundary layer decreases with increasing reaction rate parameter (β) and Schmidt number.

(iv) The skin friction at the surface increases for increasing parameter values of M and f_w but decreases for increasing B.

Conflict of Interest

The authors have not declared any conflict of interest.

REFERENCES

Alireza R, Farzaneh-Gord M, Varedi SR, Ganji DD (2013). Analytical solution for magnetohydrodynamic stagnation point flow and heat transfer over a permeable stretching sheet with chemical reaction. J. Theor. Appl. Mech. 51(3):675-686.

Arthur EM, Seini YI (2014). MHD thermal stagnation point flow towards a stretching porous surface. Math. Theory Model. 4(5):163-169. ISSN 2224-5804 (Paper) ISSN 2225-0522 (Online)

Ellahi R (2013). The effects of MHD and temperature dependent viscosity on the flow of non-Newtonian nanofluid in a pipe: Analytical solutions. Appl. Math. Modell. 37(3):1451-1467.

Ellahi R, Aziz S, Zeeshan A (2013). Non-Newtonian nanofluid flow through a porous medium between two coaxial cylinders with heat transfer and variable viscosity. J. Porous Media. 16(3):205-216.

Ellahi R, Mubashir Bhatti M, Vafai K (2014). Effects of heat and mass transfer on peristaltic flow in a non-uniform rectangular duct. Int. J. Heat Mass Transfer. 71:706-719.

Hameed M (2012). Numerical analysis of steady non-Newtonian flows with heat transfer analysis, MHD and nonlinear slip effects. Int. J. Num. Methods Heat Fluid Flow. 22(1):24-38.

Ibrahim SY, Makinde OD (2010a). Chemically reacting MHD boundary layer flow of heat and mass transfer over a moving vertical plate with suction. Sci. Res. Essays. 5(19):2875-2882.

Ibrahim SY, Makinde OD (2010b). On MHD boundary layer flow of chemically reacting fluid with heat and mass transfer past a stretching sheet. Int. J. Fluid Mech. 2(2):123-132.

Ibrahim SY, Makinde OD (2010b). On MHD boundary layer flow of chemically reacting fluid with heat and mass transfer past a stretching sheet. Int. J. Fluid Mech. 2(2):123-132.

Ibrahim SY, Makinde OD (2011a). Radiation effect on chemically reacting MHD boundary layer flow of heat and mass transfer past a porous vertical flat plate. Int. J. Phys. Sci. 6(6):1508-1516.

Ibrahim SY, Makinde OD (2011a). Radiation effect on chemically reacting MHD boundary layer flow of heat and mass transfer past a porous vertical flat plate. Int. J. Phys. Sci. 6(6):1508-1516.

Ibrahim SY, Makinde OD (2011b). Chemically reacting MHD boundary layer flow of heat and mass transfer past a low-heat-resistant Sheet moving vertically downwards. Sci. Res. Essays. 6(22):4762-4775.

Ibrahim SY, Makinde OD (2011b). Chemically reacting MHD boundary layer flow of heat and mass transfer past a low-heat-resistant Sheet moving vertically downwards. Sci. Res. Essays. 6(22):4762-4775.

Jafar K, Ishak A, Nazar R (2011). MHD stagnation point flow over a nonlinearly stretching/shrinking sheet. J. Aeros. Eng doi: 10.1061/ (ASCE) AS.1943-5525.0000186.

Javed T, Abbas Z, Sajid M, Ali N (2012). Heat transfer analysis for a hydromagnetic viscous fluid over a nonlinear shrinking sheet. Int. J. Heat Mass Transfer, doi:10.1016/ j. ijheat mass transfer .2010.12 .025 (in press).

Mahapatra TR Nandy SK, Gupta AS (2011). Momentum and heat transfer in MHD stagnation-point flow over a shrinking sheet, Transaction to ASME. J. Appl. Mech. 78:021015.

Mahapatra TR, Nandy SK (2011). Momentum and heat transfer in MHD axisymmetric stagnation-point flow over a shrinking sheet. J. Appl. Fluid Mech. 6(1):121-129. ISSN 1735-3572, EISSN 1735-3645.

Makinde OD, Charles WM (2010). Computational dynamics of hydromagnetic stagnation flow towards a stretching sheet. Appl. Comput. Math. 9(2):243-251.

Muhammad A, Shahzad A (2011). Radiation effects on MHD boundary layer stagnation point flow towards a heated shrinking sheet. World Appl. Sci. J. 13(7):1748-1756.

Rana P, Bhargava R (2012). Flow and heat transfer of a nanofluid over a nonlinearly stretching sheet: A numerical study. Comm. Nonlinear Sci. Num. Simulat. 17:212-226.

Seini YI (2013). Flow over unsteady stretching surface with chemical reaction and non-uniform heat source. J. Eng. Manufactur. Technol. JEMT 1:24-35.

Seini YI, Makinde OD (2013). MHD Boundary Layer Flow due to Exponential Stretching Surface with Radiation and Chemical Reaction. Math. Problems Eng. Volume 2013, Article ID163614, 7 pages, http://dx.doi.org/10.1155/2013/163614

Seini YI, Makinde OD (2014). Boundary layer flow near stagnation-points on a vertical surface with slip in the presence of transverse magnetic field. Int. J. Num. Methods Heat Fluid Flow 24(3):643-653.

Sheikholeslami M, GorjiBandpy M, Ellahi R, Hassan M, Soleimani S (2014). Effects of MHD on Cu-water nanofluid flow and heat transfer by means of CVFEM. J. Magnet. Magnetic Mat. 349:188-200.

Sheikholeslami M, Ellahi R, Ashorynejad HR, Domairry G, Hayat T (2014). Effects of heat transfer in flow of nanofluids over a permeable stretching wall in a porous medium. J. Comput. Theoret. Nanosci. 11(2):486-496.

Zeeshan A, Ellahi R, Siddiqui AM, Rahman HU (2012). An investigation of porosity and magnetohydrodynamic flow of non-Newtonian nanofluid in coaxial cylinders. Int. J. Phys. Sci. 7(9):1353-1361.

15

Analytical investigation of convective heat transfer of a longitudinal fin with temperature-dependent thermal conductivity, heat transfer coefficient and heat generation

D. D. Ganji[1] and A. S. Dogonchi[2]

[1]Department of Mechanical Engineering, Babol Noshirvani University of Technology, P. O. Box 484, Babol, Iran.
[2]Department of Mechanical Engineering, Mazandaran Institute of Technology, P. O. Box 747, Babol, Iran.

In this article, the heat transfer through a longitudinal fin is studied. The heat transfer coefficient, thermal conductivity and heat generation are variables and supposed to be temperature-dependent. The temperature distribution in fin with longitudinal rectangular profile was carried out by using the differential transformation method (DTM) which is an analytical solution technique. For validation of the analytical solution, the heat equation is solved numerically. The temperature distribution is shown for different values of the embedding parameters. The DTM results indicate that the fin tip temperature increases with an increase in the heat generation gradient. Results reveal that DTM is very effective and convenient. Comparison of the results (DTM and numerical) was shown that the analytical method and numerical data are in a good agreement with each other.

Key words: Fins, temperature dependent thermal properties, heat generation, analytical solutions, differential transformation method (DTM).

INTRODUCTION

Extended surfaces (also known as fins) are used to augment heat dissipation from a hot surface through its convective, radiative, or convective-radiative surface. In particular, fins are used extensively in various industrial applications such as the cooling of computer processors, air conditioning and oil carrying pipe lines. Several studies were performed on heat transfer using fins. Domairry and Fazeli (2009) solved the nonlinear straight fin differential equation to evaluate the temperature distribution and fin efficiency. Also, temperature distribution for annual fins with temperature-dependent thermal conductivity was studied by Ganji et al. (2011). The effects of temperature-dependent thermal conductivity of a moving fin and added radiative component to the surface heat loss have been studied by Aziz and Khani (2011). They applied the homotopy analysis method (HAM) to solve governing equations. Hatami et al. (2014) studied the temperature distribution for a fully wet, semi-spherical porous fin. Heat transfer and temperature distribution for circular convective-radiative porous fins was studied by Hatami and Ganji (2013). Hatami and Ganji (2014a) studied temperature distribution and refrigeration efficiency for fully wet circular porous fins with variable sections. Ghasemi et al. (2014) solved the nonlinear temperature distribution equation in a longitudinal fin with temperature-dependent internal heat generation and thermal conductivity using differential transformation method (DTM). Heat transfer and temperature distribution equations for longitudinal convective-radiative porous fins are solved by Hatami and Ganji (2014b). Heat transfer through porous fins with temperature-dependent heat generation was studied by Hatami et al. (2013). They employed DTM, collocation method (CM) and least square method (LS) for solving governing equations. Sharqawy and Zubair (2007) applied the analytical method for the annular fin with combined heat and mass transfer. Arslanturk (2005) and Rajabi (2007) obtained efficiency and fin temperature distribution by Adomian decomposition method (ADM) and the homotopy perturbation method (HPM) with temperature-dependent thermal conductivity. An analytical method for determining the optimum thermal design of convective longitudinal fin arrays is presented by Franco (2009). Lin and Lee (1999) investigated boiling on a straight fin with linearly varying thermal conductivity.

The concept of DTM was first introduced by Zhou (1986) and it was used to solve both linear and nonlinear initial value problems. This method can be applied directly to linear and nonlinear differential equation without requiring linearization, discretization, or perturbation and this is the main benefit of this method. Abbasov and Bahadir (2005) employed DTM to obtain approximate solutions of the linear and nonlinear equations related to engineering problems and they showed that the numerical results are in good agreement with the analytical solutions. Rashidi et al. (2010) solved the problem of mixed convection about an inclined flat plate embedded in a porous medium by DTM; they applied the Pade approximant to increase the convergence of the solution. Ghafoori et al. (2011) used the DTM for solving the nonlinear oscillation equation. Abdel-Halim (2008) has applied the DTM for different systems of differential equations and discussed the convergency of this method in several examples of linear and nonlinear systems of differential equations. Joneidi et al. (2009) used DTM for the analytical solution of convective straight fins with temperature-dependent thermal conductivity and comparing results with exact and numerical one. Their results reveal the capability, effectiveness, convenience and high accuracy of this method.

Moradi and Ahmadikia (2010) applied the DTM to solve the energy equation for a fin with three different profiles and temperature-dependent thermal conductivity. Balkaya et al. (2009) applied the DTM to analyze the vibration of an elastic beam supported on elastic soil. Borhanifar and Abazari (2011) employed DTM on some partial differential equations (PDEs) and their coupled versions.

The DTM is used to solve a wide range of physical problems. This method provides a direct scheme for solving linear and nonlinear deterministic and stochastic equations without linearization and yield convergent series solution rapidly.

The goal of this study is obtaining an analytical solution for temperature distribution of a fin with temperature-dependent thermal conductivity, heat transfer coefficient and heat generation. The effect of the range of values of heat transfer parameters on the temperature distribution is shown. Also, the DTM is applied to solve nonlinear problem analytically. To validate analytical results, the obtained DTM results are compared with numerical data.

PROBLEM DESCRIPTION

Consider a one-dimensional longitudinal fin, with an arbitrary profile $F(X)$ and cross-section area A_c as shown in Figure 1. The periphery of the fin is denoted by *P* and its length by *L*. The fin is attached to a fixed base surface of temperature T_b and extend to an ambient fluid of temperature T_a. The fin thickness is given by δ and the base thickness is δ_b. The $1\text{-}D$ steady state energy equation for the fin with internal heat generation can be expressed as:

$$\frac{\partial}{\partial X}\left(\frac{\delta_b}{2}F(X)K(T)\frac{\delta T}{\delta X}\right)-\frac{P}{A_c}H(T)\left(T-T_a\right)+q^*=0, \quad 0\le X\le L. \tag{1}$$

Where K, H and q^* are the non-uniform thermal conductivity, heat transfer coefficients and heat generation depending on the temperature, *T* is the temperature distribution and X is the spatial variable. An insulated fin at one end with the base temperature at the other implies boundary condition which is given by (Kraus, 2001):

$$T(L)=T_b \qquad , \frac{\partial T}{\partial X}\Big|_{X=0}=0. \tag{2}$$

Because of the heat generation varying with temperature, so we have:

$$q^*=q_a^*\left(1+\varepsilon\left(T-T_a\right)\right) \tag{3}$$

For simplifying the above equations, some dimensionless parameters are introduced as follows:

$$x=\frac{X}{L},\quad \theta=\frac{T-T_a}{T_b-T_a},\quad h=\frac{H}{h_b},\quad k=\frac{K}{k_a},\quad M^2=\frac{Ph_bL^2}{A_ck_a},\quad R=\frac{q_a^*A_c}{h_bP(T_b-T_a)}, \tag{4}$$

$$\varepsilon_R=\varepsilon(T_b-T_a),\quad f(x)=\frac{\delta_b}{2}F(X)$$

Equation 1 reduces to:

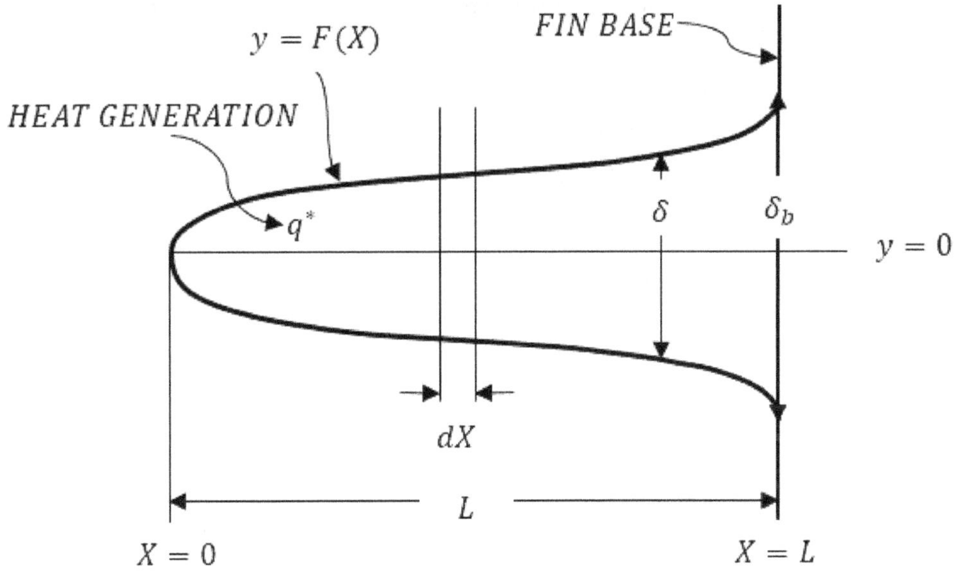

Figure 1. Schematic representation of a longitudinal fin with arbitrary profile *F(X)*.

$$\frac{\partial}{\partial x}\left(f(x)k(\theta)\frac{\delta\theta}{\delta x}\right)-M^2h(\theta)\theta+M^2R(1+\varepsilon_R\theta)=0, \quad 0\le x\le 1. \tag{5}$$

And the boundary conditions become

$$\theta(1)=1 \qquad , \frac{\partial\theta}{\partial x}\Big|_{x=0}=0. \tag{6}$$

Where *M* is the thermogeometric fin parameter, θ is the dimensionless temperature, x is the dimensionless spatial variable, q_a^* is the internal heat generation at temperature T_a, *k* is the dimensionless thermal conductivity, k_a is the thermal conductivity of the fin at ambient temperature, h_b is the heat transfer coefficient at the fin base. For most industrial application, the heat transfer coefficient maybe given as a power law (Unal, 1987):

$$H(T)=h_b\left(\frac{T-T_a}{T_b-T_a}\right)^n \tag{7}$$

Where *n* and h_b are constants. The constant n may vary between -6.6 and 5. However, in most practical applications it lies between -3 and 3 (Unal, 1987). The exponent n represents laminar film boiling or condensation when $n=-1/4$, laminar natural convection when $n=1/4$, turbulent natural convection when $n=1/3$, nucleate boiling when $n=2$, radiation when *n* = 3 and *n* = 0 implies a constant heat transfer coefficient. Exact solutions may be constructed for the steady-state one-dimensional differential equation describing temperature distribution in a straight fin when the thermal conductivity is a constant and the exponent of the heat transfer coefficient is given by $n=-1,0,1$ or 2 (Unal, 1987).

In dimensionless variables we have $h(\theta)=\theta^n$. Also, for many engineering applications, the thermal conductivity may depend linearly on temperature, that is;

$$K(T)=k_a\left[1+\gamma(T-T_a)\right] \tag{8}$$

The dimensionless thermal conductivity given by the linear function of temperature is $k(\theta)=1+\beta\theta$, where the thermal conductivity gradient is $\beta=\gamma(T_b-T_a)$.

As such, the governing equation is given by:

$$\frac{\partial}{\partial x}\left(f(x)(1+\beta\theta)\frac{\delta\theta}{\delta x}\right)-M^2\theta^{n+1}+M^2R(1+\varepsilon_R\theta)=0, \ 0\le x\le 1 \tag{9}$$

FUNDAMENTAL OF DIFFERENTIAL TRANSFORMATION METHOD (DTM)

Let $x(t)$ be analytic in a domain *D* and let $t=t_i$ represent any point in *D*. The function $x(t)$ is then represented by one power series whose center is located at t_i. The Taylor series expansion function of $x(t)$ is in form of:

$$x(t)=\sum_{k=0}^{\infty}\frac{(t-t_i)^k}{k!}\left[\frac{d^k x(t)}{dt^k}\right]_{t=t_i} \qquad \forall t\in D \tag{10}$$

The particular case of Equation 10 when $t_i=0$ is referred to as the Maclaurin series of $x(t)$ and is expressed as:

Table 1. The fundamental operations of the differential transform method.

Original function	Transformed function
$w(t)=\alpha u(t)\pm\beta v(t)$	$W(k)=\alpha U(k)\pm\beta V(k)$
$w(t)=\dfrac{d^m u(t)}{dt^m}$	$W(k)=\dfrac{(k+m)!}{k!}U(k+m)$
$w(t)=u(t)v(t)$	$W(k)=\sum_{l=0}^{k}U(l)V(k-l)$
$w(t)=t^m$	$W(k)=\delta(k-m)=\begin{cases}1, & if\ k=m\\ 0, & if\ k\neq m\end{cases}$
$w(t)=\exp(t)$	$W(k)=\dfrac{1}{k!}$

$$x(t)=\sum_{k=0}^{\infty}\frac{t^k}{k!}\left[\frac{d^k x(t)}{dt^k}\right]_{t=0} \qquad \forall t\in D \tag{11}$$

As explained in Zhou (1986) and Abdel-Halim (2004), the differential transformation of the function $x(t)$ is defined as follows:

$$X(k)=\sum_{k=0}^{\infty}\frac{H^k}{k!}\left[\frac{d^k x(t)}{dt^k}\right]_{t=0} \tag{12}$$

Where $x(t)$ is the original function and $X(k)$ is the transformed function. The differential spectrum of $X(k)$ is confined within the interval $t\in[0,H]$, where *H* is a constant. The differential inverse transform of $X(k)$ is defined as follows:

$$x(t)=\sum_{k=0}^{\infty}\left(\frac{t}{H}\right)^k X(k) \tag{13}$$

It is clear that the concept of the differential transformation is based upon the Taylor series expansion. The values of function $X(k)$ at values of argument *k* are referred to as discrete, that is, $X(0)$ is known as the zero discrete, $X(1)$ as the first discrete, etc. the more discrete available, the more precise it is possible to restore the unknown function. The function $x(t)$ consists of *T*-function $X(k)$, and its value is given by the sum of the *T*-function with $(t/H)^k$ as its coefficient. In real applications, at the right choice of constant *H*, the larger values of argument k the discrete of spectrum reduce rapidly. The function $x(t)$ is expressed by a finite series and Equation 13 can be written as:

$$x(t)=\sum_{k=0}^{n}\left(\frac{t}{H}\right)^k X(k) \tag{14}$$

Mathematical operations performed by differential transform method are listed in Table 1.

ANALYTICAL SOLUTION

By 1-D transform of Equation 9 considered by using the related definition in Table 1, we have the following:

(1) Rectangular profile ($f(x)=1$), case $n=1$

$$\begin{aligned}&(K+1)(K+2)\Theta(K+2)+\beta\left(\sum_{i=0}^{K}\Theta(i)(K+1-i)(K+2-i)\Theta(K+2-i)\right)+\beta\left(\sum_{i=0}^{K}(i+1)\Theta(i+1)(K+1-i)\Theta(K+1-i)\right)\\&-M^2\left(\sum_{i=0}^{K}\Theta(i)\Theta(K-i)\right)+M^2R\big(\delta(K)+\varepsilon_R\Theta(K)\big)=0,\end{aligned} \tag{15}$$

(2) Rectangular profile ($f(x)=1$), case $n=2$

$$\begin{aligned}&(K+1)(K+2)\Theta(K+2)+\beta\left(\sum_{i=0}^{K}\Theta(i)(K+1-i)(K+2-i)\Theta(K+2-i)\right)+\beta\left(\sum_{i=0}^{K}(i+1)\Theta(i+1)(K+1-i)\Theta(K+1-i)\right)\\&-M^2\left(\sum_{j=0}^{K}\Theta(K-j)\sum_{i=0}^{j}\Theta(i)\Theta(j-i)\right)+M^2R\big(\delta(K)+\varepsilon_R\Theta(K)\big)=0,\end{aligned} \tag{16}$$

In the above equations $\Theta(K)$ is transformed function of $\Theta(x)$. The transformed boundary condition takes the form:

$$\Theta(1)=0 \tag{17}$$

$$\sum_{i=0}^{\infty}\Theta(i)=1 \tag{18}$$

Supposing that $\Theta(0)=a$ and using Equations 17 and 18, another value of $\Theta(i)$ can be calculated.

The value of a can be calculated using Equation 18. Thus, we end up having the following:

(1) Rectangular profile, case $n=1$

$$\begin{aligned}&\Theta(2)=-\frac{1}{2}\frac{M^2\left(-a^2+R+R\varepsilon_R a\right)}{1+\beta a}\\&\Theta(3)=0\\&\Theta(4)=-\frac{1}{24}\frac{M^4\left(-a^2+R+R\varepsilon_R a\right)\left(2a-\beta a^2+3\beta R+2\beta R\varepsilon_R a-R\varepsilon_R\right)}{(1+\beta a)^3}\end{aligned} \tag{19}$$

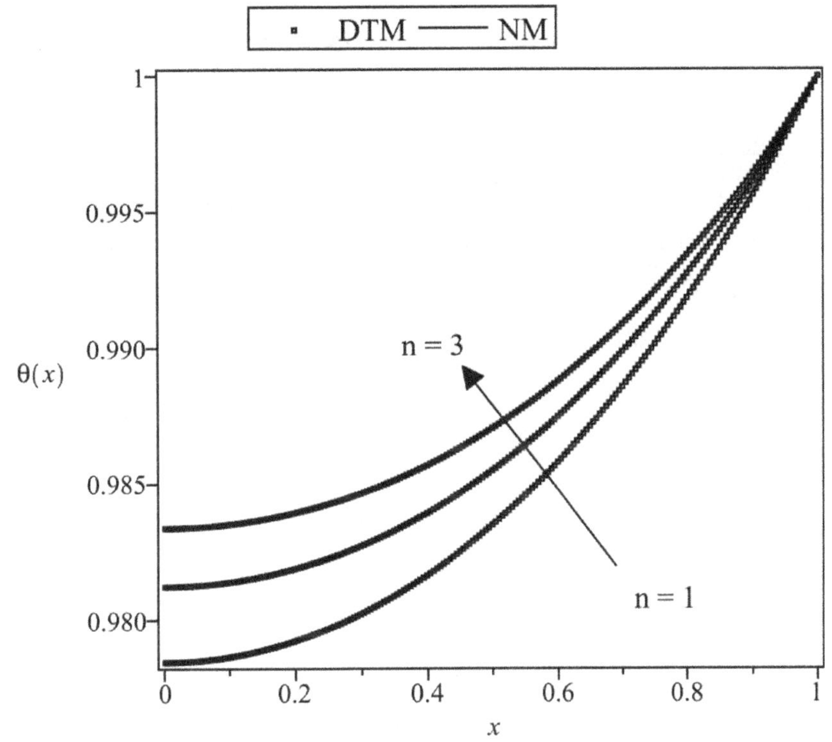

Figure 2. Comparison of (x) obtained by the DTM with numerical solution in a longitudinal rectangular fin when M = 1, β = 1, R = 0.8, ε_R = 0.1.

(2) Rectangular profile, case $n=2$

$$\Theta(2)=-\frac{1}{2}\frac{M^2\left(-a^3+R+R\varepsilon_R a\right)}{1+\beta a}$$
$$\Theta(3)=0$$
$$\Theta(4)=-\frac{1}{24}\frac{M^4\left(-a^3+R+R\varepsilon_R a\right)\left(3a^2+3\beta R+2\beta R\varepsilon_R a-R\varepsilon_R\right)}{(1+\beta a)^3} \quad (20)$$

From the above continuing process, substituting Equation 19 in Equation 14 for *H* = 1, we can obtain the closed form of the solution:

(1) Rectangular profile, case $n=1$

$$\theta(x)=a-\frac{1}{2}\frac{x^2M^2\left(-a^2+R+R\varepsilon_R a\right)}{1+\beta a} \quad (21)$$
$$-\frac{1}{24}\frac{x^4M^2\left(-a^2+R+R\varepsilon_R a\right)\left(2a-\beta a^2+3\beta R+2\beta R\varepsilon_R a-R\varepsilon_R\right)}{(1+\beta a)^3}+\dots$$

In order to obtain the value a, we used Equation 18. Then, we will have:

$$\theta(1)=a-\frac{1}{2}\frac{M^2\left(-a^2+R+R\varepsilon_R a\right)}{1+\beta a} \quad (22)$$
$$-\frac{1}{24}\frac{M^2\left(-a^2+R+R\varepsilon_R a\right)\left(2a-\beta a^2+3\beta R+2\beta R\varepsilon_R a-R\varepsilon_R\right)}{(1+\beta a)^3}+\dots=1$$

Solving Equation 22 by Maple software gives the value of a. For other cases the same process is used to obtain the value of a and temperature distribution.

RESULTS AND DISCUSSION

In this paper, the steady-state heat transfer in a longitudinal rectangular fin was studied. The dependence of the thermal conductivity, heat transfer coefficients and heat generation on the temperature rendered the problem highly nonlinear. The effects of the thermogeometric fin parameter (M), thermal conductivity gradient (β), heat generation gradient (ε_R) and *R* are investigated on the temperature distribution. To validate the analytical results, the temperature distribution through the longitudinal rectangular fin is compared with the numerical solution. The results are well matched with the results carried out by numerical solution as shown in Figure 2 and Table 2. In this table, error is introduced as follows:

$$\%Error=\left|\frac{\theta(x)_{NM}-\theta(x)_{DTM}}{\theta(x)_{NM}}\right|\times 100.$$

This accuracy gives high confidence in the validity of this

Table 2. Comparison between DTM and numerical results of θ(x) for rectangular profile when M=1, β=1, R=0.8 and $\varepsilon_R = 0.1$

$f(x)=1$		$n=1$			$n=2$			$n=3$	
x	**NM**	**DTM**	**Error (%)**	**NM**	**DTM**	**Error (%)**	**NM**	**DTM**	**Error (%)**
0	0.9784318	0.9784798	0.0048984	0.9812073	0.9812627	0.0056476	0.9833487	0.9834069	0.005912
0.1	0.9786395	0.9786799	0.0041301	0.9813846	0.9814303	0.0046542	0.9835028	0.9835497	0.0047697
0.2	0.9792382	0.9792822	0.0044861	0.9818847	0.9819353	0.0051584	0.983928	0.983981	0.0053895
0.3	0.9802475	0.9802919	0.0045252	0.982733	0.9827848	0.005271	0.9846536	0.9847085	0.0055756
0.4	0.9816774	0.981718	0.0041318	0.9839427	0.9839903	0.004838	0.985695	0.9857458	0.0051467
0.5	0.9835368	0.9835732	0.0036963	0.9855253	0.9855683	0.0043627	0.9870656	0.9871118	0.0046793
0.6	0.9858406	0.9858738	0.0033713	0.9875004	0.9875404	0.0040514	0.988788	0.9888318	0.0044207
0.7	0.9886144	0.9886403	0.0026174	0.9899023	0.9899337	0.0031732	0.9909028	0.9909375	0.0034959
0.8	0.9918745	0.991897	0.0022701	0.9927527	0.9927809	0.0028364	0.9934361	0.993468	0.003207
0.9	0.9956481	0.9956725	0.0024481	0.9960891	0.9961211	0.0032124	0.9964326	0.9964702	0.0037711
1	1	1	0	1	1	0	1	1	0

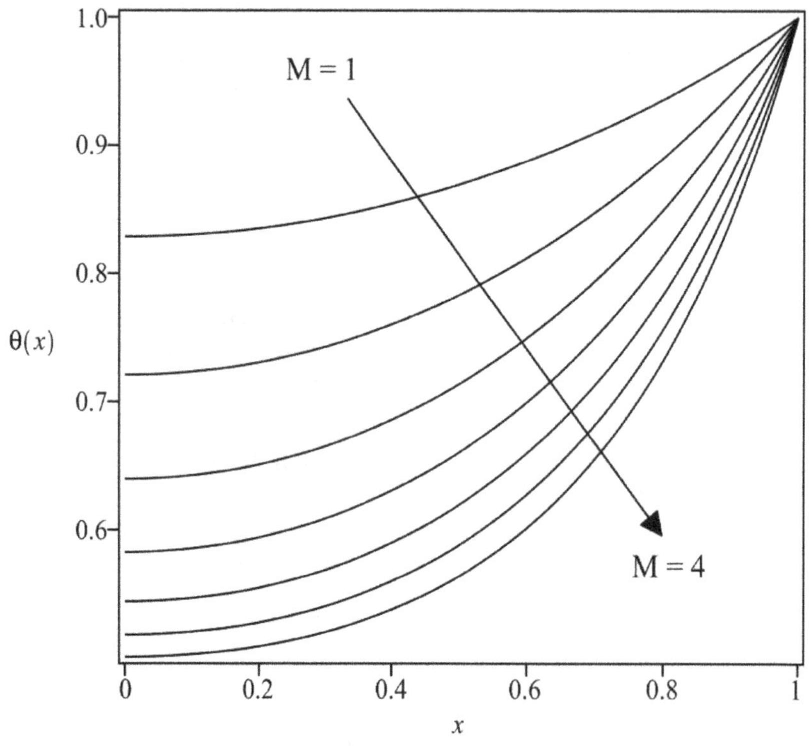

Figure 3. Temperature distribution in a longitudinal rectangular fin for varying values of thermo-geometric parameter (M) when n = 1, β = 0.5, R = 0.2, ε_R = 0.2.

problem, and reveals an excellent agreement in engineering accuracy.

The effect of thermo-geometric parameter (M) on temperature distribution is shown in Figure 3. It is illustrated that the magnitude of temperature is increased with decreasing the thermo-geometric parameter (M). Note that the thermo-geometric fin parameter $M=(Bi)^{1/2}E$, where $Bi=h_b\delta/k_a$ is the Biot number and $E=L/\delta$ is the aspect ratio or the extension factor. Undoubtedly, small values of M correspond to the relatively short and thick fins of high conductivity and high values of M correspond to longer and thin fins of poor conductivity (Mills, 1995). A fin is an excellent squanderer at small values of M. As M increases the convective heat loss increases and the temperature profile becomes steeper reflecting high base heat flow rates. In Figure 4,

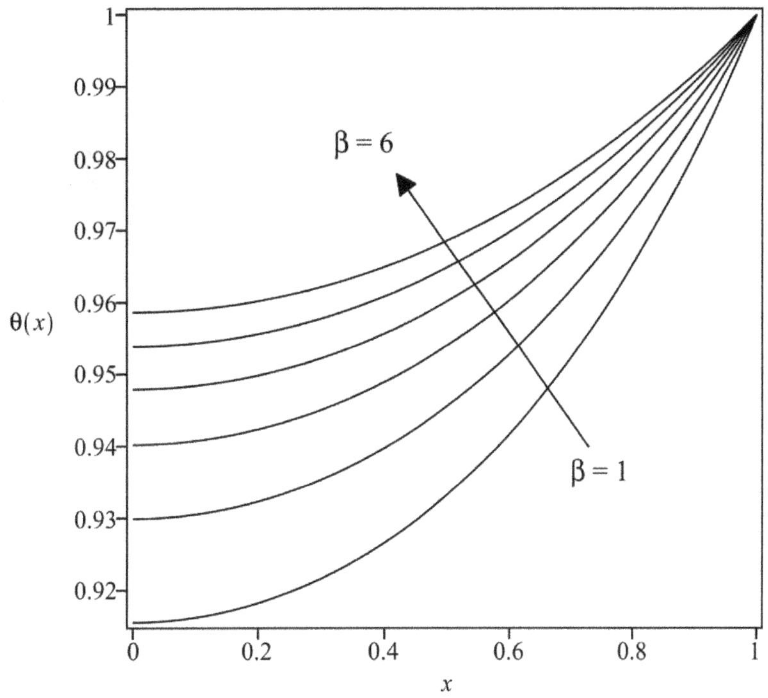

Figure 4. Temperature distribution in a longitudinal rectangular fin for varying values of β when n = 1, M = 2, *R* = 0.5, ε_R = 0.6.

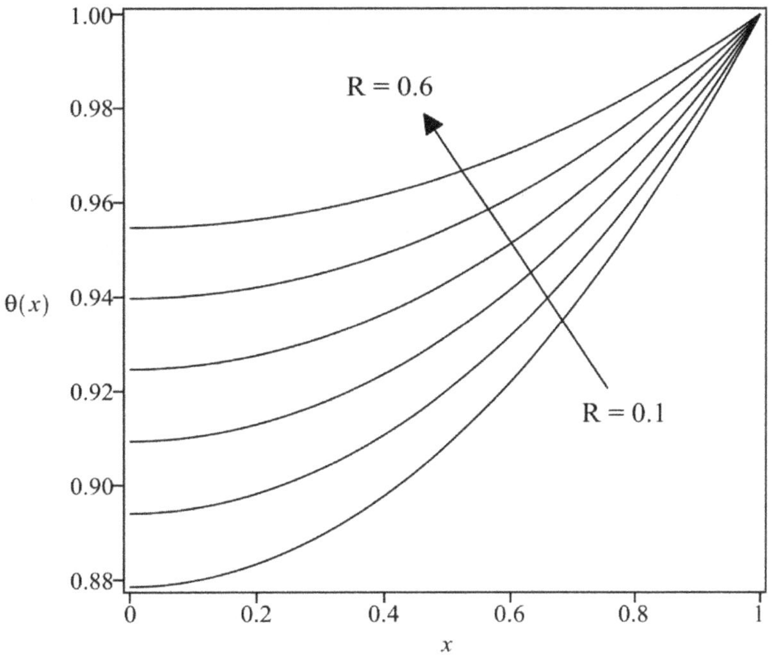

Figure 5. Temperature distribution in a longitudinal rectangular fin for varying values of *R* when n = 1, M = 1, β = 2, ε_R = 0.1.

we observe that the temperature in the fin increases with the increasing values of the thermal conductivity gradient (β). Figure 5 illustrates the variation of the temperature distribution with R. In addition, the effect of the heat generation gradient on the temperature distribution is shown in Figure 6. With a decrease in the ε_R, the losing

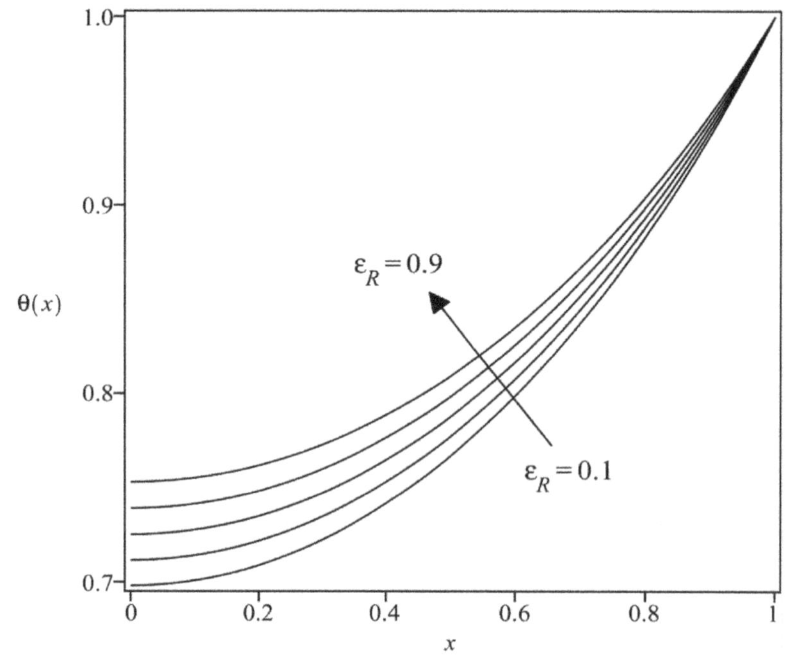

Figure 6. Temperature distribution in a longitudinal rectangular fin for varying values of ε_R when n = 1, M = 2, β = 1.5, R = 0.2.

heat from fin surface becomes stronger, thus the fin temperature decreases.

Conclusion

In this study, the DTM was applied to solve the heat transfer problem in a rectangular fin with temperature-dependent thermal conductivity, heat transfer coefficient and heat generation. Also, this problem is solved by a numerical method and some conclusions were summarized as follows:

(a) The DTM is a powerful approach for solving nonlinear differential equation, such as this problem. Also, it can be observed that there is good agreement between the present and numerical results.
(b) Increasing thermo-geometric parameter leads to a decrease in temperature distribution.
(c) By increasing thermal conductivity and heat generation gradient, temperature distribution increase.

Conflict of Interest

The author(s) have not declared any conflict of interest.

REFERENCES

Abbasov A, Bahadir AR (2005). The investigation of the transient regimes in the nonlinear systems by the generalized classical method. Math. Probl. Eng. 5:503-519. http://dx.doi.org/10.1155/MPE.2005.503

Abdel-Halim HIH (2004). Differential transformation technique for solving higher-order initial value problems, Appl. Math. Comput. 154(2):299-311. http://dx.doi.org/10.1016/S0096-3003(03)00708-2

Abdel-Halim HIH (2008). Application to differential transformation method for solving systems of differential equations. Appl. Math. Model. 32(12):2552-2559. http://dx.doi.org/10.1016/j.apm.2007.09.025

Arslanturk C (2005). A decomposition method for fin efficiency of convective straight fins with temperature-dependent thermal conductivity. Int. Commun. Heat Mass Transfer. 32(6):831-841. http://dx.doi.org/10.1016/j.icheatmasstransfer.2004.10.006

Aziz A, Khani F (2011). Convection-radiation from a continuously moving fin of a variable thermal conductivity. J. Franklin Inst. 348(4):640-651. http://dx.doi.org/10.1016/j.jfranklin.2011.01.008

Balkaya M, Kaya MO, Saglamer A (2009). Analysis of the vibration of an elastic beam supported on elastic soil using the differential transform method. Arch. Appl. Mech. 79:135-146. http://dx.doi.org/10.1007/s00419-008-0214-9

Borhanifar A, Abazari R (2011). Exact solutions for non-linear Schrdinger equations by differential transformation method. J. Appl. Math. Comput. 35:37-51. http://dx.doi.org/10.1007/s12190-009-0338-2

Domairry G, Fazeli M, (2009). Homotopy analysis method to determine the fin efficiency of convective straight fins with temperature-dependent thermal conductivity. Comm. Nonlinear Sci. Numer. Simulat. 14(2):489-499. http://dx.doi.org/10.1016/j.cnsns.2007.09.007

Franco A (2009). An analytical method for the optimum thermal design of convective longitudinal fin arrays. Heat Mass Transfer 45(12):1503-1517. http://dx.doi.org/10.1007/s00231-009-0526-5

Ganji DD, Ganji ZZ, Ganji HD (2011). Determination of temperature distribution for annual fins with temperature-dependent thermal conductivity by HPM. Therm. Sci. 15(1):111-115.

Ghafoori S, Motevalli M, Nejad MG, Shakeri F, Ganji DD, Jalaal M (2011). Efficiency of differential transformation method for nonlinear oscillation: comparison with HPM and VIM. Curr. Appl. Phys. 1:965-971. http://dx.doi.org/10.1016/j.cap.2010.12.018

Ghasemi SE, Hatami M, Ganji DD (2014). Thermal analysis of

convective fin with temperature-dependent thermal conductivity and heat generation. Case Studies in Therm. Eng. 4:1-8. http://dx.doi.org/10.1016/j.csite.2014.05.002

Hatami M, Ganji DD (2013). Thermal performance of circular convective-radiative porous fins with different section shapes and materials. Energy Convers. Manage. 76:185-193. http://dx.doi.org/10.1016/j.enconman.2013.07.040

Hatami M, Ganji DD (2014a). Investigation of refrigeration efficiency for fully wet circular porous fins with variable sections by combined heat and mass transfer analysis. int. J. Refrig. 40:140-151. http://dx.doi.org/10.1016/j.ijrefrig.2013.11.002

Hatami M, Ganji DD (2014b). Thermal behavior of longitudinal convective-radiative porous fins with different section shapes and ceramic materials (SiC and Si3N4). Ceram. Int. 40(5):6765-6775. http://dx.doi.org/10.1016/j.ceramint.2013.11.140.

Hatami M, Hasanpour A, Ganji DD (2013). Heat transfer study through porous fins (Si3N4 and AL) with temperature-dependent heat generation. Energy Convers. Manage. 74:9-16. http://dx.doi.org/10.1016/j.enconman.2013.04.034

Hatami M, Mehdizadeh Ahangar GHR, Ganji DD, Boubaker K (2014). Refrigeration efficiency analysis for fully wet semi-spherical porous fins. Energy Convers. Manage. 84:533-540. http://dx.doi.org/10.1016/j.enconman.2014.05.007

Joneidi AA, Ganji DD, Babaelahi M (2009). Differential transformation method to determine fin efficiency of convective straight fins with temperature dependent thermal conductivity. Int. Commun. Heat Mass Transf. 36(7):757-762. http://dx.doi.org/10.1016/j.icheatmasstransfer.2009.03.020

Kraus AD, Aziz A, Welty J (2001). Extended Surface Heat Transfer. Wiley. New York.

Lin WW, Lee DJ (1999). Boiling on a straight pin fin with variable thermal conductivity, Heat Mass Transf. 34(5):381-386. http://dx.doi.org/10.1007/s002310050273

Mills AF (1995). Basic heat and mass transfer. Irwin INC. Chicago.

Moradi A, Ahmadikia H (2010). Analytical solution for different profiles of fin with temperature dependent thermal conductivity. Math. Prob. Eng. 10:1-15. http://dx.doi.org/10.1155/2010/568263

Rajabi A (2007). Homotopy perturbation method for fin efficiency of convective straight fins with temperature-dependent thermal conductivity. Phys. Lett. A. 364(1):33-37. http://dx.doi.org/10.1016/j.physleta.2006.11.062

Rashidi MM, Laraqi N, Sadri SM (2010). A novel analytical solution of mixed convection about an inclined flat plate embedded in a porous medium using the DTM-Padé. Int. J. Therm. Sci. 49(12):2405-2412. http://dx.doi.org/10.1016/j.ijthermalsci.2010.07.005

Sharqawy MH, Zubair SM (2007). Efficiency and optimization of an annular fin with combined heat and mass transfer-an analytical solution. Int. J. Refrig. 30(5):751-757. http://dx.doi.org/10.1016/j.ijrefrig.2006.12.008

Unal HC (1987). An analytical study of boiling heat transfer from a fin. Int. J. Heat Mass Transf. 30(2):341-349. http://dx.doi.org/10.1016/0017-9310(87)90122-0

Zhou JK (1986). Differential transformation method and its application for electrical circuits. Wuhan (China), Hauzhang University press.

Hurricanes and cyclones kinematics and thermodynamics based on Clausius-Clapeyron relation derived in 1832

Mbane Biouele Cesar

Laboratory of Earth's Atmosphere Physics, University of Yaoundé I, Cameroon. E-mail: cesar.mbane@yahoo.fr.

Juxtaposing Clausius-Clapeyron relation derived in 1832 with Hydrodynamic concept of atmosphere parcel of air leads to the discovery of an essential property of the troposphere that will deeply ameliorates information contained in literature and audiovisual productions on tropical weather. It is indeed a rectification of the use of an ideal gas principle which enshrined the idea that, hot air is lighter than cold air throughout the atmosphere. In other words, contrary to what has been taught in schools and universities of the world, hot air is not lighter than cold air in all parts of the troposphere. Taking into account this troposphere thermodynamics reality improves our understanding of complex weather phenomena such as cyclones and hurricanes. The two equal level surfaces of water vapor and temperature rating respectively at 6.11 mb and 0.0098°C separate without any ambiguity parts of the troposphere where ideal gas assumption can be applied to parts of the troposphere where this assumption is banned.

Key words: Clausius-Clapeyron relation, Hydrodynamic concept of air parcel, equal level surfaces, rating at 6.11 mb and 0.0098°C.

INTRODUCTION

A number of questions regarding Kinematics and thermodynamics of hurricanes and cyclones remain unanswered despite the quality and quantity of ground-or space-based observations. Instead these weather phenomena are combinations of complex troposphere physical processes that occur under the accuracy of temperature and humidity conditions. In this study, regardless of the manner in which hurricanes and cyclones are consider (Arakawa and Suda, 1953; Ballenzweig,1957; Bangs, 1929; Beerbower, 1926; Cline, 1926; Conner et al., 1957; Duane, 1935; Dunn, 1956; Fassig, 1913; Fletcher, 1955; Gentry, 1955; Haurwitz, 1935; Hoover, 1957; Hughes, 1952; Jordan, 1952; Klein and Winston, 1947; Malkin and Galwaym, 1953; Malkus, 1958; McDonald, 1942; Miller, 1958; Riehl and Palmen, 1957; Rossby, 1949; Schoner and Molansk, 1956; Tannehill, 1936), we want to make a contribution to a better understanding of kinematics, and thermodynamics governing these weather events with high destructive power. Our results are obtained from effectiveness of Clausius-Clapeyron equation that leads to the slope of the equilibrium curves in the *pT*-plane (Figure 1) whose show precisely that, unlike the dry water vapor that can be assimilated to the ideal gas at all times, saturated water vapor at low temperatures (temperature below 0.0098°C) in the presence of high humidity of air (vapor pressure above 6.11 mb), has thermoelastic properties diametrically opposed to those of ideal gases (including dry water vapor). In tropical regions, saturated water vapor occupies the middle and top of the troposphere to more than 90% (it should be noted that, saturated water vapor is known as the birth place, home or bed of weather events such as hurricanes and cyclones or clouds related). Therefore, it was necessary to take account of this characteristic property of the saturated water thermodynamics to successfully draw new and unique profiles of hurricanes and cyclones.

Names assigned to tropical disturbances and related precipitating systems vary from one community to another. "*Tornade*" in French is used to refer to *cold-disturbances* while "*Tornado*" in Anglo-Saxon community is used to refer to *hot-disturbances*. In the translation of literature and audiovisual productions, this distinction is not often made and leads to inconsistencies and confusion. Indeed, vertical profiles of cold disturbances are diametrically opposite directions vertical profiles of hot disturbances. Hurricanes or tornadoes will be referred to hot-disturbances while cyclones will be referred to cold-disturbances in this work.

TROPOSPHERE DYNAMIC BALANCE

Atmosphere dynamics uses precise concept of air particle (Batchelor, 1967; Riegel, 1992). Especially:

a) Few exchanges on molecular scale: it is easy to follow quantity of air which preserves certain properties.
b) Quasi-static equilibrium: at any moment there is dynamic balance, that is, the particle has the same pressure as its environment ($P = P_{ext}$).
c) No thermal balance: heat transfers by conduction are very slow and neglected. One can have $T \neq T_{ext}$.
d) The horizontal sizes of the air particle can go from a few cm to 100 km according to the applications.
Taking into account the fact that, the atmosphere is mainly composed of dry air and water vapor, the Dalton's law connects the pressure (P) with the partial pressure of dry air (P_a) and water vapor (e):

$$P = P_a + e \qquad (1)$$

In deriving P with respect to the temperature, one has

$$\frac{dP}{dT} = \left(\frac{\partial P}{\partial T}\right)_V + \left(\frac{\partial P}{\partial V}\right)_T \cdot \left(\frac{dV}{dT}\right) \qquad (2)$$

According to the quasi-static equilibrium (or dynamic balance), the pressure in the air particle must be the same as that of the air around, including during its water contains changes in phases. In other words, P in the air particle remains constant during individual changes in phases. Hence:

$$\frac{dP}{dT} = 0 \qquad (3)$$

Equations 2 and 3 lead to the derivative of V compared to T:

$$\frac{dV}{dT} = -\frac{(\frac{\partial P}{\partial T})_V}{(\frac{\partial P}{\partial V})_T} \qquad (4)$$

Introducing (χ) the coefficient of thermal expansion of moist air at constant temperature:

$$\chi = -\frac{1}{P}(\frac{\partial P}{\partial V})_T \qquad (5)$$

Then the equation of atmosphere dynamic balance:

$$\frac{dV}{dT} = \frac{1}{\chi} \bullet \frac{1}{P}(\frac{\partial P}{\partial T})_V \qquad (6)$$

Using Clausius-Clapeyron estimations of $(\frac{\partial P_a}{\partial T})_V$ and $(\frac{\partial e_w}{\partial T})_V$, Equation 6 of troposphere (or birth place of weather events) dynamic balance become:

$$\frac{dV}{dT} = \frac{1}{\chi} \bullet \frac{1}{P}(\frac{\partial e_w}{\partial T})_V \qquad (7)$$

Equations 6 and 7 lead to a very important atmosphere dynamics statement; at any moment and throughout the atmosphere, one can use Equations 6 or 7 and Clausius-Clapeyron slope of the equilibrium curves in the pT-plane (Figure1) to predict in which direction the air parcel will move (up or down) if its temperature increases or decreases. Table 1 and Figure 2 provide an overview of possible situations in the troposphere.

BASIC KINEMATICS AND THERMODYNAMICS OF HURRICANES AND CYCLONES

Vertical profiles of hurricanes

Hurricanes are triggered by passive deep convection generated by a hot source located at the surface of the

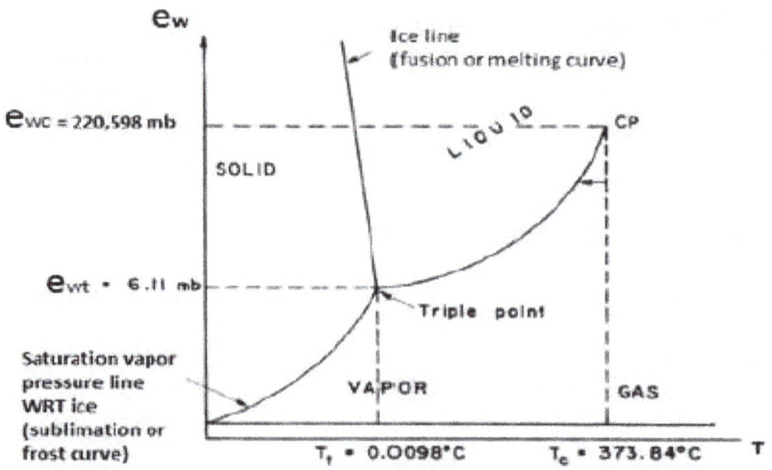

Figure 1. Saturation curves for water substance onto the p_T-plane. (p_T and T_T are called triple-point data): e_{wT} = 6.11 mb; T_T = 0.0098°C.

Table 1. Changes in volume of the moist air particle depending on temperature within a specific range of temperature and humidity.

Range of temperature coupled with range of humidity	T < 0.0098°C e_w < 6.11 mb	T < 0.0098°C e_w > 6.11 mb	T > 0.0098°C e_w > 6.11 mb
$(\frac{\partial P}{\partial T})_V$	+	-	+
$\frac{dV}{dT}$	+	-	+

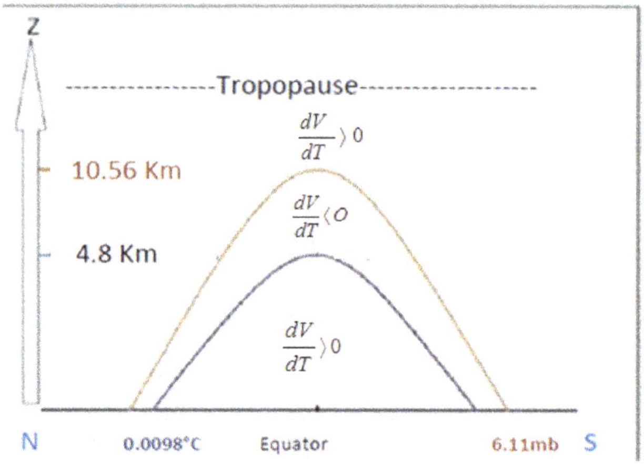

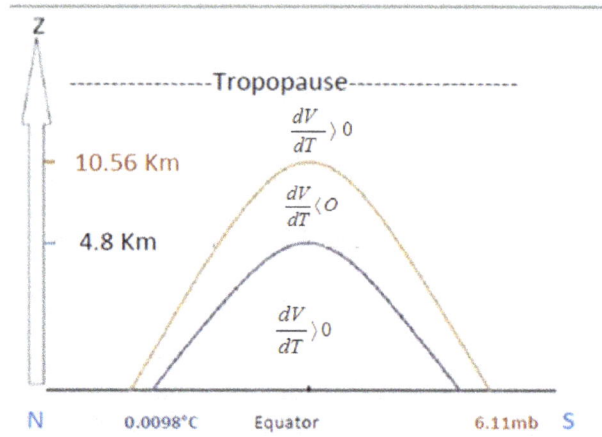

Figure 2. Troposphere specific regions depending on the manner in which V changes with T (V and T are respectively volume and temperature of an air parcel): $\frac{dV}{dT}>0$; the particle swells when its temperature increases (so it becomes lighter). $\frac{dV}{dT}<0$; the particle shrinks when its temperature increases (so it becomes less light). -10.56 Km = maximum elevation (statistic value) of 6.11 mb pseudo-isobar, -4.8 Km = maximum elevation (statistic value) of 0.0098°C isotherm, -V (air percel volume), T (air parcel temperature).

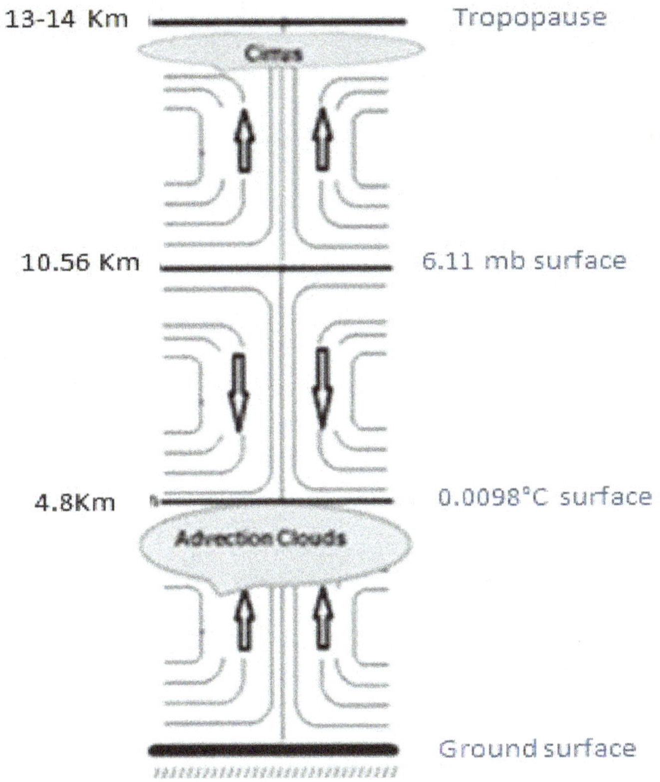

Figure 3. Hurricanes are triggered by passive deep convection generated by a heat hot source located at the surface of the Ground. They appear as high towers consisting of three floors: warm updrafts occupying floors 1 and 3 while warm downdrafts occupy floor 2.

Figure 4 (a-d). Hurricanes or tornadoes trigger dust clouds whose base is thin compared to the peak which is very broad. Hurricanes (or Tornadoes) can also electrify (Mbane, 2012) the troposphere column in which it is formed (Figure 4c). Its broadest peak indicates the presence of hot downdrafts that prevent the progression of warm updrafts.

earth and appear (Figure 3) as very high towers (from 0 to about 9 km) consisting of 3 floors: warm updrafts occupying the first and third while warm downdrafts occupies the second floor. According to ground based observations of Figure 4(a to d), over-land hurricanes (or tornadoes) trigger clouds whose base is thin compared to the peak which is very broad. Hurricanes vertical drafts can also produce electrical charges (Mbane, 2009) in the troposphere colum in which it is formed (Figure 4c). The broadest peaks of the related clouds indicate the presence of the second floor warm downdrafts that prevent the progression of the first floor warm updrafts (Figure 3). Considering the molecular scale, our model (Figure 4e) based on Clausius-Clapeyron's relation (1832) suggests, unlike ideal gas cumulonimbus model (Figure 4f), blocking of hot updrafts by hot down drafts which means installation of an additional greenhouse effect, which causes the accumulation of cloud formation latent heat with earth's surface radiate heat R_T ($R_T = \varepsilon_s \sigma T_S^4$) at the ground surface. This is consistent with based observations and explains high surface temperatures that accompany the formation of clouds in the sunny sky.

Vertical profiles of cylones

Cyclones are triggerd by very deep and passive convection generated by a cold source (squall-line) located at the summit of the troposphere and appear (Figure 5) as very high towers (from 0 to about 14 km) consisting of 3 floors: cooler downdrafts occupying 1st and 3rd while cooler updrafts occupy the 2nd floor. There is good agreement between aircraft-based observations and related cyclones second floor updrafts convective clouds whose base has to be located above 0.0098°C isotherm surface. Cyclones vertical drafts can also produce electrical charges (Mbane, 2012) in the troposphere colum in which it is formed.

Horizontal profiles of hurricane and cyclone

Observed pressure near the eyes of cyclones (or hurricane) is very low and concentrates a rapid decrease in a short distance so that the momentum of particles of air, the frictional force and the tidal force are (from surface of the earth to tropopause) negligible compared to the coriolis and pressure-gradient forces. When pressure gradient and coriolis forces are the only two factors acting, geostrophic winds (rotative in the Northern hemisphere and contra-rotative in the Southern hemisphere) immediately take place (Figures 6) within deep and passive convections. The impact of hurricanes footprint (less than a dozen kilometers in diameter) is much lower than that of cyclones (several tens of kilometers in diameter).

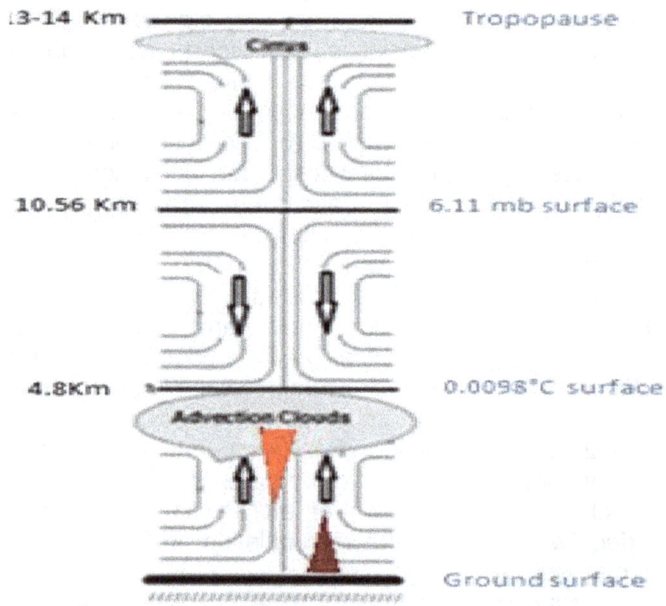

Figure 4e. Our model based on Clausius-Clapeyron's relation (1832) suggests blocking of hot updrafts by hot down drafts: then install an additional greenhouse which triggers the superposition of cloud formation latent heat (red color) with earth's surface radiate heat $\varepsilon_s\sigma$ T^4 (brown color). This is consistent with observations and explains higher temperature and humidity of atmosphere lower layers and tropical regions moderate climate.

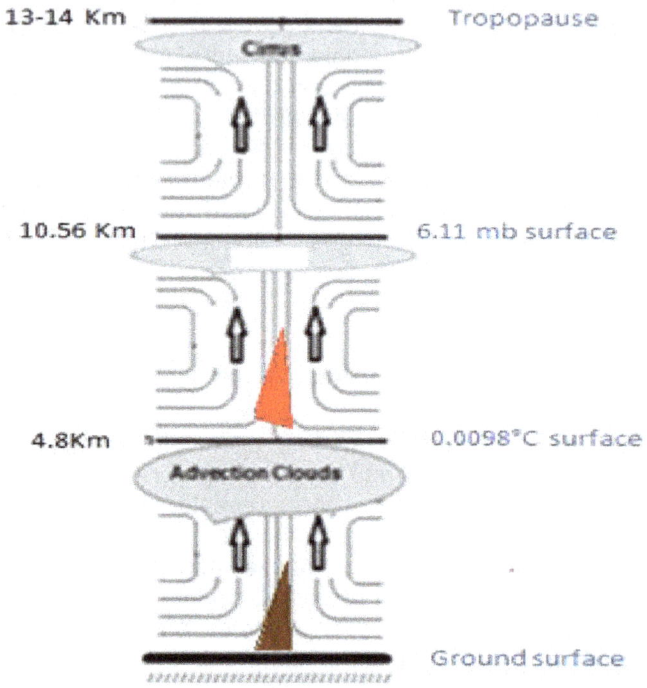

Figure 4f. Ideal gas properties suggest passive convection transfer from earth's surface to tropopause and automatically excluded the spread of cloud formation latent heat to that earth's surface.There is therefore no possibility of accumulation of heat or water vapor in lower troposphere. This is not consistent with earth's atmosphere physics.

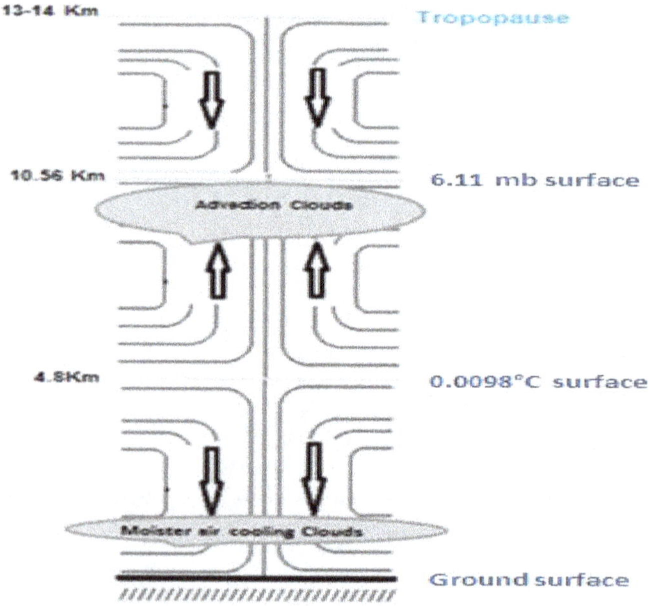

Figure 5. Cyclones are triggered by passive deep convection generated by a cold hot source located at the mid- troposphere. They appear as high towers consisting of three floors: Cold updrafts occupying floors 2 while cold downdrafts occupy floor 1 and 3

TROPOSPHERE GRAVITY WAVES AND LOCAL PRESSURE FLUCTUATIONS

In fluid medium, movement from point A to point B of local pressure fluctuations (δ P) are provided by gravity waves. For incompressible fluids (including the atmosphere because of its elasticity), these waves travel at phenomenal speeds. In the troposphere, the propagation is essentially straight and parallel to the direction of gravity acceleration ($\vec{g}$). Therefore, when a surface depression (D) occurs, it is almost spontaneously covered by altitude low pressure (L).

Reciprocally, when L occurs, it is immediately relayed by a D. This is same for Anticyclone (A) and high pressure (H). In the troposphere, D and L appear without any indication of which the two came first. Hurricanes (or tornadoes) are caused by warming ground surface areas, while cyclones are caused by mid-troposphere cooling domains. Due to gravity waves, satellites in their current configuration cannot differentiate between hurricanes and cyclones. This is a very embarrassing situation that leads to numerous confusion and makes ineffective "*Weather Alerts Systems* ".

Conclusions

The troposphere, generally known as birth place of weather phenomena, is a huge thermodynamic engine driven by the energy received from the sun. All winds,

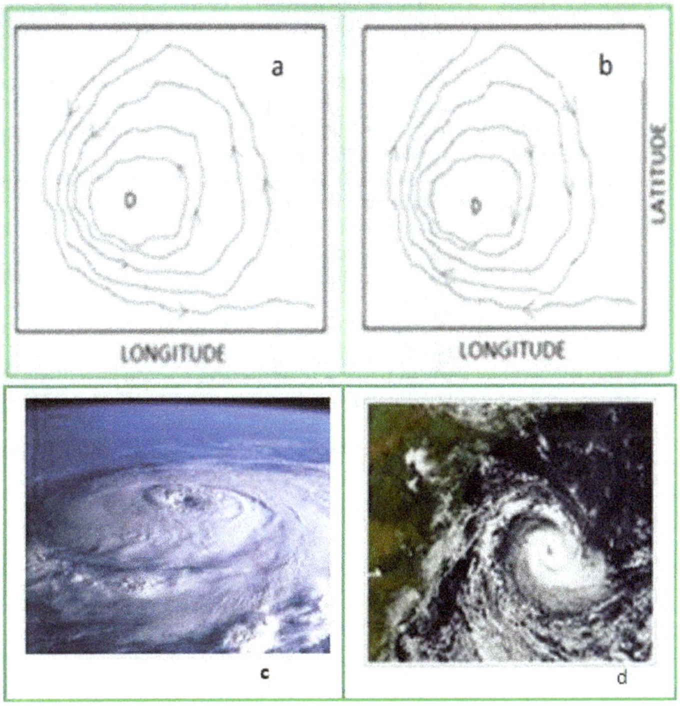

Figures 6. Streamlines of geostrophic winds triggered by cyclones around their low pressure groove. Direct rotation is observed in the northern hemisphere (e.g. a and c) while indirect rotation is observed in the southern hemisphere (e.g. b and d): based on trigonometric considerations.

storms and clouds result from the differences in the amount and utilization of this energy. Since the radiant energy appears principally as heat, it was necessary to resort the thermodynamics properties of saturated water vapor in order to better understand how the moist air reacts to heat changes in any portion of the troposphere. New and unique kinematic profiles of hurricanes and cyclones can now be easily plotted. It should be noted that, all natural meteorological phenomena included hurricanes and cyclones can be traced to the manner in which the energy from the sun is received over different parts of the earth.

Since the troposphere is a medium in which mass motions are easily started, convection is found to be one of the chief ways in which heat is transferred there. This transfer may be accomplishing either by vertical or by horizontal motions. According to our results: warmer disturbances that occur in lower-troposphere are dissipated (that is, Atmosphere is a force-restore engine or dissipative system) by a typical mass motion usually called hurricanes (or tornadoes) while cooler disturbances that occur in mid-troposphere are dissipated by another typical mass motion called cyclones. Knowing that coriolis force act to west on updrafts, everyone can now understand why hurricane and cyclone move preferentially from East to West due to the localization in updrafts of their heat sources.

Cyclones' heat source is made of huge and cooler fogs (those observed temperatures are less than -45°C) which can travel even increase (over hot oceans) in the troposphere while hurricanes' heat source is fixed on the Ground: that's why cyclone lives and travels longer than hurricane. Moreover, contrary to a widespread idea in meteorology that warm air is less dense than cold air, this work shows that: between the two equal level surfaces of water vapor and temperature respective rating 6.11 mb and 0.0098°C, hot air is less light than cold air. This explains the presence of cold upwelling in cyclones and hot downdrafts in hurricanes. We cannot conclude this work without strongly highlighting the fact that hot air, contrary to what is taught until now, is not lighter than cold air in all parts of the troposphere. This troposphere's thermodynamic property is unfortunately unknown to public and has led to numerous errors and inconsistencies that abound in many scientific books and publications (including audiovisuals productions). In the next investigation on climate, using only certainties (e.g. ideal gas approximation imposes a partition of the troposphere into 3 regions, vertical temperature gradient in the troposphere has a negative sign, moist air condenses as it cools, pressure in the air parcel is the same everywhere and is equal to that of its immediate neighbors in all circumstances (that is, state of atmosphere quasi static equilibrium) instead of wrong presumptions (e.g. ideal gas approximation is valid throughout the atmosphere, warm air is lighter than cold air throughout the atmosphere, updrafts are necessarily associated with warming, downdrafts are necessarily associated with cooling) would allow:

a) To deeply ameliorate our view of troposphere phenomena regardless of their spatial and temporal scales (e.g. Rossby's suggested representation of general circulation),

b) To greatly exorcise our fears (sometimes ridiculous) generally triggered by a lack of explanations devoid of ambiguity,

c) To better protect ourselves from disasters caused by these devastating events, those life cycles until now unfortunately escapes the human control.

Knowing physics processes behind devastating event makes it less daunting (a good example is experienced in Mexico where building's architecture is gradually adapted to the local soil structure which paradoxically amplifies seismic waves which comes from far away, instead of reducing their intensity).

REFERENCES

Arakawa H, Suda K (1953). Analysis of winds, wind waves, and swells over the sea to the east of Japan during the typhoon of September 26, 1935. Mon. Wea. Rev. 81:31-37.

Ballenzweig EM (1957). Seasonal Variation in the Frequency of North Atlantic Tropical Cyclones Related to the General circulation. National Hurricane Research Project. Report N°9:32.

Bangs NH (1929). Effects of the 1926 Florida Hurricane Upon Engineer Designed Buildings. Bull. Amer. Met. Soc. 10:46-47.

Beerbower GM (1926). Hurricanes Effects on Buildings at Hollywood, Florida. Engineering News Record. 97:752.

Cline IM (1926). Tropical cyclones. MacMillan Company. New York, N.Y. P. 301.

Conner WC, Kraft RH, Harris DL (1957). Empirical Methods for forecasting the maximum storm tide due to hurricanes and other Tropical storms. Mon. Wea. Rev. 85:113-116.

Duane JE Jr (1935). The Hurricane of September 2, 1935, at Long Key, Florida. Bull. Amer. Met. Soc. 16:238-239.

Dunn GE (1956). Areas of Hurricane Development. Mon. Wea. Rev. 84:47-51.

Fassig OL (1913). Hurricanes of the West Indies. Bull. N° 13. U.S. Weather Bureau. Washington D.C. P. 28.

Fletcher RD (1955). Computation of Maximum Winds in Hurricanes. Bull. Am. Met. Soc. 36:246-250.

Gentry RC (1955). Wind Velocities During Hurricanes. Paper N° 2731. Trans. Am. Soc. Of Civil Engineers. 120:169.

Haurwitz B (1935). The Height of Tropical Cyclones and of Eye of the Storm. Mon. Wea. Rev. 63:45-49.

Hoover RA (1957). Empirical relationships of the Central Pressures in Hurricanes to the Maximum Surge and Storm tide. Mon. Wea. Rev. 85:167-174.

Hughes LA (1952). On the Low-level Wind Structure of Tropical Storms. J. Met. 9:422-428.

Jordan E (1952). An Observational Study of the Upper Wind Circulation Around Tropical Storms. J. Met. 9:340-346.

Klein WH, Winston J.S. (1947). The Path of the Atlantic Hurricane of September 1947 in Relation to the Hemispheric Circulation. Bull. Am. Met. Soc. 28:447-452.

Malkin W, Galway JG (1953). Tornadoes Associated with Hurricanes. Mon. Wea. Rev. 81:299-303.

Malkus J (1958). On the Structure and Maintenance of the Mature Hurricane Eye. J. Met. 15:337- 349.

McDonald WF (1942). On a Hypothesis Concerning Normal Development and Disintegration of Tropical Hurricanes. Bull. Amer. Met. Soc. 23:73-78.

Miller BI (1958). The Three Dimensional Wind Structure around a Tropical Cyclone. National Hurricane Research Project. Report N° 15:41.

Riehl H, Palmen E (1957). Budget of Angular Momentum and energy in Tropical Cyclones. J. Met. 14: 150-159.

Rossby CG (1949). On the Mechanism for Release of Potential Energy in the Atmosphere. J. Met. 6:163-180.

Schoner RW, Molansky S (1956). Rainfall Associated With Hurricanes. National Hurricane Research Project. Report N° 3:305.

Tannehill IR (1936). Sea Swells in Relation to the Movement and Intensity of Tropical storms. Mon. Wea. Rev. 64:231-238.

Batchelor GK (1967). An Introduction to Fluid Dynamics. Cambridge University Press. P. 468.

Riegel CA (1992). Fundamentals of Atmospheric Dynamics and Thermodynamics. Word Scientific Publishing Co. Pte. Ltd. P. 496.

Mbane BC (2009). Vertical profiles of winds and electric fields triggered by tropical storms- under the hydrodynamic concept of air particle. Int. J. Phys. Sci. 4(4):24-246.

Mbane BC (2012). Physics of Atmosphere Dynamic or Electric Balance Processes Such as Thunderclouds and Related Lightning Flashes. Geosciences. 2(1):6-10.

17

Relation between classical mechanics and physics of condensed medium

V. M. Somsikov, A. B. Andreyev and A. I. Mokhnatkin

Institute of Ionosphere, Almaty, 050020, Kazakhstan.

An equilibrium system of potentially interacting material points in a nonhomogeneous field is studied by numerical simulations, as well as criteria which are needed in order to switch from the quantified model to the thermodynamic one. This paper studies the system passing a potential barrier and factors which affect the change of the system's internal energy. These factors include width of the barrier, number of material points in the system, as well as initial conditions. The dependence between change in fluctuations of energy parameters of the system and the number of material points is also shown. The amount of dynamic entropy is estimated. The paper also defines and describes two critical values (N1 and N2). If the system includes more than N1 material points, then its dynamics becomes irreversible. If the number of material points is greater than N2, then thermodynamic model can be used. The results obtained by numerical simulations verified the theoretical conclusions.

Key words: Nonlinearity, classical mechanics, energy, thermodynamics, Lagrange equations, non-holonomic constraints, irreversibility.

INTRODUCTION

The possibility to derive the laws of thermodynamics, statistical physics and kinetics using the laws of classical mechanics is still considered as an open question. First of all, this is due to the fact that in the nature all processes are irreversible and dissipative. However, in accordance to the formalism of classical mechanics, the dynamics of systems are reversible. It is subject of one of the key problems of physics (Ginzburg, 2007; Bohr, 1958; Klein, 1961; Prigogine, 1980; Lebowitz, 1999; Zaslavsky, 1984). Up to the present time, clear criteria for the transition to the physics of continuous media are not set. For example, it is not possible to determine the exact number of elements a system should consist of, in order to become thermodynamically applicable, as well as the accuracy of any estimation which is done using the laws of thermodynamics. The abovementioned problems are not solved yet since until recently there was no rigorous explanation of irreversibility in the framework of the laws of classical mechanics, without introducing probabilistic hypotheses (Prigogine, 1980; Lebowitz, 1999; Zaslavsky, 1984; Smoluchowski, 1967; Cohen, 1998; Klimontovich, 1995; Kadomtsev, 1995; Sidharth, 2008). Indeed, today irreversibility is usually explained by the property of exponential instability of Hamiltonian systems and the hypothesis of the existence of fluctuations. The main idea of this explanation is as follows. According to the Poincare theorem on the reversibility of Hamiltonian systems, a system's coordinates in the phase space will be close enough to their initial values in a finite (although it can be very large) period of time (Prigogine, 1980; Zaslavsky, 1984; Smoluchowski, 1967). But if averaged over an arbitrarily small neighborhood of the phase space, the system will not return back to its initial position because of exponential instability. Arbitrarily small fluctuations in the system are equivalent to such averaging. Therefore, the existence of fluctuations in Hamiltonian systems is a sufficient condition for the occurrence of irreversibility. But the hypothesis of a roughening of the phase space due to fluctuations actually involves statistical laws which are alien to determinism of classical mechanics. Therefore, the question of justifiability of this hypothesis is still open.

A lot of attempts to explain the second law of thermodynamics on the basis of the laws of classical mechanics without the use of statistical hypotheses failed, so most probably it is impossible to solve this problem in the framework of the existing formalism of classical mechanics (Prigogine, 1980; Zaslavsky, 1984). This means either that there is no explanation in the framework of classical mechanics at all or that the formalism of classical mechanics requires expansion, for example by removing the limitations under which it was built (Prigogine, 1980).

In order to find an approach to solve the problem of irreversibility the dynamics of a system of hard discs has been studied firstly. Unlike other authors who studied the system of billiards, provided that they are Hamiltonian systems (Bird, 1976; Sinai, 1995), we considered the billiards in the approach of pair collisions of disks (Somsikov et al., 1999; Somsikov, 2004). As a result the irreversibility of the system of hard disks was found. It turned out that the exchange of momentum between the discs plays a key role in irreversibility. The Hamilton's and Lowville's equations describing equilibration in a system were obtained basing on these results (Somsikov, 2004). But a number of problems still remained. For example, the pair interaction in a real system is a special case which is valid only for fairly rarefied systems. Moreover, the interactions between the system's elements are potential. Therefore, the next step was the study of systems of potentially interacting material points (Somsikov, 2005).

It followed from the study of systems of hard discs that the mechanism of irreversibility cannot rely on the formalism of classical mechanics. Therefore, it was decided to look for an answer based on the energy equation for systems of potentially interacting elements, provided that Newton's laws are valid for each individual material point. As a result of these studies it was found that the system's dynamics is irreversible. Irreversibility was caused by the mutual transformation between the system's motion energy and its internal energy (Somsikov, 2014). But one problem still remains unsolved: Why formalisms of classical mechanics are reversible, while the dynamics of systems of material points is irreversible. This contradiction should be explained; it was found that in classical mechanics irreversibility has been lost while obtaining the Lagrange equations for systems of material points because of usage of the hypothesis of holonomic constraints (Somsikov, 2014). It turned out that this hypothesis excludes the possibility of describing the nonlinear transformation of energy between different degrees of freedom, which is responsible for the irreversibility.

It turned out that it is possible to explain irreversibility using Newton's laws by removing some of the limitations of classical mechanics formalism. The mechanics of structured particles (SP) was developed and a mechanism to explain irreversibility was offered. This mechanism was named deterministic as it uses the mechanics of SP without any probabilistic principles. The essence of this mechanism is as follows (Somsikov, 2014a; Somsikov, 2014b).

Newton built the mechanics for an abstract structureless material point (MP). Therefore, the external forces change only the position of MP. This fact defines Newton's second law. But all real bodies are not structure-less. Thus the external forces change not only the position of a body, but its internal energy as well. The internal energy is accounted for the motion of elements of the body with respect to its center of mass. This means that the dynamics of the system is defined by the principle of dualism of symmetry due to the fact that the system is not structure-less. The essence of the principle is that the dynamics of real bodies is determined by two types of symmetry: The symmetry of space and the symmetry of the body itself. The principle of dualism of energy follows from the principle of dualism of symmetry. The principle of dualism of energy claims that the dynamics of a real body is defined by splitting the body's total energy into its internal energy and the energy of motion. The equation of the body's motion follows from the principle of dualism of energy. This equation takes into account both the work needed to change the position of the body, and the work than changes the body's internal energy (Somsikov, 2014b).

Thus, the total energy of the body is an invariant which determines the body's dynamics; the energy of motion itself is not an invariant, so the motion of the body's mass center should be determined based on the invariance of the total energy.

According to classical mechanics, a body can be represented by a set of potentially interacting MPs (Goldstein, 1975; Landau, 1976). This means that the equation of motion of the body should be derived upon the condition that the motion of each MP is described by the laws of Newton. In general a body is an open non-equilibrium dynamic system. Such system in many cases could be considered as it is in a local thermodynamic equilibrium state (Klimontovich, 1995; Landau, 1976; Rumer and Rivkin, 1977). In this case, the body can be represented by a set of moving equilibrium subsystems;

each subsystem is in turn a set of potentially interacting MPs (Landau, 1976; Rumer and Rivkin, 1977). Let us name such a subsystem as a structured particle (SP). This means that the dynamics of an open non-equilibrium system should be defined based on the assumption that the system consists from a lot of SPs. Therefore, first of all, the mechanics of SPs should be built in order to describe the dynamics of real bodies.

As it turned out, the equation of motion of a SP allows explaining the mechanism of irreversibility without the use of probabilistic hypothesis of fluctuations (Somsikov, 2014a; Somsikov, 2014b). This is because the energy of motion is no longer an invariant for a SP, as it was for a MP. It has been found that the energy of the SP's motion is transformed into its internal energy in the non-homogeneous field of external forces. Increase in the internal energy is proportional to the gradient of the external forces. If the external field is weak enough, then the SP remains in equilibrium along its path. At the same time its internal energy can increase only, but it cannot be transformed into the energy of motion. This is the essence of deterministic irreversibility that is such irreversibility which follows from the deterministic laws of classical mechanics; unlike the irreversibility explained within the canonical framework of classical mechanics (Prigogine, 1980; Zaslavsky, 1984), the deterministic irreversibility does not need the hypothesis of fluctuations. Thus, the deterministic mechanism of irreversibility is caused by transformation of energy of a system's motion into its internal energy when the system is in a non-homogeneous field. The fact that in this case the energy of motion is not constant means that the time symmetry is broken. However, the sum of the energy of motion and the internal energy is constant. Hence it is clear that in order to explain the nature of deterministic irreversibility for a non-equilibrium system, the following should be done:

1. Represent a non-equilibrium system by a set of moving relative to each other equilibrium subsystems.
2. Represent the energy of these subsystems as the internal energy and the energy of motion.
3. Obtain the equation of motion of subsystems directly from the dual form of energy, thus preserving the nonlinear terms which are responsible for energy exchange different degrees of freedom.

Unlike the deterministic mechanism of irreversibility, the traditional statistical mechanism refers irreversibly in Hamiltonian systems to the hypothesis of the existence of arbitrarily small fluctuations. The point is that since the Hamiltonian systems are exponentially unstable, then the presence of such fluctuations results in irreversible dynamics. The presence of fluctuations in these systems or in the external limitations is a sufficient condition for the irreversibility.

The deterministic irreversibility is a strong argument in favor of determinism of nature. This is very important; just recall the fundamental debates of Bohr and Einstein's on determinism and randomness, which took place during creation of quantum mechanics (Ginzburg, 2007; Bohr, 1958;).

Due to the dualism of energy, the equation of a SP's motion is given by independent micro and macro variables. Moreover, micro variables define the motion of the MPs relative to the center of mass of the SP, while macro variables define the motion of the SP's center of mass itself. Deriving equation of motion of the SP in this way takes into account possible transformation of the energy of the SP's motion into its internal energy and requires no use of the hypothesis of holonomic constraints, which is the basis for deriving the canonical equation of Lagrange. Unlike the canonical equation of Lagrange, the equation of the SP's dynamics describes the nonlinear transformation of the SP's motion energy into its internal energy; this transformation breaks the symmetry of time (Somsikov, 2014a).

An oscillator passing through a potential barrier was studied, and it was found that the condition of holonomic constraints eliminates nonlinear terms in the equation, which are responsible for breaking of time symmetry (Somsikov and Denisenya, 2013). This issue is considered in more detail subsequently.

Existence of deterministic irreversibility leads to the concept of "deterministic entropy" (D-entropy) (Somsikov, 2014a; Somsikov, 2014b). D-entropy is a deterministic one, because it strictly follows from the laws of classical mechanics without use of the hypothesis of fluctuations in a system. It is defined as the relative change in the system's internal energy. Unlike the thermodynamic entropy, D-entropy for a system consisting of small number of MPs can be both positive and negative.

The equation of the SP's motion is nonlinear. It is almost impossible to do any analytical analysis in order to check the theoretical conclusions following from the equation. Hence numerical simulations are the only way to check the theoretical conclusions. Numerical simulations allow determining the criteria for switching from classical mechanics to thermodynamics, statistical physics, kinetics, as well as identifying the cases when the system is irreversible, depending on the properties of the system, etc.

So, the objective of the paper is to determine the basic dynamic properties of a system of potentially interacting MPs in a non-homogeneous external field using numerical simulations. The estimation of fluctuations of the system's internal energy depending on the number of MPs, as well as initial parameters of the system and the barrier has been done, dependence of change in the system's internal energy on the barrier's width, as well as D-entropy have been studied. This study made it possible to verify the theoretical conclusions about the dynamics of a SP, define the criteria needed to switch from the deterministic to the thermodynamic model for a system.

In addition, it made it possible to show how the important statistical laws of physics may result from the strict laws of classical mechanics, as well as define some cases when the mechanism of irreversibility is applicable.

AN OSCILLATOR PASSING A POTENTIAL BARRIER

Initially, the simplest system, more specifically a one-dimensional oscillator of two MPs connected by a spring, has been considered (Somsikov and Denisenya, 2013). The oscillator's total energy includes the energy of motion and the internal energy. These two types of energy are given in the independent micro and macro variables. Micro variables describe the oscillations, while macro variables determine the motion of the oscillator's mass center.

It was found that the presence of a non-homogeneous external field made the micro and macro variables dependent. As a result, mutual nonlinear transformation of the energy of motion into the internal energy was taken into account. This nonlinear transformation was lost when deriving the Lagrange equation because of use of the hypothesis of holonomic constraints (Somsikov, 2014b). This means that it is impossible to obtain the effects resulting from a nonlinear relation between the degrees of freedom within the formalism of Lagrange.

Let us explain on an example of two-body problem, how the uses of the hypothesis of holonomic constraints excludes the possibility of describing the energy exchange between different degrees of freedom. There are two ways of obtaining the motion equations of a system of two MPs. The first way is a traditional one. The motion equation for two MPs in the external field is given by Sinai (1995):

$$\dot{v}_1 = -F_{12} - F_1^0;\ \dot{v}_2 = -F_{21} - F_2^0 \qquad (1)$$

Here F_1^0, F_2^0 are the external forces, acted on a first and second MP; F_{12} is the force of interaction of MPs

Let us add and subtract these equations. As a result, we obtain:

$$\dot{v}_1 + \dot{v}_2 = -(F_1^0 + F_2^0);\ \dot{v}_1 - \dot{v}_2 = -2F_{12} - F_{12}^0 \qquad (2)$$

Where $F_{12}^0 = F_1^0 - F_2^0$

Equation (2) can be rewritten as:

$$2\dot{V} = -(F_1^0 + F_2^0);\ \dot{v}_{12} = -2F_{12} - F_{12}^0 \qquad (3)$$

Where $V = \frac{1}{2}\left(\sum_{i=1}^{2} v_i\right);\ v_{12} = v_1 - v_2$

Thus, according to the Equation (3), the motion of the system in the external field is an independent motion of the system's center of mass and the relative motion of MPs. The motion of the center of mass is defined by the sum of the external forces applied to it. Relative motion of MPs is determined by their interaction forces and the difference between the external forces acting on each MP. According to these equations, there are two invariant of motion. The first invariant is the energy of motion of the system as a whole, and the second invariant is the energy of the relative motion of MPs.

Now consider the derivation of the motion equation for a system basing on the energy equation. By differentiating the system's energy with respect to time, we get:

$$v_1(\dot{v}_1 + F_{12} + F_1^0) + v_2(\dot{v}_2 + F_{21} + F_2^0) = 0 \qquad (4)$$

In general case the variables in Equation (4) cannot be separated. The sum is equal to zero not only when each term is equal to zero (as it is postulated by the hypothesis of holonomic constraints), but each term itself could be different from zero while their sum is zero. By regrouping the terms of equation (4), we obtain:

$$(v_1\dot{v}_1 + v_2\dot{v}_2) = -F_{12}v_{12} - v_1F_1^0 - v_2F_2^0 \qquad (5)$$

By comparing this equation to Equation (3), using as the variables the velocity of the mass center and the relative velocities of MPs, the result obtained is:

$$2V\dot{V} + \frac{1}{2}v_{12}\dot{v}_{12} \dot{=} -V(F_1^0 + F_2^0) - \frac{v_{12}}{2}(F_{12} + F_{12}^0) \qquad (6)$$

It is equivalent to the following equation:

$$V\left[2\dot{V} + (F_1^0 + F_2^0)\right] + \frac{v_{12}}{2}\left[\dot{v}_{12} + 2F_{12} + F_{12}^0\right] \qquad (7)$$

The terms in Equation (6) are grouped so that the first term determines the motion of the mass center, while the second term determines the change in the internal energy.

Equation (6) is equal to Equation (3) in the next cases: When $F_{12}^0 = 0$, when $v_{12} = 0$, and when external forces are linear. The first case is equivalent to the rigid connection between the MPs. The second case is equivalent to the homogeneity of the external field. The third case is associated with a linear dependence of external forces on the coordinates. Only in these cases the variables are separated. In general case of a non-homogeneous external field the variables of the Equation (6) cannot be separated and the hypothesis of holonomic constraints is not valid, that is, Equations (3) and (6) are not equivalent. Numerical simulations of an oscillator in a non-homogenous field have confirmed this conclusion (Somsikov, 2014b).

As it turned out, in some cases the oscillator can pass the potential barrier even if its energy of motion is less than the height of the barrier (Figure 1).

In some cases the oscillator can also reflect, even if its

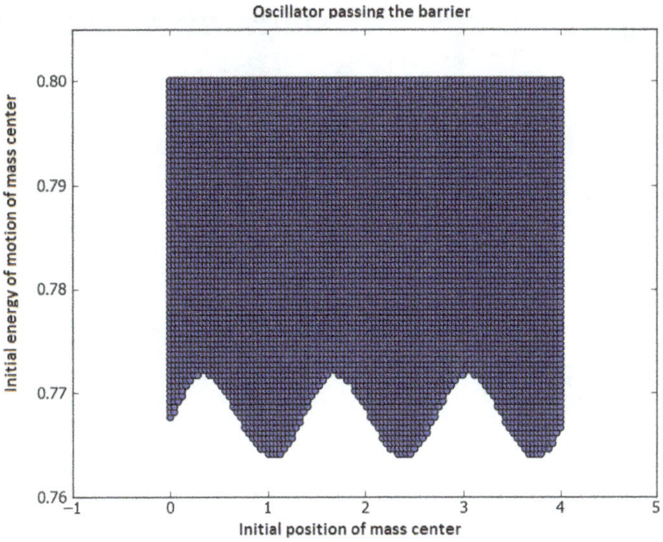

Figure 1. The filled area corresponds to the cases when the oscillator passes the barrier (depending on its initial energy of motion and phase).

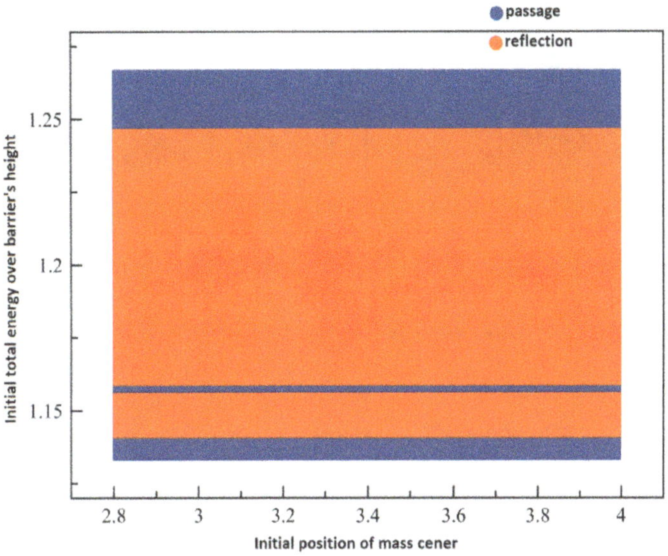

Figure 2. Oscillator's reflection / passage areas depending on its initial energy and position (red fill corresponds to reflection).

energy of motion is greater than the height of the barrier. Moreover, while gradually change the initial phase, it is possible to get interchangeable areas where the oscillator passes the barrier and where it is reflected (Somsikov and Denisenya, 2013) (Figure 2). These effects disappear if one neglects the nonholonomic constraints, that is, excludes consideration of the nonlinear mutual transformation of the oscillator's energy of motion into its internal energy.

At the moment when the oscillator is near the potential barrier, it is its phase that determines the sign of change in the internal energy (Figure 3). The result also depends on the height of the barrier, the oscillator's energy of motion, and other parameters. Thus, the calculation of oscillator's motion shows the important role of nonlinear effects in the dynamics of a system in a non-homogeneous field. These nonlinear effects can be studied only by using the principle of duality of symmetries, taking into account the transformation of the system's energy of motion into the energy of motion of the MPs relative to the system's center of mass.

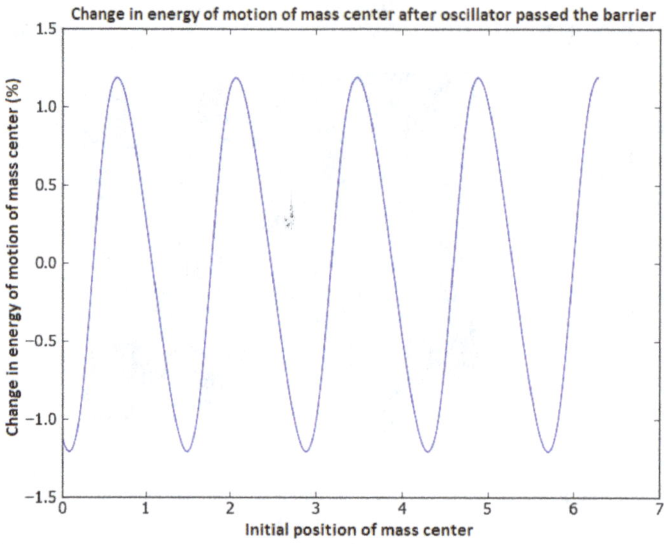

Figure 3. Change in the oscillator's energy of motion (% of its initial value) depending on its initial phase.

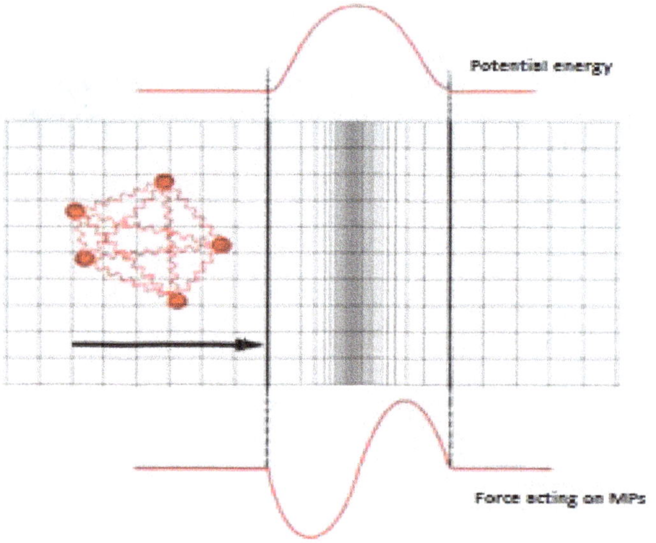

Figure 4. The scheme of numerical simulations of the system passing the potential barrier.

FORMULATION OF THE PROBLEM OF A SYSTEM PASSING A POTENTIAL BARRIER

A system of potentially interacting MPs in a nonhomogenous field has been considered (Figure 4). The initial parameters include the system's internal energy and the energy of motion of its center of mass. Coordinates and velocities of the MPs were set randomly. Their sums are equal to zero in the center of mass. The system is represented by a ball with a certain radius at which the potential energy of interaction of the MPs is equal to the total kinetic energy of the MPs.

The dual system of coordinates is used, that is the independent variables are the micro and macro variables. Microvariables determine the motion of each MP in the center of mass, while macro variables determine the motion of the center of mass itself. The barrier's height is chosen so that the system passes through it. Change in the system's internal energy, the system's motion energy, as well as D-entropy and other parameters of the problem are computed; the independent parameters include the number of MPs, the barrier's height and width, as well as the initial conditions. The results obtained are compared with the statistical laws and with the theoretical conclusions obtained on the basis of the equations of motion of a SP (Somsikov, 2014a; Somsikov, 2014b):

$$M_N \dot{\vec{V}}_N = -\vec{F}^{env} - \frac{(\Phi^{env} + \dot{E}_N^{ins})\vec{V}_N}{V_N^2} \quad (8)$$

Where $\vec{V}_N = \dot{\vec{R}}_N$ is velocity of the center of mass, $i = 1,2,3,...,N$ - MP's count, $\vec{R}_N = \frac{1}{N}\sum_{i=1}^{N}\vec{r}_i$; $M_N = Nm$; $\vec{F}^{env} = \sum F_i^{env}(R_N, \tilde{r}_i)$, $\dot{E}_N^{ins} = \sum_{i=1}^{N} \tilde{v}_i(m\dot{\tilde{v}}_i + F(\tilde{r})_i)$ - change in the system's internal energy, $\Phi^{env} = \sum \tilde{v}_i F_i^{env}(R_N, \tilde{r}_i)$, $\vec{F}_i^{env}(R_N, \tilde{r}_i)$ - external force acting on i^{th} MP, $r_i = R_N + \tilde{r}_i$, $\tilde{r}_i$ - MP's coordinates relative to the center of mass.

MPs interaction forces are given by Hooke's law. The external field is specified in the form of one period of a cosine $U(x_i) = U_b[\cos(2\pi(x_i - R_b)/a) + 1]$, provided $\left(R_b - \frac{a}{2}\right) < x_i < \left(R_b + \frac{a}{2}\right)$. Hence the forces acting on each MP are given by:

$$F_i(x_i) = U_b \sin(2\pi(x_i - R_b)/a) \quad (9)$$

Where U_b is the barrier's height; R_b is the position of the barrier's max height; a is the barrier's width; x_i is the distance between i^{th} MP and the center of mass; i - MP's count. According to Equation (9), the force is proportional to the barrier's height, and inversely proportional to its width.

The numerical simulations are done for various initial distributions of MPs and parameters of the problem in order to determine the nature of the changes in the energy of motion and the internal energy of the system, depending on the number of MPs and initial parameters.

CHANGE IN THE SYSTEM'S INTERNAL ENERGY AS A FUNCTION OF THE INITIAL PARAMETERS AND THE NUMBER OF MPS

According to the theoretical results, the dynamics of a conservative nonequilibrium system, represented by a set of equilibrium subsystems in a nonhomogeneous external field, should be irreversible due to the transformation of the system's kinetic energy into its internal energy (Somsikov, 2014b). A similar conclusion follows from the statistical methods of analysis of nonequilibrium systems (Landau, 1976; Rumer and Rivkin, 1977). Let us consider how much MPs a system should consist of in order to be described in terms of empirical equations of thermodynamics and statistical laws.

If the theoretical results derived from the equations of motion of systems are valid, then the numerical simulations should reveal that there is a certain number of MPs a system should consist of, such that the system's internal energy can increase only. This number (*N1*) can be taken as a first criterion for the system to be equilibrium. Obviously, this number should depend on the relative values of the internal energy, the energy of motion of the system, the potential barrier's height. Simulations have been done in order to verify existence of *N1* and study its behavior depending on the parameters of the system; the simulations estimated the change in the system's internal energy depending on the number of MPs.

Figure 5 shows the results of 400 experiments for different number of MPs. Number of MPs correspond to a power of two (4, 8, 16, 32, 64, 128, 256, 512, 1024). Initial macro parameters are constant: the mass of the system equals 1 kg, the mass of each MP equals to $1/N$, the kinetic energy of the system's center of mass E_s equals 150 J, the system's velocity is directed along the coordinate axis X.

The potential barrier is located in the YZ plane and has a width along the X axis equal to 0.2 m, the barrier's height equals 130 J, the system's internal energy equals 100 J, links rigidity coefficient U_o equals 300000 N/m. Initial micro parameters, such that coordinates and velocities of the MPs, are set randomly. Each dot in the Figure 5 corresponds to the ratio of change in the system's internal energy to its initial kinetic energy ($\Delta E^{ins}/E_0^{ins}$).

The figure shows that if the number of MPs is greater than 64, then the change in the internal energy can be positive only. This means that for the given parameters of the problem and $N \geq 64$, the system's dynamics is irreversible. This conclusion is made based on the fact that the impossibility of transformation of the system's internal energy into its kinetic energy can be considered as the test for irreversibility. Let us name this number as the first critical number (*N1*). It is obvious that *N1* depends on the parameters of the problem, for example, on the barrier's width.

AREA OF APPLICABILITY OF D-ENTROPY

In accordance with the law of conservation of momentum, the internal energy of a system cannot be transformed into its kinetic energy, since it is not possible to change the system's momentum. This means that the system is irreversible.

The concept of D-entropy was introduced into the mechanics because of this irreversible energy transformation fora system in a nonhomogeneous external field. D-entropy equals the ratio of the increment of the system's internal energy to its initial value, as well as the entropy of Clausius. Consider a non-equilibrium system that can be represented by a set of SPs in the

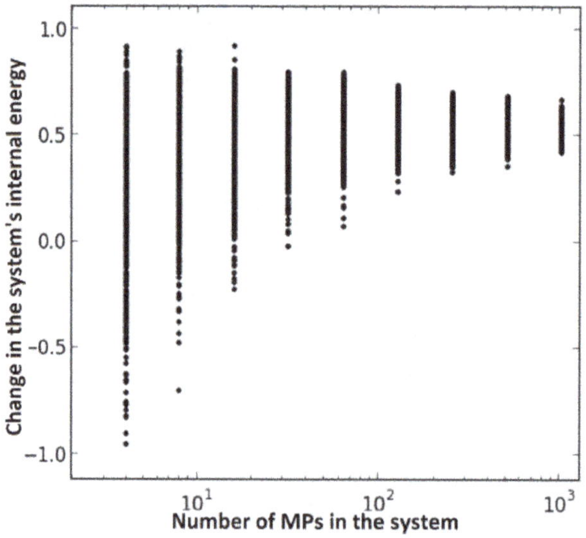

Figure 5. Fluctuations of the system's internal energy subject to the number of MPs.

approximation of local thermodynamic equilibrium; the increment of the D-entropy of this system is proportional to the energy of the relative motion of SPs, which is transforming into their internal energy. In this case, the change in the D-entropy is given by Somsikov (2014a; b):

$$\Delta S^d = \sum_{L=1}^{R}\{N_L \sum_{k=1}^{N_L}[\int \sum_s F_{ks}^L v_k dt]/E_L\} \qquad (10)$$

where E_L is the internal energy of Lth SP; N_L is the number of MPs inside the Lth SP; L=1,2,3…,R- number of SPs; S - external MPs interacting with k^{th} MP in the Lth SP; F_{ks}^L- the force acting on k^{th} MP of one SP by S^{th} MP of another SP, and v_k- velocity of kth MP.

Equation (10) is obtained based on the equation of motion of a SP, and determines the increment of the SP's internal energy comparing to its initial value. It is valid if the SP is in equilibrium along its path.

In our case L=1, so the Equation (10) is determined by a simple formula: $\Delta S^d = \Delta E^{int}/E_0^{tr}$. This formula can be verified by numerical calculations of ΔE^{int} for a system passing the potential barrier.

Figure 6 shows average values of change in the system's internal energy ΔE^{int} over its initial kinetic energy (100 J), as well as confidence intervals for these values. Each confidence interval corresponds to the confidence level of 0.99 (400 experiments) and is calculated as standard deviation of the value multiplied by Student coefficient of 2.6. The values on Figure 6 are in fact the changes in D-entropy ΔS^d up to a constant factor. The calculations show that the value will be positive at $N \geq 8$ with the probability of 0.99; for smaller number of the MPs the value can be negative. As the number of MPs goes up, the fluctuation tends to zero, and even when $N \geq 10^3$, it becomes approximately equal to 0.1 of the absolute value of ΔS^d.

A further increase in the number of MPs does not change the increment of the internal energy, that is ΔE^{int} reaches its limit at $N \approx 10^3$. If $N \geq 10^3$, then : $\Delta S^d = \Delta E^{int}/E_{11}^{int} \approx 0.55$. Since a further increase in the number of MPs does not affect the change in the thermodynamic parameters of the system, then $N = 10^3$ can be named as the second critical number (*N2*). This number determines the transition to the thermodynamic description for the problem.

CHANGE OF THE ENERGY OF THE SYSTEM PASSING THE BARRIER

Let us compare fluctuations of the system's internal energy depending on the number of MPs and their distribution function, with the law of statistical fluctuations of its mean square value. This comparison is a convincing proof of the possibility of justification of statistical laws based on the laws of mechanics. Let us recall the way it is usually proved that the relative fluctuation of any additive parameter of a system is inversely proportional to $\sqrt{N}$, where N is the number of elements in the system, on the basis of statistical laws (Landau, 1976).

The internal energy of the system (E^{int}) is an additive value. If the system is divided into N subsystems, then the average value of its internal energy is equal to the sum of the average values of the internal energies of all subsystems, that is, $\langle E^{int}\rangle = \sum_{i=1}^{N}\langle E_i\rangle$. Let us start from

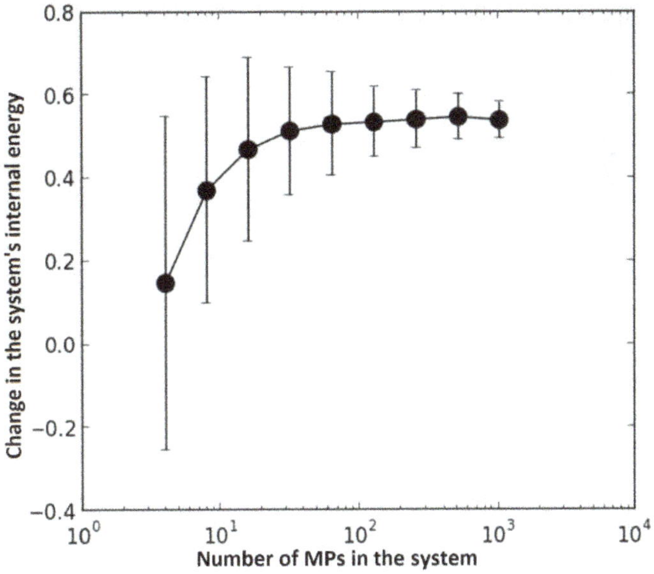

Figure 6. Change in the system's internal energy subject to the number of MPs.

the fact that the internal energy increases in proportion to the number of MPs. Then the mean square value of the fluctuations of the internal energy equals $|(\Delta E)^2| = |(\sum_{i=1}^{N} \Delta E_i)^2|$. If the fluctuations in the subsystems are independent, then $|(\Delta E)^2| = \sum_{i=1}^{N} |(\Delta E_i)^2|$. Hence the well-known law is obtained: $|(\Delta E)^2|^{1/2} \propto 1/N^{1/2}$.

Thus, if the calculated value of the relative fluctuations of E^{int} varies inversely to $\sqrt{N}$, then this fact will serve as a proof of both the law of the fluctuations, and the possibility of the justification of this law by the laws of classical mechanics.

Figure 7 shows that dots, corresponding to the fluctuations of the internal energy, fit the curve corresponding to the statistical law of decrease of fluctuations in the system with the increase in the number of its elements (Landau, 1976).

This means, firstly, that the numerical simulations of the system passing through the barrier are correct, secondly, that the dualism of energy is reflected in the statistical laws, and thirdly, that the laws of classical mechanics are suitable not only for justification of the statistical laws, but also for determining the scope of their application depending on the parameters of the system.

The slight difference between the calculated fluctuations of ΔE^{int} and the approximating line can be explained by the fact that the increase in the number of MPs results in a change of other parameters of the system that affect the value of ΔE^{int} (for example, size of the system). Another reason for a certain deviation from the statistical law may be the fact that for a given number of MPs the system cannot be strictly considered as an equilibrium one. In general, the study of these deviations may be useful to identify the areas of applicability of statistical laws in the specific problems of dynamics.

CHANGE IN THE INTERNAL ENERGY SUBJECT TO THE WIDTH OF THE BARRIER

According to the equation of motion of the system, the change in the system's internal energy ΔE^{int} nonlinearly depends on the micro and macro variables and is different from zero only when the scale of nonhomogeneity of the external field is about the scale of the system. The value of ΔE^{int} should increase as the difference between the forces acting on different areas of the system goes up (Landau, 1976; Rumer and Rivkin, 1977). This conclusion is checked by calculating the dependence of ΔE^{int} on the barrier's width. Figure 8 shows the results of these calculations.

The ordinate axis is the ratio of change in the internal energy to the initial energy of motion of the system's center of mass. The solid vertical line represents the standard deviation of coordinates of MPs (a measure of the system's size). The dotted line represents the maximum size of the system (the maximum distance between the MPs during the numerical experiment).

According to Figure 8, there is a decrease in the efficiency of transformation of the system's kinetic energy into its internal energy, with the increase in the barrier's width, that is, as gradient of the external field goes down,

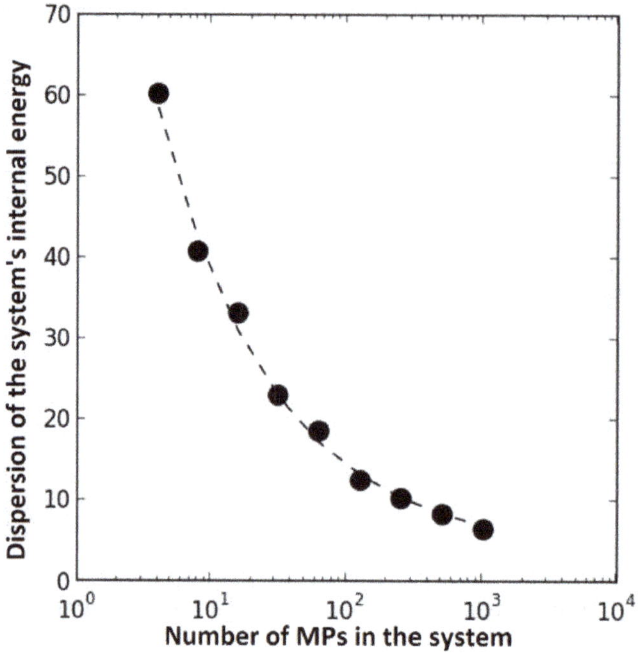

Figure 7. Dots: max amplitude of fluctuation of ΔE^{int} subject to number of MPs. Approximating line is given by $p + r/\sqrt{N}$, where p=3.5, r=110.

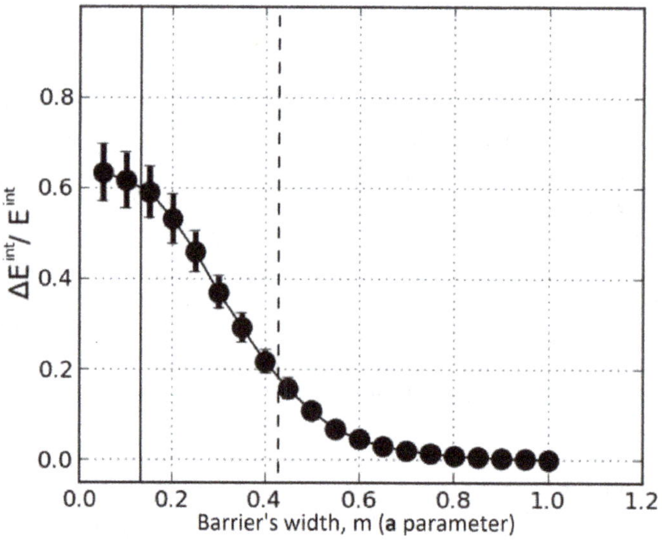

Figure 8. Change in the internal energy subject to the barrier's width, *a*.

then the change in the internal energy tends to zero. The law of diminishing is close to power-law dependence. The dependence of the internal energy on the gradient of the external forces follows from the Equation (8), if we expand the external force in the small parameter (Somsikov, 2014b).

All of the results of numerical simulations are obtained here only because the calculations were based on the concept of deterministic mechanism of irreversibility. Previously proposed D-entropy is calculated in accordance with this mechanism based on the rigorous equations of motion of systems (Somsikov, 2014b). The

D-entropy can be determined only if one uses the principle of symmetry dualism and consequent principle of energy dualism. The sum of the system's kinetic energy and its internal energy was constant during the calculations; this verified compliance with the law of conservation of energy.

CONCLUSION

Numerical simulations of dynamic parameters of motion of an equilibrium system of potentially interacting MPs in a nonhomogenous external field revealed the following patterns. A critical number of MPs is determined, such that the system's internal energy cannot go down for any given initial state of the system. In our case, this number, *N1* ~10^2. That is, when the number of SPs is greater than *N1*, then the system's dynamics becomes irreversible. This number specifies the minimum number of MPs in the system which is needed in order to apply a concept of D-entropy S^d.

Let us note that the value of $N > 10^2$ is obtained for a specific model. In general, the value of *N1* should differ for a system with different parameters. But the main thing is not the exact value of *N1*, but the fact that it can be determined using the laws of Newton. This fact is a strong argument for an idea that the laws of thermodynamics can be obtained within the frameworks of classical mechanics without use of statistical hypotheses. Moreover, it supports the idea that the statistical laws themselves can be obtained using the laws of classical mechanics

The second critical value (*N2*) has been found. In our case *N2*=10^3. The increase in the system's internal energy ΔE^{int} stabilizes if the number of its elements is greater than 10^3. In our case asymptotic value of $\Delta E^{int} \sim 0.55$. The value of *N2* determines the transition to the thermodynamic description for the system.

It is shown that the relative fluctuations of E^{int} goes down when the number of MPs goes up. The rate of this decrease is inversely proportional to $\sqrt{N}$. This relation for a systemis obtained on the basis of the equations of dynamics and this fact is an argument in favor of the idea that the statistical laws should be justifiable under the laws of classical mechanics. Since the law of fluctuations is the basis of statistical physics (Landau, 1976), this fact indicates the possibility of justification of statistical laws based on the laws of physics.

The efficiency of transformation of the system's kinetic energy into its internal energy goes down while the gradient of the external forces decrease. This dependence shows that the change in the internal energy is due to non-potential forces which themselves are proportional to the gradients of the external forces.

The numerical simulations carried out on the basis of the equation of SP's motion, allowed defining numerical criteria for the transition from a dynamic description to the thermodynamic model depending on the number of MPs. This opens the way for the justification of the laws of thermodynamics under the laws of classical mechanics.

Overall, the results confirm the need to describe the dynamics of systems in accordance with the principle of dualism of symmetry and use of the equation of SP's motion. Given that the real bodies are structured, it is dualism that allows identifying and study the connection between the laws of classical mechanics and the empirical laws of thermodynamics and statistical physics.

Conflict of Interest

The authors have not declared any conflict of interest.

REFERENCES

Bird GA (1976). Molecular Gas Dynamics. Clarendon press, Oxford.

Bohr N (1958). Debate with Einstein about the problem of knowledge in atomic physics. V. LXVI(4):571-598.

Cohen EG (1998). Boltzmann and statistical mechanics, dynamics: Models and kinetic methods for non-equilibrium many body systems. NATO Sci. Series E: Appl. Sci. pp. 371- 223.

Ginzburg VL (2007). Special session of Editorial board of the Journal of Physics-Uspekhi, honoring the 90th anniversary of VL Ginzburg. Adv. Phys. Sci. 177(4):345-346.

Goldstein G (1975). Classical mechanics. Moscow P. 416.

Kadomtsev BB (1995). Classical and quantum irreversibility. Adv. Phys. Sci. 165(8):895-973.

Klein MJ (1961). Max Planck and the beginning of quantum theory. Adv. Phys. Sci. 92(4):679-700.

Klimontovich YL (1995). The statistical theory of open systems. Moscow. Janus. P. 292.

Landau LD, Lifshitz EM (1976). Statistical physics. Moscow. P. 583.

Lebowitz JL (1999). Boltzmann's entropy and time's arrow. Phys. Today. pp. 32-38.

Prigogine I (1980). From being to becoming. Moscow. Nauka. 1980. P. 342.

Rumer Yu B, Rivkin M Sh. (1977). Thermodynamics, Stat. physics and Kinematics. Moscow. Nauka. P. 532.

Sidharth BG (2008). Modelling Time. arXiv:0809.0587v1 [physics.gen-ph] 3 Sep 2008.

Sinai YG (1995). Modern problems of ergodic theory. M,: FIZMATLIT, P. 208.

Smoluchowski M (1967). Boundaries of validity of the second law of thermodynamics. Uspehi Fizicheskih Nauk. 93(4):724-737.

Somsikov VM (2004). The equilibration of an hard-disks system. IJBC. 2004. November. V 14(N11):4027-4033.

Somsikov VM (2005). Thermodynamics and classical mechanics. J. physics: Conference series. 23:7-16.

Somsikov VM (2014a). From the Newton's mechanics to the physics of the evolution. Almaty 2014. P. 272.

Somsikov VM (2014b). Why It Is Necessary to Construct the Mechanics of Structured Particles and How to do it. Open Access Library Journal, 2014, 1PP. 1-8, DOI: 10.4236/oalib.1100586.

Somsikov VM, Denisenya MI (2013). Features of the oscillator passing through potential barrier. Proceedings of the universities. Series Physics. 3:95–103.

Somsikov VM, Merzlyakov MA, Malkov EA (1999). Numerical analysis of the matrix describing the system hard drive. Problems evolution of open systems, Issue 1, Alma-Ata, pp. 52-62.

Zaslavsky GM (1984). Stochasticity of dynamic systems. Moscow. Nauka, P. 273.

Analytical and numerical solution of heat generation and conduction equation in relaxation mode: Laplace transforms approach

Lawal M.[1,3], Basant K. Jha[2], Patrick O. Akusu[3], Ahmad R. Mat Isa[1] and Nasiru R.[3]

[1]Department of Physics Faculty of Science, Universiti Teknologi Malaysia, Johor, Malaysia.
[2]Department of Mathematics, Ahmadu Bello University, Zaria, Nigeria.
[3]Department of Physics, Ahmadu Bello University, Zaria, Nigeria.

In this article analytical solution of one-dimensional heat equation in relaxation mode of heat generation and conduction using Laplace transforms method is presented. The model adopted takes into account finite velocity of heat propagation, and relaxation of heat source capacity. The properties of heat source terms in four different cases are incorporated in the model and investigated. Temperature distributions and variations with conduction mode and relaxation time are analyzed. High relaxation time is observed to lowers the temperature profile, whereas enhanced temperature distribution changes at particular values of α, and τ_g for source capacity proportional to temperature. How the steady state solution is achieved for some selected values of coefficients is also discussed.

Key words: Relaxation time, conduction mode, pulsed heat source, Laplace transforms.

INTRODUCTION

Cattaneo was the first to build an explicit mathematical theory to correct unacceptable properties of Fourier theory of heat diffusion. The arguments used were based on the kinetic theory of gases and second-order correction to propose modification of Fourier law (Cattaneo, 1948), which gives rise to the well-known hyperbolic model of heat conduction. This also leads to suitable heat conduction models that permit the finite speed of heat flow (Ozisik and Tzou, 1994; Joseph and Preziosi, 1990). In most studies of heat propagation in systems the hyperbolic model of heat conduction is used (Jose and Juan, 2011; Al-Nimr et al., 2004; Malinowski, 1993a, Saleh and Al-Nimr, 2008; Cai et al., 2006). For instance in (Malinowski, 1993b) the analytical solutions for the relaxation equation in bodies with low heat resistance, by neglecting temperature gradient were presented. It is shown that differences between parabolic and relaxation solution fluctuate as time elapses. Differences in heat generation and conduction were reported (Lewandowska, 2001) to arise due to the time characteristics of the heat source capacity. For example when the heat is of constant strength this differences slowly decrease for long times. Furthermore, solutions of both hyperbolic and parabolic heat conduction equation for temperature dependent heat source is reported to be use in analyzing normal zones in superconductors

(Lewandowska and Malinowski, 2002), in which the amount of energy that is dissipated in the zone affect heat production by the heat source capacity which depends on temperature.

Although a lot of works has been done on the hyperbolic and parabolic heat conduction equation under different conditions, yet nobody as far as we know investigate the solutions for one-dimensional relaxation model of heat conduction taking into account the finite velocity of heat propagation, and relaxation of heat source capacity. Matlab program is one of the robust and most widely used program in areas of science and technology (Hübner et al., 2011). Also has many application in script design (Valipour et al., 2012), and in model development (Mohammad and Ali, 2012a; Mohammad et al., 2013). For example in irrigation engineering (Mohammad and Ali, 2012b) used the genetic coding in Matlab environment to determine the effective infiltration parameters in Furrow Irrigation. We use in this paper Matlab environment to write a script to compute the temperature profile in physical domain.

MODEL

By using the modified Fourier law Equation (1), which physically agree for a very short laser pulses and non-infinite speed of heat transport.

$$t_k \frac{\partial q}{\partial t} = -k\nabla\Theta - q, \quad (1)$$

Where t_k, k, q and Θ are the relaxation time of the heat flux, thermal conductivity, heat flux vector, and temperature respectively. Hyperbolic equation of heat conduction is obtained by substitution of Equation (1) into the energy conservation equation.

$$\rho c_p \frac{\partial \Theta}{\partial t} = -\nabla . q + g, \quad (2)$$

In Equation (2) ρ is the density; c_p is specific heat at constant pressure, and g is the capacity of the internal heat source. In this paper we adopt the notion of inert heat source and transient capacity of heat source g_t as seen in Equation (3).

$$t_g \frac{\partial g_t}{\partial t} + g_t = g \quad (3)$$

For the relaxation heat conduction equation that account for both finite speed of heat propagation and the relaxation of heat source capacity Equation (4) is use.

$$t_k t_g \frac{\partial^3 \Theta}{\partial t^3} + (t_k + t_g)\frac{\partial^2 \Theta}{\partial t^2} + \frac{\partial \Theta}{\partial t} = t_k a \frac{\partial \nabla^2 \Theta}{\partial t} + \mu\nabla^2\Theta + \frac{1}{\rho c_p}\left(t_k \frac{\partial g}{\partial t} + g\right) \quad (4)$$

Where μ is thermal diffusivity, t_g is relaxation time of source capacity, the length of which depends on nature of the source. The dimensionless forms of Equations (1) to (3), are given below respectively as adopted in Lewandowska (2001), which is necessary to ensure temperature variation as a function of dimensionless displacement.

$$\frac{\partial \Phi}{\partial \tau} = -\nabla . \Phi + 2\Psi_t, \quad (5)$$

$$\frac{\partial \theta}{\partial \tau} + 2\theta = -\nabla\Phi, \quad (6)$$

$$\tau_g \frac{\partial \Psi_t}{\partial \tau} + \Psi_t = \Psi. \quad (7)$$

Transformation of Equations (5) to (7) yields the equation of heat conduction below, which permits a finite speed of heat propagation and relaxation of heat source capacity.

$$\tau_g \frac{\partial^3 \Phi}{\partial \tau^3} + (2\tau_g + 1)\frac{\partial^2 \Phi}{\partial \tau^2} + 2\frac{\partial \Phi}{\partial \tau} = \tau_g \frac{\partial}{\partial \tau}\nabla^2\Phi + \nabla^2 + 2\frac{\partial \Psi}{\partial \tau} + 4\Psi \quad (8)$$

Where $\tau_g = \frac{t_g}{2t_k}$, and t_k is relaxation time due to delay of heat flux as a result of temperature gradient. Equation (8) is treated, considering the temperature gradient as a function of a dimensionless Cartesian co-ordinate, for $\frac{\partial^2 \Phi}{\partial Y^2} = \frac{\partial^2 \Phi}{\partial Z^2} = 0$. Thus, we obtained Equation (9).

$$\tau_g \frac{\partial^3 \Phi}{\partial \tau^3} + (2\tau_g + 1)\frac{\partial^2 \Phi}{\partial \tau^2} + 2\frac{\partial \Phi}{\partial \tau} = \tau_g \frac{\partial}{\partial \tau}\left(\frac{\partial^2 \Phi}{\partial X^2}\right) + \frac{\partial^2 \Phi}{\partial X^2} + 2\frac{\partial \Psi}{\partial \tau} + 4\Psi. \quad (9)$$

Equation (9) can be reduced to the classical hyperbolic equation of heat conduction for one dimensional case, when the relaxation time of heat capacity is set to zero.

$$\frac{\partial^2 \Phi}{\partial \tau^2} + 2\frac{\partial \Phi}{\partial \tau} = \frac{\partial^2 \Phi}{\partial X^2} + 2\frac{\partial \Psi}{\partial \tau} + 4\Psi. \quad (10)$$

We take into account the finite speed of heat propagation, relaxation of heat source capacity and heat conduction equations. We also consider temperature gradient to be a function of dimensionless displacement, and assumed high heat resistance.

MATHEMATICAL ANALYSIS

The boundary value problem of Equation (9) was solved after including four different source terms namely: (i) source with constant capacity, (ii) source capacity proportional to temperature, (iii) Dirac delta energy pulse, and (iv) source capacity proportional to time. This gives respectively,

$$\tau_g \frac{\partial^3 \Phi}{\partial \tau^3} + (2\tau_g + 1)\frac{\partial^2 \Phi}{\partial \tau^2} + 2\frac{\partial \Phi}{\partial \tau} = \tau_g \frac{\partial}{\partial \tau}\left(\frac{\partial^2 \Phi}{\partial X^2}\right) + \frac{\partial^2 \Phi}{\partial X^2} + 2\frac{\partial \Psi}{\partial \tau} + 4\Psi_c \quad (11)$$

$$\tau_g\frac{\partial^3\Phi}{\partial\tau^3}+(2\tau_g+1)\frac{\partial^2\Phi}{\partial\tau^2}+2\frac{\partial\Phi}{\partial\tau}=\tau_g\frac{\partial}{\partial\tau}\left(\frac{\partial^2\Phi}{\partial X^2}\right)+\frac{\partial^2\Phi}{\partial X^2}+2\alpha\frac{\partial\Phi}{\partial\tau}+4\alpha\Phi \quad (12)$$

$$\tau_g\frac{\partial^3\Phi}{\partial\tau^3}+(2\tau_g+1)\frac{\partial^2\Phi}{\partial\tau^2}+2\frac{\partial\Phi}{\partial\tau}=\tau_g\frac{\partial}{\partial\tau}\left(\frac{\partial^2\Phi}{\partial X^2}\right)+\frac{\partial^2\Phi}{\partial X^2}+2\beta\frac{\partial(\tau)}{\partial\tau}+4\beta\partial(\tau) \quad (13)$$

$$\tau_g\frac{\partial^3\Phi}{\partial\tau^3}+(2\tau_g+1)\frac{\partial^2\Phi}{\partial\tau^2}+2\frac{\partial\Phi}{\partial\tau}=\tau_g\frac{\partial}{\partial\tau}\left(\frac{\partial^2\Phi}{\partial X^2}\right)+\frac{\partial^2\Phi}{\partial X^2}+2\beta\frac{\partial(\sin\tau)}{\partial\tau}+4\beta\sin(\tau) \quad (14)$$

By using the Laplace transforms technique, with boundary conditions $\Phi(X,0)=\frac{\partial\Phi}{\partial\tau}(X,0)=\frac{\partial^2\Phi}{\partial\tau^2}(X,0)=0$, for the four different source terms respectively, solutions of Equations (11) to (14) yields:

$$\frac{d^2\bar{\Phi}}{dX^2}-\frac{A(S)}{B(S)}\bar{\Phi}+\frac{4\Psi_c}{(\tau_g S^2+S)}=0 \quad (15)$$

$$\frac{d^2\bar{\Phi}}{dX^2}-\frac{A(S)}{B(S)}\bar{\Phi}+\frac{4\alpha}{B(S)S^2}=0 \quad (16)$$

$$\frac{d^2\bar{\Phi}}{dX^2}-\frac{A(S)}{B(S)}\bar{\Phi}+\frac{2\beta}{B(S)}(S+2)=0 \quad (17)$$

$$\frac{d^2\bar{\Phi}}{dX^2}-\frac{A(S)}{B(S)}\bar{\Phi}+2\beta(1+2\tau)=0 \quad (18)$$

Where $A(S)=\tau_g S^3+2\tau_g S^2+S^2+2S$, and $B(S)=\tau_g S+1$ for Equations (11) to (14). However, in Equation (12), $A(S)=\tau_g S^3+2\tau_g S^2+S^2+2S-2\alpha S$. The α and β in Equations (16) to (18) are dimensionless coefficients that corresponding to the corresponding sources term. The solution of Equations (15) to (18) satisfying the conditions $\bar{\Phi}=0$ for $x=1$ and $x=0$ is:

$$\bar{\Phi}(X,S)=K_i\left[\frac{\cosh\sqrt{\frac{A(S)}{B(S)}}}{\sinh\sqrt{\frac{A(S)}{B(S)}}}-\frac{1}{\sinh\sqrt{\frac{A(S)}{B(S)}}}\right]\sinh X\sqrt{\frac{A(S)}{B(S)}}-K_i\cosh X\sqrt{\frac{A(S)}{B(S)}}+K_i \quad (19)$$

Where i = 1, 2, 3, and 4, $A(S)$ and $B(S)$ are as defined above with the same condition and

$$K_1=\frac{4\Psi_c}{A(S)S} \quad (20)$$

$$K_2=\frac{4\alpha}{A(S)S^2} \quad (21)$$

$$K_3=\frac{2\beta(S+2)}{A(S)} \quad (22)$$

$$K_4=\frac{2\beta}{A(S)S}\tau \quad (23)$$

The Equation (19) in Laplace transformed field is inverted for values of i = 1, 2, 3, and 4 in order to determine the temperature in physical time domain. Riemann-sum approximation (Basant and Clement, 2013) is used for the inversion of the sets of Equation (19). It involves a single summation for the numerical process. In this case the function in $\bar{\Phi}(X,S)$ is inverted to the time field.

$$\Phi(X,\tau)\cong\frac{e^{\epsilon\tau}}{\tau}\left[\frac{1}{2}\bar{\Phi}(X,\epsilon)+Re\sum_{n=i}^{N}\bar{\Phi}\left(X,\epsilon+\frac{in\pi}{\tau}\right)(-1)^n\right] \quad (24)$$

Where Re is the real part, i = $\sqrt{-1}$ is the imaginary number, N is the number of terms used in the Riemann-sum approximation. The accuracy of this method depends on the value of ϵ and the truncation error dictated by N. The ϵ is real part of Bromwich contour that is used in inverting Laplace transforms, its value must be selected so that the Bromwich contour encloses all the branch points (Tzou, 1997; Karniadakis and Beskok, 2002). For faster convergence, and reasonable results the quantity $\epsilon\tau$ should be approximately 4.7 (Vernotte, 1961). This shortens the computational time as compared to other tested values. The numerical solution is validated by considering steady state solution of Equation (25) for the first case, compared with the solution of Equation (15) for i=1 and the two results satisfy the boundary conditions x(0)= $\Phi(0)$, and x(1)= $\Phi(0)$. The pulsed energy source shows quasi-steady state behavior as the Dirac delta tends to unity. This indicates the flow is partly driven by buoyancy. This also agrees with source capacity that is proportional to time.

$$\frac{d^2\Phi}{dX^2}+4\Psi_c=0, \quad (25)$$

$$\frac{d^2\Phi}{dX^2}+4\alpha\Phi=0, \quad (26)$$

$$\frac{d^2\Phi}{dX^2}+4\beta\delta=0, \quad (27)$$

$$\frac{d^2\Phi}{dX^2}+4\beta\tau=0. \quad (28)$$

RESULT AND DISCUSSION

The results of the calculations are shown in (Figures 1 to 6). Figures 1 to 4 show the temperature profiles for the source terms that follow $\Psi=\Psi_c$, $\Psi=\alpha\Phi$, $\Psi=\beta\delta(\tau)$, and $\Psi=\beta\tau$. Using the solutions of Equations (11) to (14) for the four source terms, we write scripts that solve Equation (19) for i=1, 2, 3, and 4 by using MATLAB program in order to compute and generate the graphs. This is necessary in order to get a clear insight into the physics of the model. Different values of Ψ_c from 0.1 to 1 are used, while higher values in some cases enhance temperature profile distribution similar to the trend observed in the semi-infinite system with a time-dependent pulse energy source (Lewandowska, 2001). The resulting values of the temperature profiles $\Phi(X,\tau)$ are observed to increase for dimensionless temperature versus dimensionless time in conduction mode for the heat source capacity of constant strength. The values of τ_g are set to 1, 3, and 6

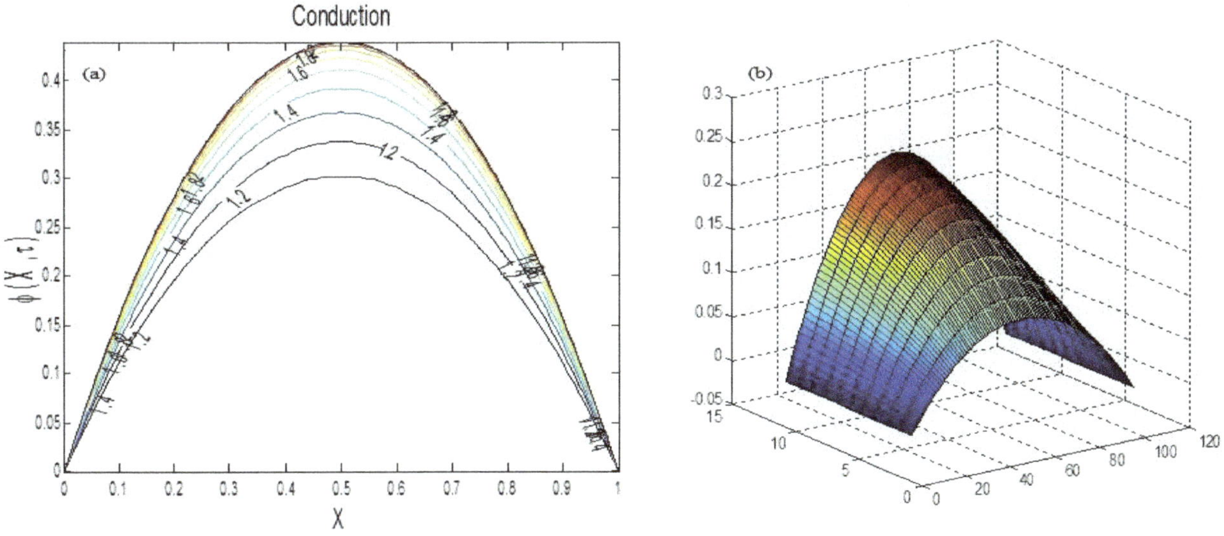

Figure 1. (a) Temperature distributions in conduction mode for constant heat source capacity, for $\tau_g = 1.$ (b) Temperature profile for $\tau_g = 3$ of the first source term.

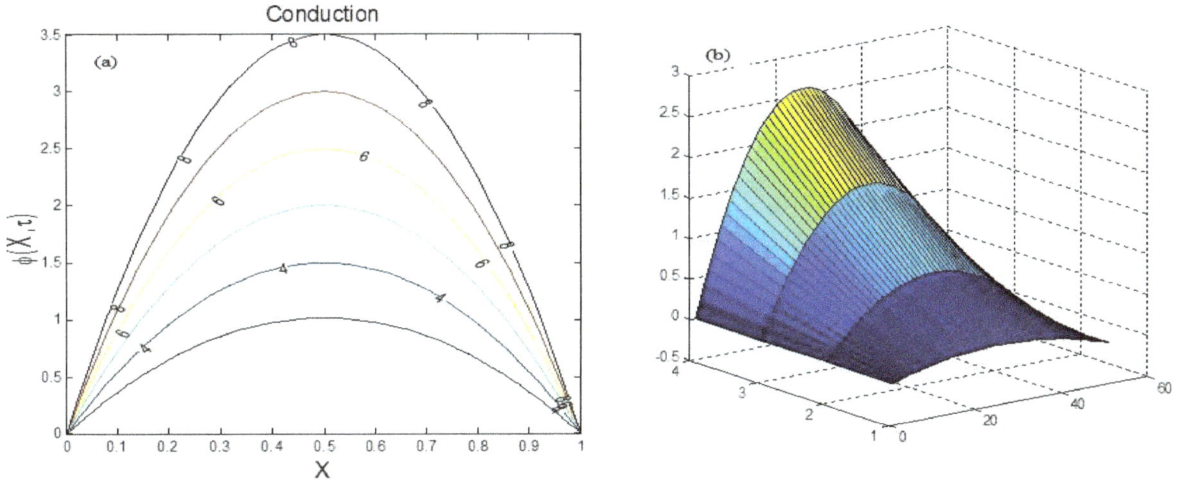

Figure 2. (a) Temperature distributions in conduction mode for source capacity proportional to temperature for for $\tau_g = 1, and\ \alpha = 1.0$, (b) Temperature profile for $\tau_g = 3$ of the second source term.

for all the four cases. In Figure 1(b), temperature profile rise as the dimensionless time slowly drop toward the direction of heat flow (Vedavarz et al., 1994), when $\tau_g = 3$ at constant source capacity. This increase of temperature in the system is caused generally by the heat generation process. Hence, dimensionless temperature distribution is indirectly proportional with the flow of heat flux as indicated in Figure 1(a). Energy is concentrated at the intermediate X for $\tau_g = 1,3,6$, in the case of conduction mode for constant heat source capacity and source capacity proportional to temperature. However, for the pulsed heat source and source capacity proportional to time shown in Figures 3(a and b) and 4(a and b), the energy is less concentrated at $\tau_g = 3$. Figure 2(a and b) display temperature distributions when the source capacity is proportional to temperature, and for $\tau_g = 3$ respectively. Our results for this case show enhanced temperature distribution at increase value of α, and τ_g. The gradual reduction in temperature along the direction of heat flow is expected to explain the well

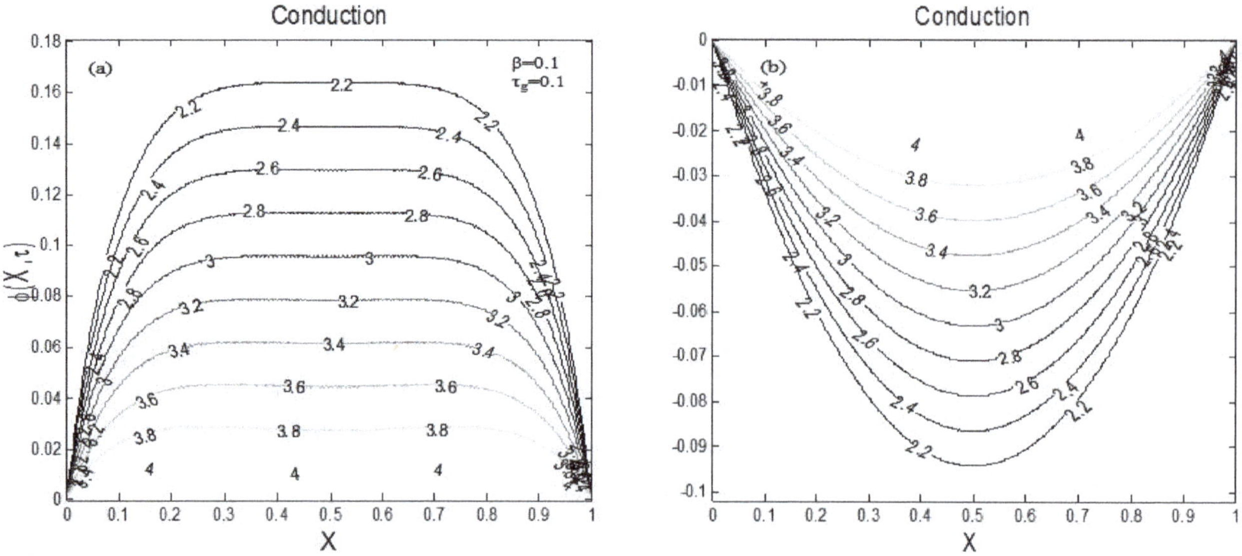

Figure 3. (a) Temperature distributions in conduction mode for pulsed heat source, (b) Temperature distributions for pulsed heat source at $\tau_g = 3$.

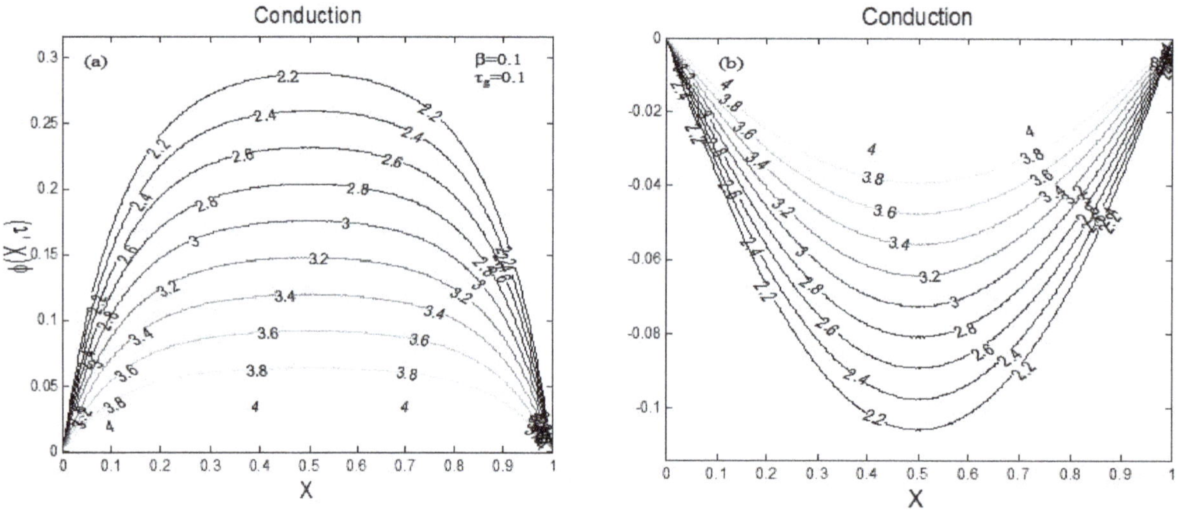

Figure 4. (a) Temperature distributions in conduction mode for source capacity proportional to time. (b) Temperature distributions for source capacity proportional to time at $\tau_g = 3$.

behavior of this model.

Figure 3(a and b) show the temperature distribution of the system, in which heat is release from a pulsed energy source. In Figure 3(a) high relaxation time lowers the temperature profile for $\tau = 0.1$, and β =1, but at higher value of τ_g the temperature profile fluctuate within the set boundary, however, the trend remains same. The temperature distribution for source capacity proportional to time in conduction mode compared to the pulsed energy source term as depicted in Figures 3(a and b) and 4(a and b). This is because when $\tau > 0$ the Dirac delta pulse approach unity, which rendered the two terms to be same at that instant that is, when $\delta(\tau) = \tau$. The little difference observed is the variation in temperature distribution, which is enhanced for source capacity proportional to time as compared to the pulsed energy source term.

Figure 5(a and b) and 6(a and b) show calculations results for the four different source terms with respect to dimensionless temperature variation versus relaxation

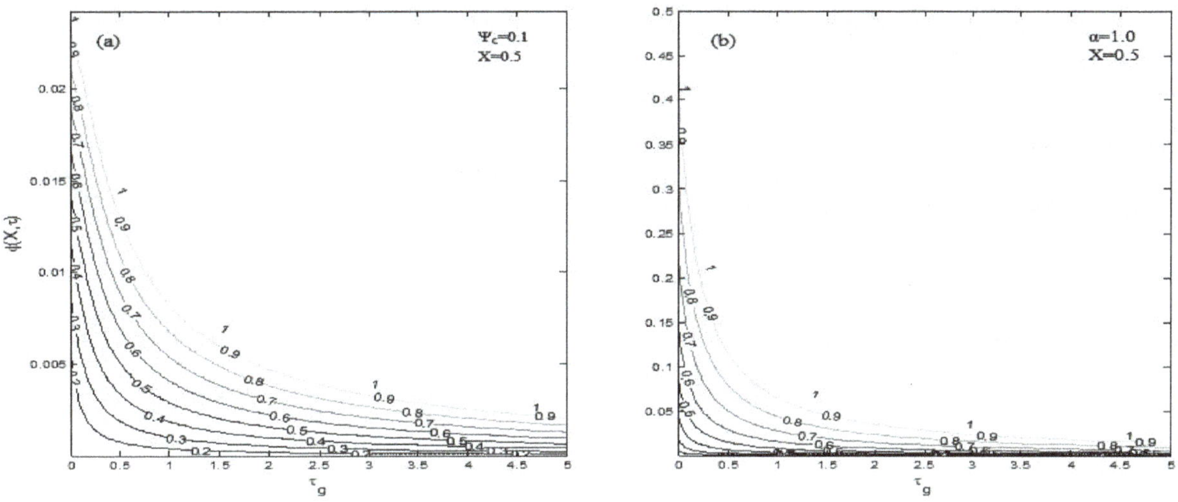

Figure 5. (a) Temperature variation with the relaxation time of source capacity calculated for the constant heat source capacity, (b) Temperature variation with the relaxation time of source capacity calculated for the source capacity proportional to temperature.

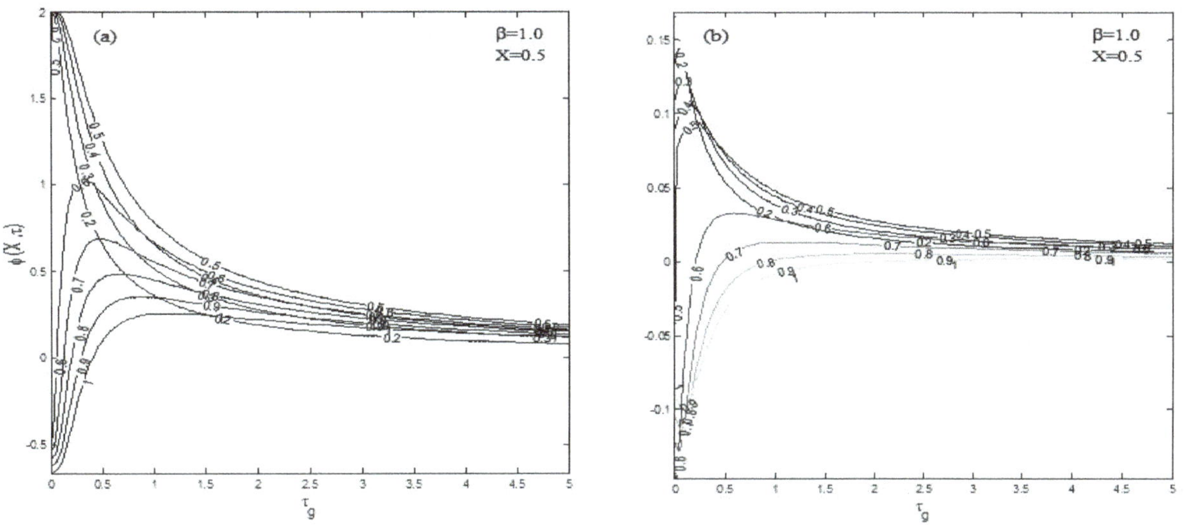

Figure 6. (a) Temperature variation with the relaxation time of source capacity calculated for the pulsed heat source, (b) Temperature variation with the relaxation time of source capacity calculated for the source capacity proportional to time.

time of source capacity. In Figure 5(a and b) uniform temperature variation occurs at shorter duration, it decrease with increase of relaxation time and stabilize at high value of relaxation time, hence the mode of conduction of heat is non-diffusive for extremely short duration. In Figure 5(b) the temperature profile is observed to depend on coefficient α, which causes oscillation at higher value of α. The overall effect is enhancement of temperature profile at high values of the coefficients that is, $\Psi_c = 1, 3, 6$. In Figure 6(a), temperature variation approaches constant value between the heat pulses of 3.5 to 5, and decrease with increase of the heat pulse from both sides of the temperature profile. The steady state solutions of Equations (26), (27) and (28) agrees with source capacity proportional to temperature, which proves the validity of the Riemann-sum approximation used in this work.

CONCLUSION

The problem of heat conduction equation for the finite

velocity of heat propagation, and relaxation of heat source capacity is solved analytically and numerically using Riemann-sum approximation. Four different expressions for dimensionless heat source capacity are considered. The effects of coefficients on temperature distribution, variation, and steady state solution are analyzed. It is observed that the temperature profile decreases when the relaxation time is high, however, at higher value of τ_g the temperature profile fluctuate within the set boundary. Furthermore, the gradual drop in temperature profile along the conduction direction agrees with the natural behavior of heat propagation.

Conflict of Interest

The authors have not declared any conflict of interest.

ACKNOWLEDGEMENT

The authors would like to thank the financial support by the Ministry of Higher Education (MOHE) Malaysia and Universiti Teknologi Malaysia (UTM) under Grant No Q.J130000.2526.06H14.

NOMENCLATURE

τ: Dimensionless time, t_k: Relaxation time of the heat flux, k: Thermal conductivity, Θ: Temperature,
Ψ: Dimensionless capacity of the internal heat source, ρ: Density, c_p: The specific heat at constant pressure, g: Capacity of the internal heat source, τ_g: Dimensionless relaxation time of source capacity, g_t: Transient heat capacity of the source, Φ: Dimensionless temperature, a: Thermal diffusivity, α: Dimensionless coefficient in expression for the Source capacity proportional to temperature, β: Dimensionless coefficient in expression for the Dirac delta pulse and Source capacity proportional to time, Ψ_c: Dimensionless coefficient in expression for the constant Source capacity, K_i: Coefficients defined by Equations (20-23),
δ: Dirac delta function, μ:Thermal diffusivity.

REFERENCES

Al-Nimr MA, Naji M, Abdallah RI (2004). Thermal behavior of a multi-layered thin slab carrying periodic signals under the effect of the dual-phase-lag heat conduction model. Int. J. Thermophysics. 25:949.

Basant KJ, Clement AP (2013). Unsteady MHD two-phase Couette flow of fluid–particle suspension. Appl. Math. Model. 37:1920.

Cai R, Chenhua G, Li H (2006). Algebraically explicit analytical solutions of unsteady 3-D nonlinear non-Fourier (hyperbolic) heat conduction. Int. J. Thermal Sci. 45:893.

Cattaneo SC (1948). Sulla conduzione de calore. Atti Semin. Mat. Fis. Univ. Modena. 3:21.

Hübner K, Sven S, Ursula K (2011). Applications and trends in systems biology in biochemistry. FEBS J. 278:2767.

Jose O, Juan JA (2011). On the stability of the exact solutions of the dual-phase lagging model of heat conduction. Nanoscale Res. Lett. 6:327.

Joseph D, Preziosi L (1990). Addendum to the paper Heat waves. Rev. Mod. Phys. 62:375.

Karniadakis G, Beskok (2002). Micro-flows Fundamentals and Simulation. Springer-Verlag, New York.

Lewandowska M (2001). Hyperbolic heat conduction in the semi-infinite body with a time-dependent laser heat source. Heat Mass Transfer. 37:333.

Lewandowska M, Malinowski L (2002). Thermal waves propagation due to localized heat inputs - the Laplace Transforms method analysis. Heat Mass Trans. 38:459.

Malinowski L (1993a). Novel model for evolution of normal zones in composite superconductors. Cryogenics. 33:728.

Malinowski L (1993b). A relaxation model for heat conduction and generation. J. Phys. D: Appl. Phys. 26:1176.

Mohammad V, Ali AM (2012a). An Evaluation of SWDC and WinSRFR Models to Optimize of Infiltration Parameters in Furrow Irrigation. Am. J. Sci. Res. P.128.

Mohammad V, Ali AM (2012b). Optimize of all effective infiltration parameters in furrow irrigation using visual basic and genetic algorithm programming. Australian J. Basic Appl. Sci. 6:132.

Mohammad V, Mohammad EB, Seyyed MRB (2013). Comparison of the ARMA, ARIMA, and the autoregressive artificial neural network models in forecasting the monthly inflow of Dez dam reservoir. J. Hydrol. 476:433.

Ozisik MN, Tzou DY (1994). On the wave theory in heat conduction. J. Heat Transfer. 116:526.

Saleh A, Al-Nimr M (2008). Variational formulation of hyperbolic heat conduction problems applying Laplace transform technique. Int. Comm. Heat Mass Trans. 35:204.

Tzou DY (1997). Macro to Microscale Heat Transfer: The Lagging Behaviour. Taylor and Francis. P. 317.

Valipour M, Banihabib ME, Behbahani SMR (2012). Monthly Inflow Forecasting using Autoregressive Artifical Network. J. Appl. Sci. 20:2139.

Vedavarz A, Kumar S, Moallemi MK (1994). Significance of non-Fourier heat waves in conduction. J. Heat Trans. 116:221.

Vernotte P (1961). Some possible complications in the phenomena of thermal conduction. Compte Rendus. 252:2191.

Permissions

All chapters in this book were first published in IJPS, by Academic Journals; hereby published with permission under the Creative Commons Attribution License or equivalent. Every chapter published in this book has been scrutinized by our experts. Their significance has been extensively debated. The topics covered herein carry significant findings which will fuel the growth of the discipline. They may even be implemented as practical applications or may be referred to as a beginning point for another development.

The contributors of this book come from diverse backgrounds, making this book a truly international effort. This book will bring forth new frontiers with its revolutionizing research information and detailed analysis of the nascent developments around the world.

We would like to thank all the contributing authors for lending their expertise to make the book truly unique. They have played a crucial role in the development of this book. Without their invaluable contributions this book wouldn't have been possible. They have made vital efforts to compile up to date information on the varied aspects of this subject to make this book a valuable addition to the collection of many professionals and students.

This book was conceptualized with the vision of imparting up-to-date information and advanced data in this field. To ensure the same, a matchless editorial board was set up. Every individual on the board went through rigorous rounds of assessment to prove their worth. After which they invested a large part of their time researching and compiling the most relevant data for our readers.

The editorial board has been involved in producing this book since its inception. They have spent rigorous hours researching and exploring the diverse topics which have resulted in the successful publishing of this book. They have passed on their knowledge of decades through this book. To expedite this challenging task, the publisher supported the team at every step. A small team of assistant editors was also appointed to further simplify the editing procedure and attain best results for the readers.

Apart from the editorial board, the designing team has also invested a significant amount of their time in understanding the subject and creating the most relevant covers. They scrutinized every image to scout for the most suitable representation of the subject and create an appropriate cover for the book.

The publishing team has been an ardent support to the editorial, designing and production team. Their endless efforts to recruit the best for this project, has resulted in the accomplishment of this book. They are a veteran in the field of academics and their pool of knowledge is as vast as their experience in printing. Their expertise and guidance has proved useful at every step. Their uncompromising quality standards have made this book an exceptional effort. Their encouragement from time to time has been an inspiration for everyone.

The publisher and the editorial board hope that this book will prove to be a valuable piece of knowledge for researchers, students, practitioners and scholars across the globe.

List of Contributors

Najma Saleem
Department of Mathematics, University of Management and Technology, CII Johar Town Lahore-54770, Pakistan

T. Hayat
Department of Mathematics, Quaid-i-Azam University Islamabad 44000, Pakistan

A. Alsaedi
Department of Mathematics, Faculty of Science, King Abdulaziz University, P. O. Box 80257, Jeddah 21589, Saudi Arabia

A. Bissessur
School of Chemistry and Physics, University of KwaZulu-Natal, Private Bag X54001, Durban, 4000 South Africa

M. Naicker
Department of Chemistry, University of South Africa, P. O. Box 392, Muckleneuk Ridge, City of Tshwane, 0003, South Africa

H. Mirgolbabaei
Young Researchers Club, Islamic Azad University, Jouybar Branch, Jouybar, Iran

M. Bozorgi
Automotive Engineering Department, Iran University of Science and Technology, Tehran, Iran

M. M. Etghani
Department of mechanical engineering, Babol Noshirvani University of Technology, Babol, Iran

Faris Mohammed Ali
Department of Physics, University Putra Malaysia, 43400 UPM, Serdang, Selangor, Malaysia

W. Mahmood Mat Yunus
Department of Physics, University Putra Malaysia, 43400 UPM, Serdang, Selangor, Malaysia

Zainal Abidin Talib
Department of Physics, University Putra Malaysia, 43400 UPM, Serdang, Selangor, Malaysia

Vipul Sharma
Department of Electronics and Communication Engineering, Gurukul Kangri University, Haridwar, India

S. S. Pattnaik
Department of ETV, NITTTR, Chandigarh, India

Tanuj Garg
Department of ETV, NITTTR, Chandigarh, India

N. Ahmed
Department of Mathematics, Gauhati University, Guwahati-781014, India

K. Kr. Das
Department of Mathematics, Gauhati University, Guwahati-781014, India

Mustafa ORDU
Dumlupinar University, Simav Technical Education Faculty, Simav-Kutahya, Turkey

Mustafa ALTINOK
Gazi University, Technical Education Faculty, Anakara, Turkey

Abdi ATILGAN
Artvin Coruh University, Faculty of Forestry, Artvin, Turkey

Murat OZALP
Dumlupinar University, Simav Technical Education Faculty, Simav-Kutahya, Turkey

Huseyin PEKER
Artvin Coruh University, Faculty of Forestry, Artvin, Turkey

Hakan Keskin
Department of Woodworking Industry Engineering, Technology Faculty, Gazi University, 06500 Beşevler, Ankara, Turkey

Neslihan Süzer Ertürk
Department of Industrial Technology, Industrial Arts Education Faculty, Gazi University, 06830 Gölbaşı, Ankara, Turkey

Mustafa Hilmi Çolakoğlu
Technology Development Foundation of Turkey (TTGV) 06800 Bilkent, Ankara, Turkey

Süleyman Korkut
Department of Forest Industrial Engineering, Forestry Faculty, Düzce University, 81620 Düzce, Turkey

M. N. Lakhoua
University of Carthage, UR: Mechatronics Systems and Signals, ESTI, Tunis, Tunisia

Jiin-Yuh Jang
Department of Mechanical Engineering, National Cheng Kung University, Tainan 70101, Taiwan

Chun-Chung Chen
Department of Mechanical Engineering, National Cheng Kung University, Tainan 70101, Taiwan

Nawaf H. Saeid
Department of Mechanical, Materials and Manufacturing Engineering, The University of Nottingham Malaysia Campus, 43500 Semenyih, Selangor, Malaysia

Yu S. Nechaev
Bardin Institute for Ferrous Metallurgy, Kurdjumov Institute of Metals Science and Physics, Vtoraya Baumanskaya St., 9/23, Moscow 105005, Russia

Nejat T. Veziroglu
International Association for Hydrogen Energy, 5794 SW 40 St. #303, Miami, FL 33155, USA

Sofiane ABERKANE
Département Energétique, Faculté des sciences de l'ingénieur, Université M'Hamed Bougara de Boumerdés-35000, Algérie

Malika IHDENE
Université de Yahia Farès, Médéa- 26000, Algérie

Mourad MODERES
Faculté des hydrocarbures et de la chimie, Université M'Hamed Bougara de Boumerdés-35000, Algérie

Abderahmane GHEZAL
Laboratoire de mécanique des fluides théorique et appliquée, Faculté de physique, Université des sciences et de la technologie de Houari Boumediene Bab Ezzouar, Alger-16111, Algérie

E. J. Christian
Department of Mathematics, Faculty of Mathematical Sciences, University for Development Studies, P. O. Box 1350, Tamale -Ghana

Y. I. Seini
Department of Mathematics, Faculty of Mathematical Sciences, University for Development Studies, P. O. Box 1350, Tamale -Ghana

E. M. Arthur
Department of Mathematics, Faculty of Mathematical Sciences, University for Development Studies, P. O. Box 1350, Tamale -Ghana

D. D. Ganji
Department of Mechanical Engineering, Babol Noshirvani University of Technology, P. O. Box 484, Babol, Iran

A. S. Dogonchi
Department of Mechanical Engineering, Mazandaran Institute of Technology, P. O. Box 747, Babol, Iran

Mbane Biouele Cesar
Laboratory of Earth's Atmosphere Physics, University of Yaoundé I, Cameroon

V. M. Somsikov
Institute of Ionosphere, Almaty, 050020, Kazakhstan

A. B. Andreyev
Institute of Ionosphere, Almaty, 050020, Kazakhstan

A. I. Mokhnatkin
Institute of Ionosphere, Almaty, 050020, Kazakhstan

M. Lawal
Department of Physics Faculty of Science, Universiti Teknologi Malaysia, Johor, Malaysia
Department of Physics, Ahmadu Bello University, Zaria, Nigeria

Basant K. Jha
Department of Mathematics, Ahmadu Bello University, Zaria, Nigeria

Patrick O. Akusu
Department of Physics, Ahmadu Bello University, Zaria, Nigeria

Ahmad R. Mat Isa
Department of Physics Faculty of Science, Universiti Teknologi Malaysia, Johor, Malaysia

R. Nasiru
Department of Physics, Ahmadu Bello University, Zaria, Nigeria

www.ingramcontent.com/pod-product-compliance
Lightning Source LLC
Chambersburg PA
CBHW080650200326
41458CB00013B/4804